ANIMAL MODELS IN
LIGHT OF EVOLUTION

ANIMAL MODELS IN LIGHT OF EVOLUTION

NIALL SHANKS, PhD   C. RAY GREEK, MD

BrownWalker Press
Boca Raton

Animal Models in Light of Evolution

BrownWalker Press
Boca Raton, Florida • USA
2009

ISBN-10: 1-59942-502-5 *(paper)*
ISBN-13: 978-1-59942-502-3 *(paper)*

ISBN-10: 1-59942-503-3 *(ebook)*
ISBN-13: 978-1-59942-503-0 *(ebook)*

www.brownwalker.com

Library of Congress Cataloging-in-Publication Data

Shanks, Niall, 1959-
  Animal models in light of evolution / Niall Shanks and C. Ray Greek.
    p. ; cm.
  Includes bibliographical references and index.
  ISBN-13: 978-1-59942-502-3 (pbk. : alk. paper)
  ISBN-10: 1-59942-502-5 (pbk. : alk. paper)
  1. Animal models in research. 2. Animal experimentation. 3.
Evolution (Biology)--Simulation methods. 4. Evolutionary genetics--
Research. I. Greek, C. Ray. II. Title.
  [DNLM: 1. Biomedical Research. 2. Models, Animal. 3.
Developmental Biology. 4. Evolution. W 20.5 S5277a 2009]

RA1199.4.A54S53 2009
616'.027--dc22

2009027846

For my dogs Brutus, Gnasher, and Lummocks
N.S.

♦

In memory of my mother
R.G.

ACKNOWLEDGEMENTS

One of us (NS) has had many fruitful conversations over the years with a wide range of biologists and biomedical researchers concerning the implications of evolutionary biology for the biomedical sciences. Particularly worthy of mention are Rebecca Pyles and Dan Johnson who have done much to shape his understanding of evolutionary biology.

RG would like to thank his wife Jean for all of her assistance, knowledge, and support in this and other endeavors.

We thank Jamie Aitchison for proofreading our unruly text.

Parts of the chapters on prediction were originally published in the journal *Philosophy, Ethics, and Humanities in Medicine* (http://www.peh-med.com/home/) under the title "Are Animal Models Predictive for Humans?" We thank them for allowing us to use that material.

The authors wish to thank: the American Association for the Advancement of Science for use of the Enard et al. figures; U.S. Department of Energy Genomes to Life Program for the gene regulation diagrams; Marshall Clemens for allowing us to use his diagrams of complex systems; the National Academies Press for the distribution of funding figure; *Xenobiotica* via Informa for use of the figure from Chiba et al.; Nature Publishing Group for reproduction of the Horrobin article; *Toxicologic Pathology* via Sage Publications for the Ulrich et al. figures; *Drug Discovery World* for the chart on immune differences between humans and mice; Elizabeth Behrman for helpful suggestions on the topic of mathematical models for complex systems; Cold Springs Harbor Press for the Varki et al. table; Ron Coleman and *Drug Discovery Today*; James Harris, PhD for the bioavailability figures; Garland Science Press for the metabolism figure; Cambridge University Press for the Phillips machine diagram; *Regulatory Toxicology and Pharmacology* via Elsevier for the Olson study; Elizabeth Behrman at Wichita State University for assistance with the physics and math; Mr. William Artz of the College of Arts and Sciences at Wichita State University for help with technical details surrounding the preparation of this manuscript.

TABLE OF CONTENTS

TABLE OF FIGURES

PART I

◆

HISTORY AND BACKGROUND

CHAPTER 1

◆

INTRODUCTION

There is a principle which is a bar against all information,
which is proof against all arguments and which cannot
fail to keep a man in everlasting ignorance—
that principle is contempt prior to investigation.

—Herbert Spencer

Why this book?

The drug phenylephrine is used in pregnant women and is similar to epinephrine in that both drugs known can be used to increase blood pressure. When women are in labor or when they are having a cesarean section, their blood pressure may decrease because of blood loss or the administration of an epidural to decrease or eliminate pain. In these instances a drug may be administered in order to keep blood pressure within normal limits.

When one of us (RG) was completing his residency in anesthesiology, phenylephrine was contraindicated for use in pregnant women. Phenylephrine was never used in women who were pregnant. Administering phenylephrine to a woman in labor or undergoing a C-section would probably have resulted in the resident changing careers or at least specialties. Phenylephrine was thought to decrease blood supply to the baby thus risking numerous complications. This view of phenylephrine was taught to anesthesiologists and obstetricians as being incontrovertibly true since the late 1960s/early 1970s. The drug given to women who needed blood pressure support was ephedrine instead of phenylephrine.

All that changed in the late 1990s/early 2000s. It was at that time that someone noticed that phenylephrine did not really harm the fetus as had been taught for decades to tens of thousands of physicians. The reason for this contradiction was that all the studies that showed phenylephrine was harmful had been done in animals, mainly sheep (Cooper et al. 2002; Lee, Ngan Kee, and Gin 2002). Humans are not sheep. People died as a result of the mistaken application of the above results from animals to humans. These examples, and other instances like them (see [(Greek and Greek 2000; Greek and Greek 2002)]) are the reason for this book.

The issues we have just alluded to have reverberations in, for example, the field of cancer research. Thus Dennis, writing in *Nature* in 2006 reports:

> It was in 1991 that Bob Weinberg first realized he had a problem with mice. He and his postdoc Tyler Jacks were trying to develop a mouse model for retinoblastoma, a childhood cancer of the retina. It results from the loss of a gene called *Rb*, so the team genetically engineered mice to lack the same gene. But the mice didn't get retinoblastoma. Instead, they developed tumours in their pituitary glands. The finding shocked Weinberg. "Up until then, *I had always believed that all mammals were biologically equivalent*," he says: "This planted the seeds of doubt in my mind."
>
> Weinberg, based at the Whitehead Institute for Biomedical Research in Cambridge, Massachusetts, is one of the pioneers of the molecular age of cancer research. He was involved in the early work on the first human cancer-causing and cancer-suppressing genes in the early 1980s. But when he saw that mutations in such genes didn't cause the same kind of cancer in mice and humans, he began to ask himself why. He became aware of other examples that challenged researchers' faith in how accurately mice could replicate human tumours, and has since sought to bring this to his colleagues' attention. "There is a laundry list of problems with mouse models of cancer," he says. (Dennis 2006) (Emphasis added.)

Just because two things are similar does make them interchangeable. The chemist Primo Levi in his autobiography *The Periodic Table* warns against using the "almost-the-same" in chemistry:

> I thought of another moral...that one must mistrust the almost-the-same...the practically identical, the approximate, the or-even, all surrogates, and all patchwork. The differences can be small, but they can lead to radically different consequences, like a railroad's switch points; the chemist's trade consists in good part in being aware of these differences, knowing them close up, and foreseeing their effects. And not only the chemist's trade. [(Levi 1984) p60]

There are two words for knowledge in the German language. *Kennen* means *knowledge by acquaintance*. This meaning can also be found in archaic English, as in "Do ye *ken* John Peel. . . ." The other German word for knowledge is *wissen*. This is not mere knowledge by acquaintance, but is that kind of knowledge that involves *understanding*, and has a conceptual component. The German word for science is *wissenschaft*. Science in this sense means much more than a passing acquaintance with facts. Science has a conceptual component involving the understanding, interpretation and analysis of (among other things) observations, experiments and data in the broad sense. A large part of this book is devoted to the question of the interpretation and understanding of what has been revealed by the biomedical sciences. In particular, we will be concerned to bring out the implications of evolutionary biology for the fruits of these activities.

This book can also be categorized under the heading of *philosophy of science* and as such it is important that we spend a moment to discuss some of the salient features of philosophy of science. Philosophy of science in the analytic tradition has traditionally studied words like *prediction, theory, law, and hypothesis* and has been concerned to elucidate and clarify the meanings of these terms. Philosophers of science have also examined methodologies employed in various branches of science (some of these investigations are comparative, some are critical). Other traditional concerns in the philosophy of science have revolved around the concept of *evidence*. Such issues concern the nature of evidence, the methods used to gain access to it, and its subsequent evaluation. One of the hallmarks associated with the dawn of modern science was the realization that not all evidence was on a par—that evidence had to be examined for quality and for relevance to problems of interest (witch-hunters had plenty of evidence to support their accusations, for example, but most of it was rubbish extracted under conditions of torture). Learning to think clearly about these and related matters involves critical thinking (and this in turn involves the study of reasoning errors that are common enough to be categorized under the heading of *fallacies*).

Given these points of interest, philosophers of science have many of the same concerns as scientists themselves. The motto for the Royal Society for the Advancement of Science (the world's oldest scientific society) is *Nullius in Verba*, which means, roughly speaking, "don't take anyone's word for it." This is as good a starting point for science as it is for philosophy of science. In a way, sound philosophy of science, like good science itself, is a rational, critical extension and elaboration of sound commonsense.

Issues surrounding critical thinking about science have been neglected in the context of science education. Thus Williams has recently pointed out in *The Scientist*:

As a science educator, I train science graduates to become science teachers. Over the past two years I've surveyed their understanding of key terminology and my findings reveal a serious problem. Graduates, from a range of science disciplines and from a variety of universities in Britain and around the world, have a poor grasp of the meaning of simple terms and are unable to provide appropriate definitions of key scientific terminology. (Williams 2008)

In an editorial accompanying the Williams article, *The Scientist* editor Gallagher stated:

You might expect that newly minted science graduates—who presumably think of themselves as scientists, and who I'd thought of as scientists—would have a well-developed sense of what science is. So it's pretty shocking to discover that a large proportion of them don't have a clue ... [Williams] found that a sizeable proportion of science graduates entering teacher training couldn't define what is a scientific fact, law or hypothesis. (Gallagher 2008)

The issue of animal experimentation in biomedical research has traditionally been of interest to experimental biologists, theoretical biologists and historians of science. Insofar as philosophers have had an interest in these matters, it has been primarily from the standpoint of the ethics of animal use in research. In this book, we hope to show that standard uses of animals in biomedical research raise a host of issues in the philosophy and methodology of science that have nothing to do with the ethical confines of traditional philosophical interest. An examination of animal experimentation from the standpoint of ideas rooted in the philosophy of science, will, we hope, illuminate issues about the nature of science itself, especially experimental science (a matter all too often neglected by philosophers of science). If successful, perhaps experimentalists may come to see their activities in a different light. There is, after all, a big difference between *doing* science on the one hand, and *making sense* of what one has actually done—and *why*—on the other.

At this point, we should explain how we will be using some terms. We will use interchangeably the terms and phrases *animal model, animal-based research*, and *animal experimentation* to mean the use of any nonhuman animal for scientific research and testing purposes. We will use the word *animal* to mean nonhuman animals even though we do of course realize humans are also members of the Animal Kingdom. We will discuss the meanings of the words *hypothesis* and *prediction* (along with allied concepts) in more detail in later chapters dealing specifically with the prediction question in biomedical research. We will be concerned thereto ask under what conditions (if any) can results in animal test subjects be extended and extrapolated to human populations of interest. In other words, when do animal experiments predict human outcomes, and how are these predictions to be tested and validated. (We recognize here, and throughout the book, that animals are used in many ways in biomedical research, and the use of animal subjects to predict human outcomes is but one use of animals in biomedical research, albeit a very important one).

In this book, we present a critical analysis of the use of animals in the context of biomedical research aimed at *predicting* human responses with respect to such matters as the study of disease, the safety of pharmaceutical products, and the effects of environmental toxins. We will raise concerns about the clinical relevance of *predictive animal modeling*. We will argue first that there is a large body of empirical evidence undergirding these concerns. Second we will argue that the concerns we raise have a solid theoretical grounding from both the standpoint of evolutionary biology and dynamical systems theory (especially its implications for the study of complexity). Third we will argue that there are serious methodological and evidential concerns raised by the practice of predictive animal modeling. These issues will be presented in ways that are relevant to professional biologists, as well as those interested in the history and philosophy of science. As is the case with any volume that crosses disciplinary lines, some will find some of the material simplified while others will appreciate the foundations explained by that same material. We beg the reader's indulgence and ask her to remember that others who are not specialists in her field will be reading the sections that fall in her domain of expertise. Our goal has been to explain the concepts so a college science major could understand the basic issues.

In September 2003, in a debate with one of the authors of this volume (RG)—a debate that took place at the Labour Party conference in Bournemouth, England—Dr Ian Gibson, MP (a biologist by training) stated in response to a question about the odds of an animal model getting the right answer in the context of drug testing: "Well, I mean Ray would say seventy: thirty [against] or something like that, I would say fifty: fifty." Even those who defend predictive animal modeling appear to be modest about the prospects of scientific fruit from those research practices.

In an August 4, 2004 article in the *New York Times* titled, "In Drug Research, the Guinea Pigs of Choice Are, Well, Human," Andrew Pollack observes of a new trend in the search for new drugs:

Drug researchers are conducting small, fast, relatively inexpensive tests on people to get a quick gauge of a drug's promise before committing to full-scale clinical trials that may involve hundreds of patients, millions of dollars and many years of study . . . In the past, many of these experiments might have been done only on animals. Often called experimental medicine, the approach is meant to reduce the huge costs of drug development and speed the most promising treatments into the marketplace . . . And scientists and industry executives, while acknowledging the potential for ethical issues, say that experiments on people are more reliable, because animal tests often fail to accurately predict whether a drug will work on people.

In an article published in *The British Medical Journal*, Pound et al. have recently observed:

Clinicians and the public often consider it axiomatic that animal research has contributed to the treatment of human disease, yet little evidence is available to support this view. Few methods exist for evaluating the clinical relevance or importance of basic animal research, and so its clinical (as distinct from scientific) contribution remains uncertain. Anecdotal evidence or unsupported claims are often used as justification—for example, statements that the need for research is "self-evident" or that "Animal experimentation is a valuable research method which has proved itself over time." Such statements are an inadequate form of evidence for such a controversial area of research. We argue that systematic reviews of existing and future research are needed. (Pound et al. 2004)

In an FDA White Paper issued in March 2004, titled, "Innovation or Stagnation: Challenge and Opportunity on the Critical Path to New Medical Products," the authors observe that a new medical compound entering Phase 1 human trials after up to a decade of preclinical screening (using animal models) has about an 8% chance of reaching the market [(FDA News 2006) p 8]. Concerning the causes of this state of affairs the authors observe:

A number of authors have raised the concern that the current drug discovery process, based as it is on *in vitro* screening techniques and animal models of (often) poorly understood clinical relevance, is fundamentally unable to identify candidates with a high probability of effectiveness. [(FDA News 2006) p. 9]

Because the use of animals as research subjects has been a source of moral controversy, there has been a growing interest in the scientific and lay communities concerning the roles played by animals in the biomedical sciences. Much of this interest has been prompted by the relatively recent (circa 1980) animal rights movement (and related social movements) in the United States and Western Europe. This is *not* our concern in the present volume where our focus is on matters of science and not morality. This point deserves emphasis. We fully realize this book has potential implications for a whole host of extra-scientific questions about the conduct of biomedical research (though not necessarily the ones you might think). However, those implications, important thought they may be for persons with relevant interests, are debates for another day.

We do not deny that our discussion may be of relevance to some limited aspects of moral debates surrounding the use of animals in biomedical research. Moreover, we do not deny that our volume has relevance to public policy debates concerning the use of increasingly scarce biomedical research funds. These research activities, insofar as they promise great value to human health and well-being, receive widespread public support. Still, there is no escaping the *animal issue*. Large numbers of animals are consumed annually in the name of predictive biomedical research. Estimates vary but even conservative estimates place the number of animals used in these research endeavors in the United States alone to be of the order of millions per year (see Appendix 1).

Certainly lurking behind our central concerns in this book is a social cost-benefit analysis of current research practices. Giles wrote of this issue in *Nature*:

> In the contentious world of animal research, one question surfaces time and again: how useful are animal experiments as a way to prepare for trials of medical treatments in humans? The issue is crucial, as public opinion is behind animal research only if it helps develop better drugs. Consequently, scientists defending animal experiments insist they are essential for safe clinical trials, whereas animal-rights activists vehemently maintain that they are useless. (Giles 2006)

It is, of course, possible to have concerns about the reality of the social benefits promised by animal investigators, their lobbyists and policy advocates, that are quite independent of any interest in animal rights. Increasingly burdened taxpayers, many of whom couldn't care less about animal rights, for example, have a strong interest in the pursuit of such matters. In the spirit of the Royal Society, they are fed up with being told to take someone else's word for it (one way or the other). We cannot settle these contentious questions and what little we do have to say has been relegated to the appendices accompanying the main body of our arguments.

Interest has also been generated because of new developments in science itself. These developments are derived in no small measure from various genome projects and their implications for the relative positions of humans and other animals in nature. The new biological discipline of genomics reflects the fruits of these inquiries. The biomedical implications here can be seen in such fields as pharmacogenomics and toxicogenomics.

The purpose of this book is to address the ability, or lack thereof, of animals to predict human response and to see what other roles they may have in research and testing. We will argue that claims concerning the great utility of animals as predictive models of human biomedical phenomena are unsupported by evidence and are compromised by both methodological issues and issues arising from basic biological theory.

In this book we will thus discuss the following propositions:

1. *When the animal model community discusses the use of animals in research they give the definite impression that such results have been and will be translated directly to humans.* (The community here includes those who use animals as models for humans, their employers, those in the press who support their activities, and so on.) Some theorists in these debates acknowledge the difference between basic and applied research but even these commentators often encourage belief in the predictive utility of animal models with respect to translational research. Interestingly enough, animal welfare activists often buy into these claims about the predictive utility of animal-based research, hoping (with varying degrees of disingenuity) that animal-based research can be replaced by non-animal methods that work *just as well*. We will argue here that they should be careful what they wish for given the actual predictive track record of animal-based research.

2. *Animal models are not predictive for humans, indeed even different humans respond differently to drugs and disease, for many reasons.* We will discuss some of these reasons and we will examine the meaning of the word *predict*—a matter that calls for attention if only because it has acquired a semantic shiftiness that makes its usage highly susceptible to equivocation in public discussions of these matters.

3. *Animals can be used in science in many endeavors that have little or nothing to do with prediction.* Animals can be used as bioreactors, for the study of other animals of the same species or strain, as an aid in learning and so forth. Clearly, one can obtain much important basic scientific knowledge that may or may not go on to be important in the study of human disease. It is here however that we again criticize the animal model community. There are indeed important connections between basic biological research on animals on the one hand, and human medicine on the other, but these connections are typically much more distant and indirect and suggestive than those engaged in predictive animal modeling tell the public and their policy makers.

    a. Organisms belonging to different species or even different strains of the same species may manifest different responses to the same stimuli due to:

        i. differences with respect to genes present, and also with respect to the versions (*alleles*) of genes present;

        ii. differences with respect to mutations in the same gene (where one species has an ortholog of a gene found in another);

        iii. differences with respect to proteins and protein activity;

        iv. differences with respect to gene regulation;

        v. differences in gene expression;

        vi. differences in protein-protein interactions;

        vii. differences in genetic networks (robustness, pleiotropy etc);

        viii. differences with respect to organismal organization (humans and rats may be intact systems, but may be differently intact);

        ix. differences in environmental exposures; and last but not least

        x. differences with respect to evolutionary histories.

These are some of the important reasons why there are species differences with respect to the response to drugs and toxins, and why different species (and strains of a given species) experience different disease states.

    b. Even nearly identical organisms (e.g., chimpanzees and humans in some debates, or monozygotic twins in other contexts) may respond differently to drugs and experience different diseases. These observations serve to sharpen the tion problem with which we are concerned.

    c.   Current biomedical research is studying disease and drug response in ways that have uncovered reasons, even at the molecular and cellular levels of description where very small differences between organisms (be they members of two different species or members of different strains of a given species) become highly significant in the generation of responses to the same stimuli. Hence, by the standards of our current best biomedical sciences, using animals (e.g. mice) as predictive models for human disease and drug testing is highly questionable from a scientific point of view.

    d.   Immense empirical evidence supports this position.

4.   *The above issues are important because there are human lives at stake.* The National Institutes of Health, for example, are funded in large measure by US taxpayers in order to improve the health of Americans. If the funding of animal models is not accomplishing this (see Appendix 1 and 2), then the basic assumption upon which the NIH operates must be addressed.

The reader is warned that this volume, dealing as it does with issues surrounding some common research practices and theoretical claims in the biomedical sciences, involves numerous examples and citations. Perhaps there are too many, but we find ourselves damned if we do and damned if we don't. If we do not dwell on these matters at length, we will be accused of arguing about positions that no one has held, and for which there is little or no evidence. If we do dwell on these matters, we run the risk of annoying some readers who may say, "enough already." We decided that it would be better to risk annoying some readers with an excess of quotation (which those who are willing to grant our claims may skip), than to run the risk of being accused of attacking a straw man that no one takes seriously. (If we are guilty of this last error, it will not be for the trivial reason that a critic simply does not like what we say and wishes to dismiss it out of hand). Readers puzzled by our cautions here are reminded that few topics in the scientific arena stir passions so vigorously—one way or the other—as discussions about the roles played by animals in biomedical research.

Now is a good time to address an issue that frequently crops up in the context of discussions of these contentious matters. Because we make extensive use of quotations we run the risk of being accused of taking the words of others out of context. This is an easy accusation to make, but without carefully reasoned backing, the accusation, in and of itself, is little more than an indication of intellectual laziness. Sometimes, of course, accusations of this kind are indeed justified and can be defended by citing additional evidence. Without credible supporting evidence, the accusation that someone has taken something out of context is largely empty. Its main function is to immunize what was quoted from scrutiny in the court of rational inquiry.

The issue here is not whether the authors we quote agree with our overall point of view (or indeed with any of it) but whether what they said was what we reported. We quote the words of animal modelers who have admitted in print, in publicly checkable sources, that there are methodological and evidential problems associated with the claim that experiments on nonhuman animals can be used to predict human responses (regardless of whatever else animals are useful for in science). That many of these same researchers believe that animals can be used to predict human responses is, with no further discussion, neither here nor there. We discuss at length the theoretical reasons given to justify predictive uses of animal models, and we criticize these justifications in due course. Our point is that the prediction issue is on the table for discussion.

Many of the scientists we quote are speaking very specifically about an area or example that falls directly within their expertise. When they comment that animal models failed in some way, in their area of expertise, their claim carries weight no matter how committed they are to the use of animals in biomedical research!

The FACTS of species differences relevant to the prediction issue are not in question. What is at issue are the INTERPRETATIONS authors place on the facts they refer to. We are justified in citing facts (and quoting researchers who do so) while rejecting interpretations. We cite many examples of the same phenomenon (species difference) to establish that it is a more or less agreed fact. Our point is not to get involved in an endless dingdong about examples but to suggest that there are other well-established ways to deal with what we see (i.e. evolutionary biology). It matters not to us that those referring to the facts of species differences are also ardent advocates of animal experimentation. What matters are the facts, along with consideration of alternative hypotheses about their significance. We will offer some alternative hypotheses about the significance of the facts we uncover. Science, after all, is as much driven by disputes about the meaning and significance of facts as it is about the facts themselves. We have nothing to apologize for in this regard.

We are about to begin a detailed analysis of the roles played by animals in biomedical research. This is a good place to make clear, once again, what we are interested in, and what we are not. There can be no doubt whatsoever that if you wish to make discoveries about rats and mice you will be forced of methodological necessity to perform careful scientific studies of R. *rattus* and M. *musculus* respectively. In fact, in writing this book, we are the beneficiaries of the results of careful scientific studies of animals. There is no doubt that careful biological studies of rats and mice can help clarify the general contours of mammalian biology. Such studies can also play a valuable heuristic role by prompting new ways of thinking about human biological problems of interest. The issue we *are* concerned with is this: notwithstanding these cautions, *are animal models predictive of human outcomes* in, say, toxicology, drug discovery, and the study of the causes and cures of human diseases?

CHAPTER 2

◆

MATTERS OF SCIENCE

There is nothing as deceptive as an obvious fact.
—Sir Arthur Conan Doyle via Sherlock Holmes

Before turning to biology, consider an issue in physics. In physics, in order to have a proper understanding of the limitations of Newtonian mechanics, you need to understand relativity theory and quantum theory, vector spaces, partial differential equations and tensor calculus. In order to understand all this, you need calculus, linear algebra and other college math courses, not to mention a host of physics courses. In order to understand college math you need to understand high school math and so on. In this book we want to examine limitations of some current uses of animal models in biomedical research and drug testing. That is to say, we are concerned with the use of animal models as predictors of human responses. Just as in the case of physics, so too here.

Thus, in order to understand the limitations of certain animal modeling practices, you need to understand many rather diverse areas of study: evolutionary biology; population genetics, developmental genetics (gene networks and gene regulation); what models are and how they are used; the implications of dynamical systems theory for the study of biological complexity; what science in general is and how it is practiced; and how the practice of medicine is influenced by the results from animal models. When all of this is understood, then you can appreciate the arguments for and against using animals in biomedical research. It is clearly no trivial matter to be settled by sound-bites and slogans.

The issue of the use of animals in biomedical research crosses many scientific disciplines and hence representatives from these disciplines must weigh in on the topic in order for the scientific community as a whole to have an understanding of the issue and for those conducting the research to proceed in the way most compatible with the best scientific knowledge of the day. Hence, in this book, we evaluate animal models not just in light of biology but also other scientific disciplines.

Animals are used in science in at least nine distinct ways: (1) as predictive models for human disease; (2) as predictive models to evaluate human exposure safety in the context of pharmacology and toxicology (e.g., in drug testing); (3) as sources of 'spare parts' (e.g., aortic valve replacements for humans); (4) as bioreactors (e.g., as factories for the production of insulin, or monoclonal antibodies, or the fruits of genetic engineering); (5) as sources of tissue in order to study basic physiological principles; (6) for dissection and study in education and medical training; (7) as heuristic devices to prompt new biological/biomedical hypotheses; (8) for the benefit of other nonhuman animals; and (9) for the pursuit of scientific knowledge in and of itself.

In this volume we are primarily concerned with the practice of using animals as predictive models of human biological and biomedical phenomena (items (1) and (2) above). Insofar as we are interested in the scientific issues raised by the use of one group of evolved biological systems (a sample of an animal population, and often one that does not reflect the variation found in the species as a whole) to draw conclusions about another group of such evolved biological systems (typically a much larger and varied population of humans), our investigation might be described as an investigation into the logic of inter-species extrapolation. For reasons that will become clearer in our discussion of evolutionary biology, this problem is related to the problem of intra-species extrapolation (for example, between distinct individuals, varieties or strains of a given species).

This book is *not* intended to be a criticism of the use of animals in the context of basic biological research. There can be no doubt that careful studies of animals have prompted important hypotheses about basic biological principles, and there can be no doubt that studies of animals have contributed greatly to our scientific understanding of life, and there is little doubt that these studies will continue to illuminate these matters in the future (items (7) and (9) above).

This book is deeply rooted in evolutionary biology and its consequences. One important consequence (to be explored in detail as the book proceeds) is the fact that evolutionary processes occurring in accord with basic biological principles have modulated the manifestation of life in different organismal lineages, and in this way they have contributed to biological diversity. A further consequence of this is that observed similarities between different species can be highly misleading.

As a foretaste of what is to come in this book, we note that both Antarctic notothenioids and Arctic cod produce antifreeze glycoproteins that serve *similar* functions in the respective organisms. Such proteins contribute to the ability of these fish to make a living in their respective (chilly) positions in the economy of nature. The antifreeze glycoproteins are similar in structure and function, but they nevertheless do not arise from a common gene. Different genes are involved and this is a case of convergent biochemical evolution [(Hochachka and Somero 2002) p413]. Similarities can be deceptive. Thus, similar functions can be achieved through very different causal pathways.

Another way to bring this point out is to consider the concept of *homology*, since this is a good way to focus attention on the meaning of similarities and differences between species. To an evolutionary biologist such as Ernst Mayr, homology refers, "…to the structure, behavior, or other character of two taxa that are derived from the same or equivalent feature of the nearest common ancestor [(Mayr 2002) p287]." The arms of humans, the wings of bats, the flippers of whales and the forelimbs of horses are homologous in this sense, yet they serve very different functions. By contrast the wings of moths and those of birds achieve a similar function—exploitation of an aerial niche—but they are not homologous in the evolutionary sense.

By contrast, in molecular biology, homology refers to sequence similarities with respect to DNA or amino acids. As Brigandt has noted, ". . . molecular homology is a statement about the similarity of genes and proteins, not about their evolutionary origin—inheritance from a common ancestor. . . Molecular homology as mere similarity of DNA or amino acid sequence is an understanding of homology that is tied to the experimental practice of molecular biology. It is effective to *organize knowledge about molecular mechanisms and direct experimental practice* [(Brigandt 2003) p12]." Matters become more complex, however, from the standpoint of evolutionary developmental biology. In this context Brigandt observes:

> . . . non-homologous genes may be involved in the production of homologous structures, and, conversely, non-homologous structures may essentially depend on the expression of the same gene. This is possible because in the course of evolution the importance of a gene for the origin of a structure may diminish and it may be-

come relevant for another character and finally acquire a new function (co-option). [(Brigandt 2003) p13]

West-Eberhard has recently elaborated on the implications of the existence of multiple developmental pathways for the homology concept. She discusses several examples where phenotypes are conserved while developmental pathways change [(West-Eberhard 2003) p494-6]. Thus West-Eberhard comments:

> Multiple pathways are important evidence that selection acts on phenotypes, not on the mechanisms producing them, and they show how evolution can work to maintain structures by tinkering with different resources at hand. They also show that developmental criteria may be unreliable indicators of common decent and reduce the kinds of evidence that can confidently be used to test for homology. [(West-Eberhard 2003) p497]

Evolution leads us to expect complex suites of similarities and differences between different species reflecting descent from common ancestors with subsequent evolutionary modification. However, the remarks above suggest, at the very least, that an assessment of the nature and significance of evolved similarities is going to be a complex matter that may not admit of simple quantitative comparisons.

The theory of evolution provides the conceptual glue that unites all the disparate branches of biological inquiry. This theory has been deeply affected by genomic studies, and the impact of such studies is seen nowhere more clearly than in evolutionary developmental biology. As we will see in the course of the present volume, these genomic studies have had enormous impacts on our understanding of the nature of the similarities and differences between species, and hence of the impact of evolutionary processes that generate the biodiversity we see around us. It should be equally obvious that an understanding of the nature of these similarities and differences will also be important when it comes to a scientific assessment of the use of nonhuman animals as models of human biomedical phenomena. In these experiments, animal subjects representing samples from nature's diversity are used to form expectations about the likely behavior of human subjects, themselves distinct parts of the same natural diversity.

Modern evolutionary biologists do not make the anthropomorphic error of assuming that humans are the most complex creatures on the planet, and that other animals can be judged relative to humans by the degree to which they differ with respect to complexity. Science has revealed no such *scala naturae* with regard to complexity. For this reason evolutionary biologists no longer talk of higher and lower mammals, for example. In this sense, animal models are unlike models found in many other branches of science, where, for example, the model system is usually considered to be a simpler system than the system modeled—an idealization, perhaps, in which needless complications can be ignored (or hidden in a host of *cateris paribus* clauses).

If there are not clear differences with respect to degree of complexity, there may well be differences with respect to the *kind* of complexity. Hence it may make evolutionary sense to say that humans and the rodent species typically used to model them are *differently* complex. An important part of this volume consists of an attempt to make sense of evolution's implications for biological complexity, and the implications of this, in turn, for the use of nonhuman animals as models of human biomedical phenomena.

As we noted earlier, we are not directly concerned here with the ethical positions that are taken by supporters and opponents of animal experimentation. The philosophical and ethical issues have been explored in numerous books and articles well known to both communities. We will be concerned to argue that the scientific issues concerning the use of animals as models of human biomedical phenomena are by no means as simple as are commonly supposed by participants in moral and public policy debates. An understanding of the scientific issues we are concerned with will involve a careful examination of the nature of science itself. This task is undertaken in the early chapters of this book since it provides the conceptual backdrop against which the practice of animal modeling can be assessed. Ours is an ahistorical age where lessons from the past are frequently ignored, if only because of ignorance and apathy. Yet scientific practices in the biomedical sciences have evolutionary histories by way of methodological analogy with the historical evolutionary factors shaping the very biological objects of those biomedical inquiries. Historical legacies in both cases are both interesting and instructive.

Matters of History

Since history is an issue that shapes the matters at hand, some remarks about where we stand with respect to the historian's distinction between *internalists* and *externalists* seem to be in order. Morrell explains externalism as:

> The view that social, political, and economic circumstances affect the pursuit of knowledge of nature . . . Externalist historians are in-

terested in scientific groups (both institutionalized and informal), the reasons for the development of certain kinds of scientific research, scientific careers and patronage of science. They claim social and economic circumstances have affected the rate and the direction of some scientific work. Committed externalist historians usually assume that the response to such circumstances has on occasion helped constitute scientific knowledge itself. [(Bynum, Browne, and Porter 1981) p145]

By contrast, Morrell explains internalism as:

The view that science is primarily an abstract intellectual enterprise insulated from social, political, and economic circumstances. Internalist historians focus on the obviously intellectual aspects of the setting and solving of problems concerned with the understanding and control of the natural world; they highlight conceptual frameworks, methodological procedures, and theoretical formulations. For these historians, often concerned to defend science as the supremely rational form of thought, changes in past science were exclusively or chiefly occasioned by the solving of inherited and abstract problems within a particular field of inquiry. [(Bynum, Browne, and Porter 1981) p211]

While we do not pretend that the distinction between internalist and externalist analyses of the history and practice of science is always straightforward or easy to draw, our approach in this volume is located primarily within the internalist framework. What does this mean?

First and foremost we are concerned with conceptual frameworks, methodological procedures and theoretical issues surrounding research practices in the biomedical sciences. Our analysis is self-referential or reflexive in that we intend to examine extant scientific research practices *using scientific concepts, methods and theories.*

Science has an important feature that is absent from much human activity. In the long run, it is *self-correcting.* This means that reasonable judges sitting in the court of rational inquiry ought to be able to overturn earlier decisions by other such judges if there is warrant to do so. Such warrant will inevitably involve evidence interpreted in the light of our current best scientific theories. Our scientific theories, fallible as they are, are the only lights we have to go by, and being themselves rooted in high-quality evidence, they are usually good lights to work with.

One potential consequence of this approach to the study of scientific activity is that if a practice, method or theory is found to be flawed, and is nevertheless resistant to change in the light of relevant data and analysis, hypotheses may have to be formulated from an externalist standpoint to explain the continuation of adherence to the practice, method or theory. Such hypotheses will refer to extra-scientific explanatory factors such as economic, social or political factors that undergird a practice, method or theory in the face of countervailing scientific considerations. Our approach is thus not reductionistic—we do not claim that internal factors are the only relevant explanatory factors in an explanation of why scientific activity takes place in the way that it does, as a matter of fact (this applies with particular force to activity in the name of science which may have little to do with science the way science is normally understood from an internalist perspective). In fact we explicitly recognize that real science involves a complex interplay, from the institutional level to the level of the individual practitioner, of internal and external factors. Our point is that the perception of the validity of scientific activity hinges on its being, or being believed to be, good science.

The importance of externalist factors as factors shaping scientific inquiry should nevertheless not be forgotten. Until the 1980s, the study of retroviruses was a comparative backwater in the field of virology. This would change dramatically with the AIDS crisis. The availability of research funding shifted the focus of scientific inquiry. Here there is a biomedical corollary to Parkinson's Law to the effect that the amount of research done in a field expands to consume available funding. It is true that availability of research funding was motivated in part by public health concerns, but there was clearly more going on here from an externalist perspective. After all, there were many diseases, some of them major killers in the developing world, that did not get the sort of attention that was devoted to research on HIV. The point here is not that we shouldn't be doing research into HIV—arguably we should be doing more. Rather, the point is that while scientists may ideally bring the best standards of scientific practice to the work they conduct, the guidance of the direction of scientific research, at the institutional level where policy evolves, transcends purely internalist scientific concerns. Having said this, we note here that implications of the present book for the conduct of animal-based HIV research will be discussed in chapter 15, where some issues about actual scientific practice will be discussed. But that lies ahead.

In looking at the ways in which externalist factors can trump purely internalist concerns, one important factor to consider is *institutional iner-*

tia—institutional resistance to change (for whatever reason). Not all inertia is bad. If the forces for change are malignant, arguably they should be resisted (for example, forces for change in school science curricula that attempt to mandate inclusion of the backward creationist perspectives of Christian fundamentalists—well-represented at many levels of government in the US—should be resisted).

The problem is that institutional inertia can slow the pace of positive change and innovation too. The virtue of inertia is that it makes institutions robust in the face of perturbing forces. Unfortunately, the vices of institutional conservatism consequent upon inertia have much the same origin as the virtues in this regard. Scientific practice has arguably never consisted of the works of the heroic "gentleman investigator," independently wealthy, pursuing scientific truth for no other reason than a pure fascination with the general disposition of things. Today the practice of science is supported and constrained by a variety of institutions that include (but are not limited to) governmental funding and regulatory agencies, corporations with extra-scientific interests in profitability, scientific societies, journals, and universities (with all their attendant administrative horrors).

In Appendix 3 we provide some examples of research proposals claiming potential human relevance for lines of research that may be meritorious in their own terms, regardless of their implications for human puzzles of interest. Sometimes there are administrative pressures to claim human relevance for research where there likely is none, or where, if there is, it is so indirect and distant, that it may as well not be there. We think that some rethinking of the administrative hoops that researchers have to jump through is long overdue.

One consequence of this is that research practices, once established, are hard to change, even when evidence points in the direction of a need for changes. The existence of institutional inertia should serve as a warning to those who see science in the light of a very naïve empiricism in which the whole enterprise is driven by, and is highly responsive to, the dispassionate discovery for evidence (jurists have long abandoned such evidential dreams with respect to legal practices). Thus, for example, what a naïve empiricist might consider to be conclusive empirical refutation of a group of hypotheses, may under the right administrative circumstances, with appropriate institutional investments, simply become *incomplete validation of an otherwise successful methodology!*

For example, as early as 1957 observers could point out, notwithstanding massive amounts of accumulated human clinical data concerning the hazards of tobacco use, that, "The failure of many investigators...to induce experimental cancers [in animals], except in a handful of cases, during fifty years of trying, casts serious doubt on the validity of the cigarette-lung cancer theory" (Northrup 1957). The handful of cases would count as incomplete validation of an otherwise successful practice of great value to *Big Tobacco*. Courts of law and policy makers responded positively for decades to the existence of healthy test animals, notwithstanding sickly, wheezing human smokers!

Perhaps, then, the conservatism flowing from institutional inertia is a reflection of the wisdom in Belloc's poem *Jim*, about a willful boy who ignored his keeper and was consumed by a lion:

> And always keep ahold of Nurse
> For fear of finding something worse.

An interesting issue here, of course, is this: how many *incomplete validations* do we need to have, before we call the success of the *otherwise successful methodology* into question? Even if we do not answer this question directly, it is one that haunts the rest of the book.

The actual practice of science, then, reflects a wide range of factors, some of which are internal in nature (scientific methods, statistical analyses, modeling techniques, and so on). It also reflects external considerations relating to public and private funding or other extra-scientific factors—externalist religious claims, for example, have led to intellectually reprehensible restrictions on embryonic stem cell research and research into reproductive technologies in the US during the Bush administration. These considerations often involve economic, social and political factors. To an internalist, religious beliefs about the *realm invisible* and its putative denizens (ranging from *God the creator* to *souls* allegedly present at the instant of conception) can, in the nature of the case, have no place in science. From an externalist perspective, religious (or moral or political or economic) restrictions on science, for good or ill, are not matters to be settled simply by looking at science itself, hence the apparent need for seemingly endless public policy debates!

These macroscopic observations about the mix of internal and external motivations behind the practice of science are also reflected in the microcosm in the minds of individual scientists. Studies are undertaken for their intrinsic scientific value, perhaps to solve an extant scientific problem, or extrinsically for the sake of tenure, benefit to humanity and so on. Whatever the motivation, it is always a good question as to whether the practice, method or theory is good science as judged by the standards of science itself. That is our interest with the issue of the use of animal models in biomedical research.

♦

CHAPTER 3

A BRIEF HISTORY OF ANIMALS IN EARLY SCIENCE

For every action, there is an equal and opposite criticism.

The central concern of this book is with the use of animal subjects as predictive models of human biomedical phenomena. This is a complex subject that will require illumination in terms of our current best biological theories. Before this task can be undertaken we believe it is important to appreciate the historical context out of which current research practices have evolved. Thus, in order to understand the complex roles played by animals in the biomedical sciences, it will be helpful to understand some of the central ways in which science has evolved through the study of animals, and how science itself has shed light on the nature of animals, human and nonhuman alike. A more detailed discussion can be found in Shanks (Shanks 2002). In this chapter we examine the ways in which animal experimentation shaped early modern science in the 16th and 17th centuries. This is important since as we shall see *physik* (i.e., medicine) had enormous implications for *physics* (i.e., study of *phusis* or nature). In the next chapter we will see that physics, in turn, had enormous implications for the methodological debates concerning the use of animals in 19th century biomedical inquiries—debates that reverberate to the present. These reverberations are modulated, however, by another aspect of 19th century biological inquiry: the theory of evolution. Consequently, before discussing the current state of affairs with respect to biomedical research, there is much to learn from the study of the past, if only because there is some truth to the old adage that the child is the father of the man!

Medicine and method: the rise of anatomy and physiology

Though human and animal dissection took place in the medieval period, the explosion of interest in anatomical and physiological inquiry—

involving widespread, systematic studies of human cadavers and live, nonhuman animals—occurs in the Renaissance. This is as good a place as any to begin our historical inquiries.

An important part of the explanation for this blossoming of anatomical and physiological inquiry lies in the way that renaissance investigators became increasingly reliant on mechanical metaphors to conceptualize the objects of their inquiries—*bodies*—in mechanical terms. Here we will see how the metaphor of *body-as-machine* came not only to guide anatomical and physiological studies, but reflected in turn a systematic method of inquiry, known as the *resoluto-compositive method* or *method of analysis and synthesis*. We will see that the combination of metaphor and method resulted in an intellectual whole greater than the sum of its parts.

The metaphors we use to understand new aspects of the world around us typically reflect prior experience with other, more familiar features of the world. So we can say, for example, it is as though X (the body revealed by dissection) is like a machine Y (perhaps a mechanical clock with the back removed to display the interacting cogs, springs and other components). The metaphor of body-as-machine did not appear all at once, but rather it evolved from crude mechanical analogies early in the Renaissance to a fully crystallized and articulated mechanical picture of human and nonhuman animal bodies by the middle of the 17th century.

We will see that the metaphor of body-as-machine had enormous implications for medical inquiries, and especially for the use of animals as research subjects. We will also see that the mechanical metaphors that fueled the growth of anatomical and physiological inquiry also had broader implications, culminating in the metaphor of *Nature-as-machine*. It is arguably no accident that a method that had proved so fruitful in the domain of *physik* should come to shape early inquiries in *physics* as well.

Since machines need designers and makers, God, as the designer of the natural machine, was just one of the ways in which science and religion came to enjoy a cooperative relationship—a relationship that would only be soured by events, forced in large measure (but by no means exclusively), by a growing understanding of the consequences of Darwin's theory of evolution. As an amusing aside (since one of us lives in Kansas, where these matters have taken on a life of their own), contemporary advocates of so-called "intelligent design" are at least 300 years behind the times.

Andreas Vesalius was perhaps the greatest of the Renaissance anatomists, and his book, *De Fabrica Humani Corporis* (The Structure of the Human Body) was published in 1543—the same year that saw the publication of Copernicus' *De Revolutionibus Orbium Caelestium* (Of the Rotation of Celestial Bodies). The *Fabrica* deserves attention partly because it corrects errors in the old Galenical tradition of anatomy—Vesalius shows that the jaw was one bone not two, and that the breastbone consisted of three bones, not seven, and that men and women had the same number of ribs. More importantly, the text also includes a discussion of the importance of *vivisection*, the dissection of live animals. This topic has been a source of much controversy ever since. The *Fabrica* thus takes us from a static study of organic structure (something that could be accomplished through the study of cadavers) to the study of organic function—a dynamical study that depends on observations of the working bodily machine, and thereby moves the focus of the discussion from anatomy to physiology.

There is also a method at work. The body was clearly a highly complicated thing. Anatomy was the first step toward an understanding of this system. In order to understand the structure of any complex system (perhaps a clock or a car), you must first *analyze* the system into parts and come to understand the structural properties of the most basic parts of the system (in Vesalius' day, organs and tissues). The second stage of anatomical inquiry then consists of a further *analysis* of the structural relationships among the parts so identified, with the *synthesis* of this information of the parts and their mutual relationships yielding knowledge of the fabric of the complex system as a whole—something conceived of as an interwoven, intertwined system of parts. The bodily fabric of interest to anatomists was described by Vesalius as " . . .a fabric not built of ten or twelve parts, as it appears to the casual observer, but of several thousand diverse parts" [(Clendening 1960) p139] but a fabric that could nevertheless be systematically unraveled to reveal its secrets.

Vesalius' anatomical method thus accords well with the *resoluto-compositive* method, a systematic method of inquiry developed by medical logicians at the medical school at Padua, beginning in the early 14th century with the work of Pietro d'Abano, and culminating in the early 16th century with the work of Zabarella (1533-1589). The *resoluto-compositive* method is a method designed to enable the investigator to understand a complex phenomenon, like disease, or a complex system, like a body.

Suppose you want to understand a complex system like a body. The first step involves the *resolution* or *analysis* of the body into its component parts. Having resolved the system into its parts, the properties of the parts must then be studied, so that their causes, and what in turn they cause, may be understood. Having understood the properties of the parts, one must go on to discern the static, structural relationships between the parts—how they are connected or stand to each other. In short, one must discern the anatomy of the system.

From knowledge of the anatomy of the system, one must then study the dynamical relationships between the parts, not only to see how changes in the parts are brought about, and themselves bring about further changes in the system, but also to discern the *functions* of the parts. In short, their physiology. This understanding of the parts and their mutual relationships can then be *composed* or *synthesized* into an understanding of the complex system of interest.

As we shall see shortly, this method was employed by Harvey in his investigations of the motions of the heart. A variant of the method was employed by Galileo in investigations of the motions of falling bodies, and also by Thomas Hobbes in the formulation of his mechanical conception of humans and nonhuman animals. It should be obvious that the resoluto-compositive method is ideally suited for the study of complex machines with many interrelated parts (e.g., clocks), and for other complex systems conceptualized as being mechanical in nature.

The method is also evident in the foundations of Vesalius' anatomy, and especially in his physiology, to which we now turn. Anatomy can teach us much about the structure of the parts of the body and their static relationships, but to understand vital function, one must look not just at the parts and their static relationships; one must also understand the dynamical interactions between the parts. This in turn means going beyond cadaver studies on human and nonhuman animals. It involves the dissection of living organisms—vivisection—to see the parts in action.

The methodological importance of vivisection of animals is explained by Vesalius as follows:

> Just as the dissection of the dead teaches well the number, position and shape of each part, and most accurately the nature and composition of its material substance, thus also dissection of a living animal clearly demonstrates at once the function itself, at another time it shows very clearly the reasons for the existence of the parts. Therefore, even though students deservedly first come to be skilled in the study of dead animals, afterward when about to investigate the action and use of the parts of the body they must become acquainted with the living animal. [(Clendening 1960) p142]

But why should studies of animals be relevant for human physiology?

The answer to this question is to be found in the observation that by the time Vesalius was writing there was an extensive body of comparative anatomical knowledge. Vesalius, like other anatomists before him, had noticed all manner of structural similarities and analogies between humans and nonhuman animals (including, but not restricted to dogs, cats, sheep, pigs and horses). For example, in the study of the arrangements of the nerves in man and beast Vesalius writes:

> They will be arranged almost in this manner: In man one of them is the third, and is carried into the forearm along the anterior side of the elbow joint; another, in fact, the fifth, runs to the elbow next to the posterior portion of the internal tuberosity of the humerus. For in this manner the nerves are also observed in the dog. These nerves having been tied somewhere before they reach the elbow joint, the motion of the muscles flexing the digits and arm will be abolished, and if thou wilt intercept with a band the nerve which in man is reckoned by me the fourth and is extended along the humerus to its external tuberosity, then the motion of the muscles extending the foreleg and digits will be abolished. [(Clendening 1960) p143]

It is these structural similarities that permit the testing of hypotheses about the function of structures found in human cadavers by experiments on live animals.

Here is an example of the kind of experiment that Vesalius performed in order to reveal information about biological function:

> ... I note the recurrent nerves lying on the sides of the rough artery [trachea] which I sometimes intercept with ligatures, at other times I cut. And first I do the same on the other side, in order that it may be clearly seen when one nerve has been tied or cut how half the voice disappears and is totally lost when both nerves are cut. And if I loosen the ligatures the voice will return again ... and it is clearly proved how the animal struggles for deep breaths without its voice when the recurrent nerves have been divided with a sharp knife.

This is all in a manner reminiscent of the way in which someone studying the function and dysfunction of a machine—such as a clock—might systematically study the parts in motion, and then remove or impede the action of selected parts to assess specific kinds of defect. It is also part and parcel of the new methods that were both shaping and being reflected in medical inquiry in the renaissance.

William Harvey and the triumph of a method

After graduating from Cambridge, William Harvey (1578- 1657) spent five years at the University of Padua working with Fabrizio di Aquapendente (1537-1619), who was Galileo's personal physician. It was at Padua that he came to understand and appreciate the importance of comparative anatomy.

Harvey's use of resoluto-compositive method, along with the use of explicit mechanical metaphors, can be found in the context of work on the motions of the heart—published in 1628 as *De Motu Cordis et Sanguinis* (Of the motions of the Heart and Blood)—work that was made possible in part through extensive experimentation of a wide variety of animal subjects. The problem confronting Harvey was the problem of understanding the complex motions of the heart. This was a daunting task for as Harvey himself notes, "I found the task so truly arduous, so full of difficulties, that I was almost tempted to think with Fracastorius, that the motion of the heart was only to be comprehended by God [(Clendening 1960) p155]."

The problem was generated by the speed with which the heart's motions occur, especially in mammals, whose hearts had been exposed to public view without benefit of anesthesia, and who consequently were in great physical distress:

> For I could neither rightly perceive at first when the systole and when the diastole took place, nor when and where dilatation and contraction occurred, by reason of the rapidity of the motion, which in many animals is accomplished in the twinkling of an eye, coming and going like a flash of lightening . . . [(Clendening 1960) p155]

The problem can be solved through an application of the resoluto-compositive method if there is some way to analyze these fast, complex motions into component motions, and to understand the causes of the component motions.

This could be achieved through experimentation on appropriate animal subjects—subjects in whom the component motions would be visible:

> These things are more obvious in the colder animals, such as toads, frogs, serpents, small fishes. . . They also become more distinct in warm-blooded animals, such as the dog and the hog, if they be attentively noted when the heart begins to flag, to move more slowly,

44

and, as it were, to die: the movements then become slower and rarer, the pauses longer, by which it is made much more easy to perceive and unravel what the motions really are, and how they are performed. [(Clendening 1960) p156]

So with the help of appropriate subjects—from the standpoint of species and physiological condition—the component motions of the heart can be resolved. Here is an early expression of the idea that for any problem of interest there is usually some animal upon which it can be conveniently studied. The convenience here is methodological in nature. It involves the search for animal subjects in which phenomena believed to occur in one subject can be seen very clearly to occur in another type of animal subject. The methodology involves the formation of analogical hypotheses: normal mammalian hearts (whose motions are fast and indistinct) operate on *similar principles* to those actually observed in frogs (for example) and in mammals whose hearts are flagging. The conclusion about the behavior of the normal mammalian heart was not, for Harvey, simply a matter of observation unsullied by analogical hypotheses

Harvey analyzes the complex motion of the heart into component motions associated with structures discernible in the heart (*ventricles* and *auricles*, the latter being the old word for *atria*). Harvey is then able to synthesize his understanding of the properties of the parts into an understanding of the complex motion of the whole system:

> These two motions, one of the ventricles, another of the auricles, take place consecutively, but in such a manner that there is a kind of harmony or rhythm preserved between them, the two concurring in such a wise that but one motion is apparent. . . *Nor is this for any other reason than it is in a piece of machinery, in which, though one wheel gives motion to another, yet all the wheels seem to move simultaneously*; or in that mechanical contrivance which is adapted to firearms, where the trigger being touched, down comes the flint, strikes against the steel, elicits a spark, which falling among the powder, it is ignited, upon which the flame extends, enters the barrel, causes the explosion, propels the ball, and the mark is attained—all of which incidents, by reason of the celerity with which they happen, seem to take place in the twinkling of an eye. [(Clendening 1960) p161] (Emphasis added.)

In this passage, we see how the fruits of animal experimentation, united with the resoluto-compositive method and the explicit use of mechanical metaphors could yield natural resolutions of problems that had hitherto been viewed as mysteries beyond the reach of human ken.

Thinking of the operation of the heart in mechanical terms—and hence as a system admitting of a rudimentary quantitative mathematical description—yielded further fruits. Even granting a large margin of error, Harvey estimated that in an hour the heart could pump more blood by weight than its human owner. Where was all this blood coming from, and where did it go? Harvey had a radical solution. There is a mystery:

> . . . unless the blood should somehow find its way from the arteries into the veins, and so return to the right side of the heart; I began to think whether there might not be A MOTION, AS IT WERE, IN A CIRCLE. Now this I afterwards found to be true; and I finally saw that the blood, forced by the action of the left ventricle into the arteries, was distributed to the body at large, and its several parts, in the same manner as it is sent through the lungs, impelled by the right ventricle into the pulmonary artery, and it then passed through the veins and along the vena cava, and so round to the left ventricle in the manner already indicated. [(Clendening 1960) p164]

Harvey thereby united his own research on the structure and function of the heart, with earlier work on pulmonary circulation, to conceptualize the conjoined system of heart and blood vessels as a closed, mechanical circulatory system.

Hooke and the coupling of natural and artificial machines

In the course of the development of the biomedical sciences in 16th and 17th centuries, humans and animals gradually came to be seen as machines. The mechanical metaphors lost their original heuristic purpose and came to take on a more literal significance. The implications of this for controversies concerning the relative places of humans and nonhuman animals in nature will be examined in the next chapter. However, here we wish to point out that the evolution of a thoroughly mechanistic view of organisms was accompanied by a view that if organisms were *natural* machines resulting from providential design, they could nevertheless be studied with the aid of machines—artifacts—of our own design. This is a fragment of a more general phenomenon in the 17th century, where the emerging conception of *Nature-as-machine* was accompanied by the view that it could be understood with the aid of machines (such as telescopes, microscopes, barometers, vacuum pumps, and so on). In biomedical investigations, this had some interesting consequences for the direction of research.

First, investigators began designing machines to gain quantitative data about medical phenomena. An early example is provided by the work of Santorio Santorii (1561-1636), a professor of medicine at Padua, and a colleague of Galileo. He is credited with the invention of a number of devices such as the pulsilogium, which was a device for measuring pulse rate, as well as a clinical thermometer. In addition, ". . . he described an experiment which laid the foundation for the modern study of metabolism. He spent days on an enormous balance, weighing food and excrement, and estimated that the body lost weight through invisible perspiration (Crombie 1959)."

Nonhuman animals could also be mechanically coupled to machines to resolve more radical physiological puzzles. In this regard there were some puzzles about the function of the motions of the lungs. With the aid of an air pump, Leonardo da Vinci had shown that air did not travel from the lungs to the heart. After the work of Servetus and Harvey, the details of pulmonary circulation and its coupling to the general circulatory system had become much clearer.

Yet there were still puzzles concerning the function of the motion of the lungs. Some physicians maintained that the motions of the lungs served to promote the circulation of the blood. Writing after Harvey, Robert Hooke (1635-1703) performed an experiment in 1667 to settle this matter—an experiment which involved the coupling of an *animal machine* to machines of human design. The report of the experiment was originally published in the *Philosophical Transactions of the Royal Society of London* under the title *Preserving Animals Alive by Blowing through their Lungs with Bellows*. Hooke's experiment was performed before members of the Royal Society. He describes his experimental set-up as follows:

> . . . I caus'd at the last meeting the same experiment to be shewn in the presence of this *Noble Company*, and that with the same success, as it had been made by me at first; the dog being kept alive by the Reciprocal blowing up of his lungs with *Bellowes*, and they suffered to subside, for the space of an hour or more, after his *Thorax* had been so display'd, and his *Aspera arteria* cut off just below the *Epiglotis* and bound on upon the nose of the Bellows. [(Zucker 1995) p22]

In other words, the experimental subject, the dog, has had his windpipe cut and then tied to the nose of a pair of bellows and has been kept alive. It is clear from other remarks Hooke makes that the dog's lungs were exposed in the course of the experiment, so that they could be observed and manipulated. This experiment gives us an opportunity to examine an early example of the use of an artificial, mechanical respirator.

The purpose of this part of the experiment is to give the investigator control over the motions of the lungs through the use of the bellows. Thus, Hooke comments:

> The Dog having been kept alive. . . for above an hour, in which time the Tryal hath often been repeated, in suffering the dog to fall into *Convulsive* motions by ceasing to blow the Bellows, and permitting the Lungs to subside and lye still, and of suddenly reviving him again by renewing the blast, and consequently the motion of the Lungs. [(Zucker 1995) p22]

This part of the experiment shows at least three things. First, it shows the dog can be kept alive with the aid of the respirator (there had apparently been some earlier failures to replicate Hooke's first results with this technique). Secondly, it shows that there is a connection between vital function and the motion of the lungs. Thirdly, it sets us up for the crucial question to be investigated: is the immediate cause of vital function the motion of the lungs in and of itself, or is it something else (something that might be *supported by* the motion of the lungs in a natural setting)?

To investigate this question, Hooke modifies the structure of his respirator in an ingenious manner:

> I caused another pair of Bellows to be immediately joyn'd to the first, by a contrivance, I had prepar'd, and pricking all the outercoat of the Lungs with the slender point of a very sharp pen-knive, the second pair of Bellows was mov'd very quick, whereby the first pair was always kept full and was always blowing into the Lungs; by which means the Lungs also were always kept very full, and without any motion, there being a continual blast of Air forc'd into the Lungs. [(Zucker 1995) p22]

Here the experimental modification of both the respirator and the lungs of the animal subject permits the introduction of continuous airflow through the lungs.

This in turn allows the observation of airflow with no motion of the lungs, as well as the cessation of airflow, again with no motion of the lungs. It permits the investigator to discern whether the motion of the lungs is the immediate cause of vital function, or whether it is something else—the airflow—which is the immediate cause of vital function (with the motion of the lungs supporting such airflow in the natural context of an unmodified experimental subject). Hooke reported his results as follows:

> This being continued for a pretty while, the dog, as I expected, lay still, as before, his eyes being all the time very quick, and his heart beating very regularly. But upon ceasing the blast, and suffering the Lungs to fall and lye still, the Dog would immediately fall into Dying convulsive fits; but be as soon reviv'd again by the renewing the fullness of his Lungs with the constant blast of fresh Air. [(Zucker 1995) p22]

In this way Hooke was able to ascertain that the immediate cause of vital function was the airflow, not the motion of the lungs.

Hooke was able to cross-check his result by some further observations of his experimental subject:

> Towards the latter end of the Experiment a piece of the Lungs was cut quite off; where 'twas observable, that the blood did freely circulate, and pass thorow the Lungs, not only when the Lungs were kept thus constantly extended, but also when they were suffered to subside and lye still. [(Zucker 1995) p22]

In this way, Hooke was able to establish that the motion of the lungs did not support circulation—that occurs whether the lungs move or are stationary. What matters for vital function is a flow of fresh air. Normally this is achieved through the motion of the lungs, but it can be achieved without such motion. What it was about fresh air that supported life was unknown at this time—since it would be more than a hundred years before oxygen was isolated, let alone named.

From physiology to physics

Correlative with the rise of modern science is the dual phenomenon of nature being conceptualized with the aid of mechanical metaphors, and nature being studied with the aid of machines. It was the incredible success of this new way of thinking and this new way of exploring nature that cemented the union between science and technology—a union that owes its existence in no small measure to the work of investigators in anatomy and physiology.

There is a method too. The resoluto-compositive method predates a fully articulated conception of nature as a machine, growing out of medical inquiries at Padua. Nevertheless, the resoluto-compositive method is ideally suited for an understanding of mechanical objects such as clocks, or of objects conceptualized in mechanical terms, for example cats and dogs. Thus, it was not just useful in the study of bodily motions in humans and nonhuman animals, it found applications in the study of motion generally. For example, a variant of the method was used by Galileo to study the motions of falling bodies. In particular he was able to show the complex motion of a projectile describing a parabolic trajectory, can be resolved into the combined effects of two independent motions (Cohen 1985).

More importantly, in the course of the 17th century, nature itself came to be seen as a complex system of interacting bodies in motion that could be understood in mechanical terms. Arguably, the crowning achievement of 17th century physics is to be found in Sir Isaac Newton's (1642-1727) great work, *Principia Mathematica*, published in 1687. The resulting system of physics—Newtonian mechanics—provides a vision of the universe itself as a giant machine whose parts are held together, and whose motions are interrelated, through gravitational interactions.

Newton's work on scientific method has enormous implications for the practice of science in domains that extend far beyond physics. Here we pause to examine aspects of Newtonian science. In the *Principia* [(Thayer 2005) p3-5], Newton provided rules of reasoning to guide the practice of science (or *natural philosophy* as it was then known). The first two rules concern the nature of causation. According to the first rule, simplicity is a virtue: "We are to admit no more causes of natural things than such as are both true and sufficient to explain their appearances." According to the second rule, "Therefore to the same natural effects we must, as far as possible, assign the same causes." Newton goes on to explain that this rule applies, ". . . to respiration in a man and in a beast, the descent of stones in Europe and America, the light of our culinary fire and of the sun, the reflection of light in the earth and in the planets."

The third and fourth rules concern evidence. According to the third rule, "The qualities of bodies . . . which are found to belong to all bodies within the reach of our experiments, are to be esteemed the universal qualities of all bodies whatsoever." In essence, the results of properly conducted experiments have implications for nature beyond the confines of the laboratory. The fourth rule tells us, "In experimental philosophy we are to look upon propositions inferred by general induction from phenomena as accurately or very nearly true, notwithstanding any contrary hypotheses that may be imagined, till such time as other phenomena occur by which they may either be made more accurate or liable to exceptions." Science thus emerges as an attempt to understand the causes and effects of natural phenomena using experiments in a process driven by evidence.

As applied to physics, Newton saw several implications. First there is *causal determinism*. This is an idea rooted in two claims. First, all events have causes; and second, for qualitatively identical systems, the same cause is followed by the same effect. Causal determinism is a presupposition of much scientific activity (notwithstanding indeterministic quantum phenomena). The idea that results in the laboratory can be extended to form expectations about qualitatively similar systems outside the laboratory is embodied in this idea, as is the claim that experiments should be replicable.

In Newton's physics, causal determinism finds expression in the use of deterministic differential equations to describe the behaviors of systems of interest. For example, consider a rocket under conditions of constant acceleration, so acceleration $a = const$. Then (where v = velocity, t = time, x = position and x_o and v_o are initial position and initial velocity respectively) we know that:

[1] $a = dv/dt$,

and

[2] $v = dx/dt$.

In the present case, integration of [1] yields:

[3] $v = v_o + at$,

and integration of [2] yields:

[4] $x = x_o + v_o t + at^2/2$.

Equations [3] and [4] tell us that if the initial position and initial velocity (x_o, v_o) are known, along with the rate of acceleration, then the position of our rocket and its velocity, (x_i, v_i), at any other time can be calculated. Two qualitatively identical rockets that begin with the same positions, velocities and accelerations, and travel for the same amounts of time, will end up in the same positions with the same velocities—something counted on by those who fire missiles with inertial guidance. In this case a deterministic mathematical model enables us to understand the behavior of a certain class of physical system, and to make predictions about the future behavior of such systems. The end-state of our physical system thus depends on two things: laws and initial conditions. Where the initial conditions are the same, the same laws will yield the same predicted end-state. Where initial conditions vary, the same laws will yield a varied range of end-states.

Newton also provided laws in terms of which inertia and acceleration could be understood.

Law 1. Inertial systems remain inertial unless acted on by net forces. (Inertial systems are systems at rest or moving at a constant velocity in a straight line.)

Law 2. Force equals mass times acceleration. (This tells us that forces can be evaluated by observations of how masses accelerate. If two identical masses accelerate in the same way when a force is applied identically to each, then the forces must be the same, if such masses accelerate differently, there must be some difference in the force applied. Importantly, it is also possible to perform quantitative adjustments: two different masses can be made to accelerate in the same way if different forces are applied in such a way as to take into account the difference in mass. Notice the role played by determinism in these remarks).

Law 3. Action and reaction are equal and opposite. (This is needed to exclude the possibility of self-acting forces that might otherwise be consistent with the first two laws).

In accord with Newton's second law, the *universal law of gravitation* states that the force of gravity between masses m_1 and m_2 is given as:

$$[\text{UG}] \; F = ma = G \, m_1 m_2 \, / \, r^2,$$

where G is a constant of proportionality known as the gravitational constant and r is the distance between the centers of mass of m_1 and m_2 respectively. The law of gravity itself thus reflects the resoluto-compositive method, for to understand the complex motions of a system of bodies in a state of mutual gravitational attraction, one must understand the relevant properties of the parts, that is, the masses of the bodies, and how they are related to each other, in particular the (squares of) the distances between them.

The power of Newton's methods may be seen in the following example. In 1846, the French astronomer Leverrier, noticed oddities in the orbit of the planet Uranus—hence the acceleration of this particular planetary mass. Part of this effect arose from the influence of the other planets, but there was a residual effect that was left unexplained. Following Newton's advice that where possible same effects should be assigned similar causes, Leverrier proposed that the orbital oddity could be explained if another planetary mass, hitherto unobserved, was also in orbit around the

Sun. The gravitational effects of this mass would explain the oddities in the orbit of Uranus. A mass now known as Neptune was found within one degree of its predicted position. Newton's laws could predict the existence of things that humans had never seen before. Experiments could confirm these predictions. Furthermore, Newton did not just influence the development of physics and astronomy.

Newton's ideas partly explain the development of chemistry in the 18th century. Chemistry bloomed after physics partly because it belonged to a different intellectual tradition, but partly because there were problems with the availability of pure reagents. The great French chemist Lavoisier was able to make enormous intellectual strides in chemistry in part because he had access to pure substances for his experiments. As we have seen, for qualitatively similar systems, same cause is followed by same effect. If the systems are chemical systems, this requires systems that are similar with respect to substances present. Different effects—hence anomalous results—will occur if two systems differ with respect to initial conditions. In the present case if different contaminants (or different concentrations of contaminants) are present. With pure reagents good recipes could be formulated, and Lavoisier's gunpowder was in much demand during the American Revolution—so much so that he sent his assistant DuPont to establish a gunpowder factory in Delaware!

Newton had a more direct influence on the development of Lavoisier's chemistry in connection with the rejection of *phlogiston theory*. In order to explain the transformation of a combustible substance during the process of combustion, as well as the extraction of metals from ores, the rusting of metals such as iron, a rich body of 18th century chemical theory focused on a hypothetical substance known as phlogiston. Even the young Lavoisier subscribed to a version of phlogiston theory.

According to phlogiston theory, when a combustible substance burned, a substance—phlogiston—was given off or emitted. In a way, phlogiston was the substance of fire. In the 18th century, many explanations of chemical phenomena were formulated in terms of phlogiston and its properties. A candle burning in a sealed jar eventually goes out. According to theory, the candle emitted phlogiston as it burned. The air in the jar absorbed phlogiston until it was saturated and could absorb no more. At that point, the candle went out. Air that could absorb phlogiston was known as dephlogisticated air. Joseph Priestley prepared particularly pure samples of dephlogisticated air—air that seemed to have a good capacity to absorb phlogiston and support combustion. We know this "air" today as oxygen.

However, there were problems that Lavoisier, with early training in accountancy and with the aid of Newtonian intuitions, was able to discern. When some things burned (today magnesium can be used for effective demonstrations), the ashes could be measured to weigh more than the original substance. How could something burn, giving off phlogiston in the process, and end up weighing more? According to phlogiston theorists, phlogiston had negative weight, so substances gained weight as they surrendered their phlogiston to the air. But phlogiston's advocates claimed phlogiston was a form of matter. To the Newtonian Lavoisier, matter had mass, and hence positive weight due to gravity. The explanation lay elsewhere. Careful experiments indicated that far from combustion involving the emission of a substance, it was a process whereby the combustible substance united with a component of atmospheric air. A substance Lavoisier called oxygen, with the process of combustion being revealed as an oxidation process that resulted in the formation of an oxide. The oxide weighs more than the unburned substance because oxygen has mass and hence, due to gravity, weight.

Newtonian ideas also had a profound influence on the biomedical sciences—this becomes clear when we examine 19th century physiology, the subject of the next chapter. These ideas will be seen to have enormous implications for the debates about animal modeling that emerge in the late 20th century.

CHAPTER 4

ANIMALS AND THE METHODS OF 19TH CENTURY PHYSIOLOGY

Every real thought on every real subject knocks
the wind out of somebody or other.
—Oliver Wendell Holmes

There are four revolutions in 19th century biology: the theory of evolution, genetics, cell biology and the germ theory of infectious disease. While all these branches of biological inquiry were of immense scientific importance, the theory of evolution represented a break with earlier ways of thinking about science, and will thus need to be treated in a separate chapter. This is especially true since the deep implications of evolutionary biology for medicine were undeveloped for most of the 20th century. Serious interest in evolutionary or Darwinian medicine is really a phenomenon of the last two decades of the 20th century (Shanks and Pyles 2007). Classical genetics, owing its origins to the work of Gregor Mendel (1822-1884), had to wait until the dawn of the 20th century to have any significant biological and biomedical impact, and will be discussed later in connection with evolutionary biology.

Cell biology, by contrast enabled investigators to follow the resoluto-compositive method one step further than it had gone in the renaissance by analyzing tissues into their component cells and cell types. Cell biology really had to wait until the 19th century, for though Hooke had coined the term *cell* in his *Micrographia* (1665), it was the refinement of optical technologies in the 19th century that permitted the construction of microscopes with sufficient resolving power to discern the different cell types, and indeed intracellular structures, found in plants and animals.

By 1858, with the publication of Rudolf Virchow's *Cellular Pathology*, the cell had been recognized as a fundamental unit in medical inquiries. States of health and disease could be analyzed from a cellular point of view, as could the phenomena of life itself. Thus, Carl von Voit could observe in 1881:

> The unknown causes of metabolism are found in the cells of the organism. The mass of these cells and their power to decompose materials determine the metabolism. . . In speaking of the power of cells to metabolize, I have not meant thereby. . . that the cells must always use energy in order to metabolize, but rather I have understood thereby the sum of the unknown causes of metabolic ability of the cells—as one speaks of the fermentative "power" of yeast cells.
> [(Clendening 1960) p598-99]

We will shortly observe that the study of the fermentative "power" of yeast cells had enormous implications for the study of medicine.

For the present, the advent of cell biology permitted a continuation of biological inquiries into the nature of life as part of that older physiological tradition extending back to the Renaissance—a tradition initially rooted in mechanical metaphors that had come to be significantly bolstered by the mechanical picture of nature that had triumphed in the form of Newtonian or classical physics. As we saw at the end of the last chapter, the Newtonian vision for science had, by the dawn of the 19th century, come to have implications for the conduct of science far beyond the confines of physics. Interest in life at the microscopic level also led to the germ theory of disease which grew out of the work of Louis Pasteur in the 1850s and 1860s, and in particular out of his work on fermentation processes.

Animals and the germ theory of disease

Pasteur's work on fermentation processes led to the important concept of *specificity*. Specific ferments—such as the lactic ferment involved in the souring of milk, the butyric ferment involved in the rotting of broth, the alcoholic ferment involved in the production of ethanol—involved the actions of specific microorganisms that could be recognized, like the cell types observed in multicellular creatures, on the basis of morphological features

Investigators such as Pasteur and Robert Koch were able to extend these ideas about microorganismal specificity to the study of the microscopic causes of infectious disease. The microbial causes of disease could be simultaneously categorized on the basis of morphology and on the basis of specificity for organs, tissues and cell types in the multicellular creatures they attacked. An understanding of the specific chemical products of these microorganisms would gradually develop as biochemistry emerged from the womb of 19th century developments in organic chemistry. Research such as Buchner's discovery in 1897 that ethanol could be catalyzed by "juice of yeast," and didn't require the presence of a living organism, was important in this regard.

The germ theory of disease represents one of the places where non-human animals came to play an important role in medicine. To see what happened, we must see how the idea of specificity found its way into Koch's postulates—a set of methodological prescriptions that continue to influence investigations into the pathology of infectious disease to this day.

Koch's postulates guide the study of the bacterial causes of disease as follows: [1] the hypothetical germ causing a disease must be discoverable in all cases of the disease, and found in the body wherever the disease lies; [2] extracted from the body, the germ must grow in a pure culture, for several bacterial generations; and [3] this culture must give the disease to a susceptible animal, be recoverable from it in a pure culture, and transmit the disease to another such animal.

As Koch himself observed in his study of the causes of tuberculosis, correlations are easy to establish, but causation requires rigorous tests:

> On the basis of my numerous observations I consider it established that, in all tuberculous affections of man and animals, there occur constantly those bacilli that I have designated tubercle bacilli and which are distinguishable from all other microorganisms by characteristic properties. However, from the mere coincidental relation of tuberculous affections and bacilli it may not be concluded that these two phenomena have a causal relation, notwithstanding the not inconsiderable degree of likelihood for this assumption . . .
> [(Clendening 1960) p399]

Koch thus continued:

> To prove that tuberculosis is a parasitic disease, that it is caused by the invasion of bacilli and that it is conditioned primarily by the growth and multiplication of the bacilli, it was necessary to isolate the bacilli from the body; to grow them in pure culture until they were freed from any disease-product of the animal organism which might adhere to them; and, by administering the isolated bacilli to animals, to reproduce the same morbid condition, which, as known, is obtained by inoculation with spontaneously developed tuberculous material . . . [(Clendening 1960) p399]

These methods provide the scientific tools, which underlay the so-called "golden age" of bacteriology—the period from the 1870s to the

First World War. While most animals used in the context of contemporary biomedical research are *not* used in the context of studies involving Koch's postulates, there are some lessons here concerning analogical reasoning the study of which is instructive. The following case is drawn from virology rather than bacteriology.

A case in point concerns investigations into the causes of polio. This is a disease with a viral cause, that as late as the 1950s, hospitalized thousands of people in the United States each year. There is no doubt that the introduction of vaccination programs in the late 1950s have had enormous benefits for public health and well-being, and there can be no doubt that animal-based research played a role in the provision of these benefits. However, the role played by animals in the war against polio is not a simple one, and there are some important cautionary notes that need to be sounded.

Crucial to the use of Koch's postulates is an idea that we examined in the last chapter in connection with Newton's methodological rules for the conduct of science: wherever possible, to the same effects we should apply the same causes, not just with respect to the falling of stones in Europe and America, but also with respect to respiration in man and beast. In the light of Koch's work, we might add to this list, also with respect to tuberculosis in man and beast! Since many infectious diseases afflicting humans have evolutionary origins in diseases affecting the animals that humans commonly associate with, this is perhaps not too surprising.

Sometimes the key to coping with an infectious disease involves simple public health measures, such as cleaning up the water supply and properly disposing of human waste—such measures were, and remain, highly effective against such human scourges as cholera and typhoid. But sometimes an infectious disease is so intractable—even in the face of such drastic measures as quarantining—that other methods must be sought. In the case of polio, investigators sought to take the first steps to combat polio by reproducing the disease in nonhuman animals.

The injunction to apply the same causes to the same effects "in man and beast" could turn out to be seriously misleading. The *nasal hypothesis,* which dominated much of the early work in the war against polio, maintained that the virus entered the human body through the nose, traveled to the brain via the olfactory nerves, and, once ensconced in the central nervous system, migrated from the brain to the spinal cord where it was capable of producing paralytic lesions. The hypothesis was arrived at by Flexner and Lewis who found that they could induce polio in rhesus monkeys via intranasal exposure. The hypothesis was bolstered by the observation that placing substances such as alum, zinc sulfate and picric acid into the nasal passages of rhesus monkeys blocked absorption of simian poliovirus.

Though nasal sprays were ineffective in humans, and though poliovirus was found in the intestines of humans afflicted with the disease (clinical observations dismissed as the result of back-swallowing nasal secretions) incautious application of the "same effect same cause" rule dominated much thinking about the pathology of polio. It turns out that the susceptible animal—in this case the rhesus monkey—was not susceptible in the same way as humans are susceptible to polio infection. Thus medical historian J. Paul has observed:

> Of great importance was Sabin's discovery by quantitative studies that the central nervous system of lower primates (rhesus and cynomolgus monkeys) was more susceptible to poliovirus than that of higher primates (chimpanzees)—and by epidemiologic analogy, man. The reverse was true for the alimentary tract particularly in rhesus monkeys; whereas the susceptible human intestinal tract was readily infected by doses of virus that were ineffective in monkeys. [(Paul 1971) p451]

As noted by Paul, important steps in the study of the pathology of human polio involved the culturing of the virus in non-neural tissue, especially human embryonic intestinal tissue.

While it is often possible to find animals that are susceptible to human diseases, they may not be susceptible in the same way. Later chapters of this book will explore why the "same effect same cause" principle breaks down so readily for man and beast. Related to this we will also want to know why causes that bring about effects in beasts have no effect (or very different effects) in humans. Some strains of Ebola virus cause fatal hemorrhagic fever in humans, others, such as the Reston strain of Ebola, though lethal in rhesus monkeys, have no obvious detrimental effects in humans. This, however, is to raise but one issue that emerges from developments in 19th century biomedical research and its influence on research practices in the 20th century. There are other 19th century ideas about the roles of animals in biomedical research that have enormous implications for contemporary research practices, and to these we now turn.

Mechanism in 19th century physiology

We have just seen how the pursuit of biomedical inquiries was enriched by studies of cells and the effects on cells of microorganisms. We should not lose sight of the fact that focus on biological events at the microscopic level had enormous implications for physiology as well as pathology.

Fundamental physiological processes came to be seen as taking place in the cellular arena. Especially with respect to the study of such phenomena as respiration "in man and beast," these inquiries in turn led back to physics and chemistry. In the minds of prominent physiologists, the hope was expressed that the proper account of the phenomena of life might be reduced to an account couched in the mechanical language of physics and chemistry. Thus Coleman has observed:

> Life, the modern physiologist argues, depends on the regular and slow release of energy derived from oxidation of ingested foodstuffs. This energy provides the temperature appropriate to chemical reactions occurring within the body, including syntheses, and it underlies bodily motion, the electrical behavior of nerves and the secretory activity of glands. The certainty of this dependency, customarily advanced as a causal connection, derived from the accomplishments of nineteenth century respiratory physiology. [(Coleman 1978) p121-22]

These investigations owed much to Lavoisier's foundational work on the chemistry of oxidation.

Indeed Lavoisier himself had observed that both candles and organisms took in oxygen from the air and replaced it with carbon dioxide, liberating heat in the process. By the 1870s it had become apparent oxygen was conveyed by blood to the cells of the body, where it was involved in the oxidative metabolism of energy rich substances, notably glucose, found within the cellular arena. In this way, studies of vital function converged with the fruits of studies in cellular biology and bacteriology to bring to the cell and its internal structure and dynamics, the central place in the study of life in both health and disease.

While we could focus on the empirical triumphs of 19th century physiology—a subject of great historical importance—the study of the concepts and methodology of 19th century physiological research is arguably of great importance too, for it shaped not only the character of 19th century inquiries, but also the contours of current inquiries in biomedical research. There is no better place to look for an understanding of the methodological issues undergirding 19th century physiology than the works of Claude Bernard. The following remarks about Bernard's work owe much to earlier work by LaFollette and Shanks (LaFollette and Shanks 1994).

There is a puzzle about biological objects. Geneticist Richard Lewontin has put it this way:

> ... biological objects are internally heterogeneous in a way that is functionally relevant. While the earth has some variation in its internal composition, the behavior of the earth as a solar system object is unaffected by that heterogeneity. Organisms, on the other hand, are affected at every level of their functioning by their internal heterogeneity. (They can even change the way they fall "freely" through space as gravitational objects, as every sky diver knows). [(Lewontin 1995) p8]

In the 19th century, when similar observations were subjected to theoretical inquiry, the significance of the behavior of biological objects could easily be misunderstood. The functional effects (seen, for example in the purposiveness of the behavior of biological systems) was sometimes seen as evidence of special, *vital* forces beyond scope of physico-chemical laws of the kind that had shown themselves to be so fruitful in physics and chemistry. This idea, known as *vitalism*, took a long time to die. In many ways, vital forces were to biologists what phlogiston had been to chemists. Claude Bernard was in the vanguard in rejecting vitalism:

> I propose, therefore, to prove that the science of vital phenomena must have the same foundations as the science of the phenomena of inorganic bodies, and that there is no difference in this respect between the principles of biological science and those of physico-chemical science. Indeed ... the goal which the experimental method sets itself is everywhere the same, it consists in connecting natural phenomena with their necessary conditions or with their immediate causes. [(Bernard 1949) p60]

Bernard compares organisms to machines and the vitalist to a superstitious person, who, seeing a machine move seemingly of its own volition, thought it must violate the usual laws of nature [(Bernard 1949) p63].

The problem arises from a failure to distinguish between two arenas of causation: the internal environment (*milieu intérieur*) from the external environment (*milieu extérieur*). Making this distinction clear was one of Bernard's great contributions to biological theory, as was his realization that the proximate causes of physiological phenomena had to be found in the complex, heterogeneous, interactive internal environment.

The causes of the motions of inorganic objects lie in the Newtonian forces impinging upon them from the outside. The motions of organisms, by contrast, reflect the external environment and the way the heterogeneous subsystems of the organism interact *in the internal environment* simultaneously in response to each other and to the influences impinging from outside. Causes external to an organism cause it to come into existence. *Organisms come from organisms.* By Bernard's day spontaneous generation was no longer taken seriously. However, the organism, once it exists, becomes in large measure the cause and effect of its internal environment, albeit in response to stimuli from without. As Bernard himself observed:

> We must therefore seek the true foundation of animal physics and chemistry in the physical-chemical properties of the inner environment. The life of an organism is simply the result of all its innermost workings. All of the vital mechanisms, however varied they may be, have always but one goal, to maintain the uniformity of the conditions of life in the internal environment. [(Cziko 1997) p51]

Organisms thus emerge as homeostatic systems, capable of physiological adaptation to changes in the external environment so that key physiological parameters can be kept within narrow boundaries.

In the language of modern thermodynamics, organisms are examples of *open-dissipative systems*—systems that have exchanges with their surrounding environments, but whose complex dynamics reflects the operation of causal mechanisms and pathways internal to them that are animated and do work as energy flows through them—flows of energy that keep the internal environment of the systems in question away from thermodynamical equilibrium with their surroundings.

Bernard was ahead of his time. It turns out, moreover, that open-dissipative systems are precisely the sorts of system that display the phenomenon of self-organization—a phenomenon at work everywhere in biology, but also in such non-biological, physical systems such as hurricanes, tornadoes, and spiral galaxies; and in non-biological chemical systems such as the famous oscillatory Belousov-Zhabotinski reaction (originally proposed as an *in vitro* model for the Krebs cycle).

Claude Bernard is without doubt one of the great figures in the history of science. Bernard made many contributions to the biomedical sciences in his own right, but for our purposes, his *Introduction to the Study of Experimental Medicine* (Bernard 1973) is an extremely important text since it lays out the philosophical and methodological rationale for the newly emerging biomedical sciences. More than a century after his death, his methods still guide experimental practices in the biomedical sciences. The American Medical Association, in its 1992 *White Paper* on animal experimentation, praised Bernard for establishing the basic principles that guide the practice of experimental medicine. No lesser figure than the late immunologist and Nobel laureate, Sir Peter Medawar commented that, "The wisest judgments on scientific method ever made by a working scientist were indeed those of a great biologist, Claude Bernard [(Medawar 1984) p73]."

Bernard denigrated clinical medicine—and at the time he was writing, it was still in a very backward state. He believed that genuine biomedical science involved carefully controlled experiments on animals, and that such experiments were of direct and predictive relevance to human biology. Humans and other mammals, though differing in size, seemed to exhibit a great deal of similarity with respect to gross anatomical internal heterogeneity (presence of livers, kidneys, brains, and so on).

In fact, Bernard wanted to do for medicine what other scientists had done for physics and chemistry:

> We cannot imagine a physicist or a chemist without his laboratory. But as for the physician, we are not yet in the habit of believing that he needs a laboratory; we think that hospitals and books should suffice. This is a mistake; clinical information no more suffices for physicians than knowledge of minerals suffices for chemists and physicists. [(Bernard 1949) p148]

Bernard did not deny that clinical observations had a place in the practice of medicine, but he did believe that science took place, not in the clinical context, but in the laboratory.

To understand the wedge that Bernard drives between clinical medicine and scientific medicine, we must examine his views about the methodology of science. The business of science is in the formulation and rigorous testing of hypotheses about phenomena of interest:

> The experimental hypothesis . . . must always be based on observation. Another essential of any hypothesis is that it must be as probable as may be and must be experimentally verifiable. Indeed, if we made an hypothesis which experiment could not verify, in that very act we should leave the experimental method and fall into the errors of the scholastics and makers of systems. [(Bernard 1949) p33]

Clinical medicine could provide the observations that prompted the formulation of hypotheses, but it could not, in the nature of the case, be the context where the hypotheses were tested.

The clinical setting does not permit the adequate control of experimental variables, and typically it is not a setting in which, for ethical reasons, those variables can be experimentally manipulated. Or as Bernard put it, "In a word, I consider hospitals only as the entrance to scientific medicine; they are the first field of observation which a physician enters; but the true sanctuary of medical science is a laboratory . . . . In leaving the hospital, a physician must . . . go into his laboratory [(Bernard 1949) p146-7]." If the biomedical researcher is to conduct controlled experiments, carefully manipulating variables of interest, there must be appropriate subjects. Experiments on humans would be immoral. So Bernard becomes one of the most influential advocates (and practitioners) of animal experimentation.

But the focus on animals is not just motivated by a moral worry about human experimentation. Nonhuman animals are also scientifically appropriate objects of study:

> Experiments on animals, with deleterious substances or in harmful circumstances, are very useful and *entirely conclusive* for the toxicology and hygiene of man. Investigations of medicinal or of toxic substances are also wholly applicable to man from the therapeutic point of view; for as I have shown, the effects of these substances are the same on man as on animals, save for differences in degree. [(Bernard 1949) p125] (Emphasis added.)

To understand Bernard's views on the applicability of animal-based research, we must examine his commitment to *causal determinism*, an idea we have examined in connection with Newtonian physics.

Causal determinism is a doctrine underlying a lot of experimental science, both within and outside of biology. Until the advent of indeterministic quantum mechanics, it lay at the heart of physics (and still does for many physical phenomena of interest). As we saw in the last chapter, it is an idea based on two principles: *the principle of causality* according to which all events have causes, and the *principle of uniformity* according to which, for qualitatively identical systems (systems identical with respect to scientifically relevant properties), same cause is followed by same effect.

The idea can be illustrated by a consideration of Newton's second law, according to which *force equals mass times acceleration*. If you have two masses that accelerate in exactly the same way when the same force is identically applied, then the masses must be equal, if they accelerate differently under these conditions, then the masses must be different. Correspondingly, if the masses are known to be equal (these are the *initial conditions*), but accelerate differently when a force is applied in the same way, there must be a difference in the magnitude of the force applied. In accord with causal determinism, accelerations are caused by forces, and for identical masses (in this case, the *mass* of the object is scientifically relevant, its *color*, for example, is not), same *cause* (force applied) is followed by same *effect* (acceleration of mass).

For Bernard, the application of this type of causal reasoning to the biomedical context was direct:

> If a phenomenon appears just once in a certain aspect, we are justified in holding that, in the same conditions, it must always appear in the same way. If, then, it differs in behavior, the conditions must be different. But indeterminism knows no laws; laws exist only in experimental determinism, and without laws there can be no science. [(Bernard 1949) p139]

So, if seemingly identical systems behave differently, there must be a difference in initial conditions to account for the observed difference. A mature science should be able to account for such differences by analogy with the way we dealt with the differences in masses above. For Bernard, experimental medicine should yield deterministic laws akin to those of Newton.

Though living systems look different, and though their masses may vary, they obey the same physiological laws. The heterogeneous internal environments of animals are essentially the same:

> Physiologists. . . deal with just one thing, the properties of living matter and the mechanism of life, in whatever form it shows itself. For them genus, species and class no longer exist. There are only living beings; and if they choose one of them for study, that is usually for convenience in experimentation. [(Bernard 1949) p111]

For Bernard, differences between species were not physiologically relevant—all species obey the same laws:

> In living bodies, as in inorganic bodies, laws are immutable, and the phenomena governed by these laws are bound to the conditions on which they exist, by a necessary and absolute determinism. . . De-

> terminism in the conditions of vital phenomena should be one of
> the axioms of experimental physicians. [(Bernard 1949) p69]

Of course, even if different species obey the same laws, we must be certain that there are no relevant differences between them that will undermine an extrapolation of results found in members of one species, for members of the other species.

Here we will discuss how Bernard conceptualized species differences, for his views have been enormously influential. The issue has become the source of much controversy, especially in the light of modern evolutionary biology—a theory whose ramifications for medicine were largely unknown in Bernard's day, but which is now starting to revolutionize not only the way in which investigators think about human health and disease, but also the ways in which they think of the similarities and differences with respect to the heterogeneous internal environments of humans and the animals of other species used to model them.

Bernard knew that physiologists must confront problems different from those encountered by physicists and chemists when they investigate inorganic systems. The study of a living system requires the study not simply of the external factors and forces acting on the system, but also of the *inner organic environment*—the organism viewed as a complex, dynamical system with many interacting parts:

> ... *a created organism is a machine* which necessarily works by virtue of
> the physico-chemical properties of its constituent elements. Today
> we differentiate three kinds of properties exhibited in the phenomena of living beings: physical properties, chemical properties and
> vital properties. But the term "vital properties" is itself only provisional; because we call properties vital which we have not yet been
> able to reduce to physico-chemical terms; but in that we shall
> doubtless succeed someday. [(Bernard 1949) p93] (Emphasis added.)

Once the inner environment is understood in appropriate physico-chemical terms, it will be possible to describe its behavior using physico-chemical laws—laws that are the same for all species. These laws of the inner environment will be the universal laws of the biological science.

Species differences—which Bernard was well aware of—were not impediments to biomedical research. Nevertheless, Bernard was aware that some investigators did think they were significant:

> Even today, many people choose dogs for experiments, not only because it is easier to procure this animal, but also because they think experiments performed on dogs can be more properly applied to man than those performed on frogs. [(Bernard 1949) p123]

For Bernard, this is a mistake. Bernard thinks the mistake is generated because some experimenters mistake quantitative differences in initial conditions for fundamental qualitative differences between species. Bernard disagreed; he thought the fundamental properties of *vital units* (biological units) were the same for all species. Livers may come in different sizes and shapes, but they all respond to stimuli in basically the same way. In so far as there are differences, these seem to consist in slightly different arrangements of essentially similar building blocks:

> Now the vital units, being of like nature in all living beings, are subject to the same organic laws. They develop, live, become diseased and die under influences necessarily of like nature, though manifested by infinitely varying mechanisms. A poison or a morbid condition, acting on a definite histological unit, should attack it in like circumstances in all animals furnished with it; otherwise these units would cease to be of like nature; and if we went on considering as of like nature units reacting in different or opposite ways under the influence of normal or pathological vital reagents, we should not only deny science in general, but also bring into zoology confusion and darkness. [(Bernard 1949) p124]

Of course there are idiosyncrasies and differences in the way members of different species behave when experimentally manipulated. Bernard thought these should be studied and brought under universal physiological laws.

Here is an analogy to explain his view. Recall from the last chapter that in the 19ᵗʰ century, idiosyncrasies had been observed in the orbit of the planet Uranus (then believed to be the outermost planet). Did this mean that Newton's law of gravity was wrong, hence that a large mass was not obeying the cosmic rules? No. In 1846, Leverrier realized that these orbital oddities could be explained, using Newton's laws, if Uranus was being acted on by the gravitational field of a hitherto unobserved planet. Calculations were done, and observations were performed, and these led to the discovery of the planet Neptune. Idiosyncrasies do not necessarily show the laws are wrong, but they do signal the need for more science! If species behaved differently when similarly stimulated, what we need to do is to seek the physiological "Neptunes"—the hitherto *hidden variables*—which explain the oddities in a law-like manner.

In Bernard's approach to biomedical science, species differences are quantitative, and not qualitative, in nature. That is to say, once you make allowances for (quantitative) differences in weight of the animals, or in doses administered, same cause will be followed by same effect in humans and the test species. Bernard illustrates this idea in his discussion of toad venom. Doses of venom that stop the hearts of frogs, do not stop the hearts of toads. Does this mean we have same cause followed by different effect? Bernard reasons as follows:

> Now, in logic, we should necessarily have to admit that the muscular fibers of a toad's heart have a different nature from those of a frog's heart, since the poison which acts on the former does not act on the latter. That was impossible: for admitting that organic units identical in structure and in physiological characteristics are no longer identical in the presence of a toxic action identically the same would prove that phenomena have no necessary causation; and thus science would be denied. [(Bernard 1949) p180]

So how do we accommodate both science and the puzzling observations? Bernard continues:

> Pursuant to these ideas, I rejected the above mentioned fact as irrational, and decided to repeat the experiments. . . I then saw that toad's venom easily kills frogs with a dose that is wholly insufficient for a toad, but that the latter is nevertheless poisoned if we increase the dose enough. So that the difference described was reduced to a question of quantity and did not have the contradictory meaning that might be ascribed to it. [(Bernard 1949) p180]

Bernard here lays down one of the basic principles of the science of toxicology: once purely quantitative differences have been allowed for (differences in metabolic rate, body weight, surface area, etc), same cause will be followed by same effect in members of a given species, or in members of different species.

Bernard experimented on many species—domestic animals such as dogs, cats, rabbits and pigs get explicit mention, but, like Harvey before him, he experimented on frogs. As John Parascandola has noted:

> Before the twentieth century, the frog played a major role in the history of research in the life sciences. The widespread use of frogs

was due in part to their general hardiness for surviving severe operations and the excellent survival capacity for their isolated tissues. Frogs were experimented on so frequently that in 1845 Hermann Helmholtz referred to them as "the old martyrs of science." His fellow physiologist, Claude Bernard, called the frog "the Job of Physiology." (Parascandola 1995)

In the 20th century, rodents (rats and mice in particular) have become the new *Jobs of physiology*, amounting to somewhere between 80 to 90 percent of the millions of animal subjects used in biomedical research in the US each year.

Bernard does not just think of organisms as matter in motion. He does not think physiology can be reduced to physics. He does think that life can be understood in physico-chemical terms. He is an intellectual heir to the grand tradition of medicine that emerged from the Renaissance:

> It is doubtless correct to say that the constituent parts of an organism are physiologically inseparable from one another, and that they all contribute to a common vital result; but we may not conclude from this that the living machine must not be analyzed as we analyze a crude machine whose parts also have their role to play in a whole. [(Bernard 1949) p89]

Though there is more to life than mere matter in motion, Bernard was aware that to understand the whole, you had to understand the properties of the parts, and the mutual relationships between them. In this sense he is a true heir to the resoluto-compositive method.

The fundamental basis of Bernard's methodology is the tightly controlled laboratory experiment. The experiments themselves involved extensive vivisection.

> [As for vivisection on animals] No hesitation is possible; the science of life can be established only through experiment, and we can save living beings from death only after sacrificing others. Experiments must be made on either man or on animals. . . the results obtained on animals may all be conclusive for man when we know how to experiment properly. [(Bernard 1949) p102]

As Bernard had pointed out, controlled experiments on humans are immoral, so animal experimentation is inevitable if medical science is to advance. The interspecific tradeoff is clear, if we are to save human lives, we must sacrifice animals.

Since we are concerned with the science behind animal modeling we must now turn to examine the issue of modeling more closely. This will be the first step in an assessment of the viability of this interspecific tradeoff.

PART II

♦

MODELS AND SCIENCE

CHAPTER 5

◆

INTRODUCTION TO MODELS IN SCIENCE

Seek simplicity but distrust it.
　　　　　　　　　　　　　—Alfred North Whitehead

The main concern of this book is the issue of predictive animal modeling in the biomedical sciences. However, concerning models in science, zoologist Robert Hinde has observed:

> However, the value of models in science lies in part in their difference from the originals. Three issues must be considered here: (i) Models are useful because, by virtue of their availability or simplicity, they pose questions, suggest relations, or can be manipulated in ways not possible with the original. (ii) If a model is very like the original, it is easy for an investigator to assume that all the properties of the model exist also in the original, and to confuse the two in arguments in which the model is employed. (iii) The third issue is of critical importance where animals are confined or experimented upon. The more closely related the experimental animal to humans, the more the infliction of suffering requires justification in terms of potential benefit to animals or humans. For these reasons, the most useful animal models are not always the most similar to humans. (Hinde 1987)

We have had to postpone an examination of these issues until now because it was important to see some of the roles played by animals in scientific research in a historical context. Modern uses of animal models are similar in some respects, but different in others to these traditional uses—they are activities that, to pursue an evolutionary metaphor, descend with modification from historical uses of animals as research subjects. Secondly, it was important to introduce some biological theory. Scientific activity—from the experiments deemed worthy of performing to the analysis and interpretation of the data generated by those experiments—is a matter richly informed by the theories that the experimenter believes to be true. However, even with all this under our belts, we are not quite ready to examine animal modeling in biomedical research, for the very issue of modeling in science, and the business of modeling complex systems in particular, must be discussed before the issue of animal models can be profitably examined.

At a fairly abstract level, we can view scientific models as devices by means of which a scientist may come to grips with some puzzling aspect of the world. Even at this abstract level of description, the connection between modeling and problem solving is already apparent.

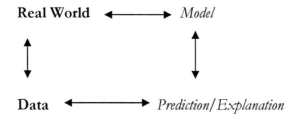

In this context, a model is a device that is intended to capture some puzzling aspect of the world. The model can be used to make predictions (i.e., enable the investigator to form expectations about the characteristics of future ranges of observation) and/or explanations (i.e., it can offer an account of ranges of observation in the past). Our observations of the world (in the context of experiments, field studies, clinical observation, and so on) generate data. Ideally the data agree or disagree with our predictions, and on that basis we can decide whether the model we have produced is a model of the real world. While this picture of scientific activity no doubt involves gross simplifications, it provides us with a starting point—a first approximation.

Consider the following example: suppose our puzzle to be solved is the gravitational attraction between two mice ten meters apart in free space. We might model the situation by idealizing our mice. In the model they will be treated as point-masses (masses located at points in space) m_1 and m_2 respectively. Our model will also incorporate Newton's laws and hence the law of gravity, according to which $F = G\, m_1 m_2\, /\, r^2$. Using the law and the *idealization* of the mice as points initially ten meters apart, we can predict the way in which the mice should accelerate toward each other. Observations could be performed to test the prediction. Notice that the model does not deal directly with mice, but with idealizations of mice in which some features—the masses, for example—are given prominence and others—the colors of the mice, for example—are ignored. The idealization of the mice *as if* they are points is a theoretical decision reflecting an estimate in the model of what is relevant to the real situation and what is not. Confirmation of the model prediction does not mean that mice *really are* point masses! Examples like this are familiar from textbook science. Idealization of the mice, as it occurs in this context, simply means avoiding needless complications. However, not all complications are needless. As we will see below, mice are internally heterogeneous and complex in ways that are relevant when biomedical phenomena are being studied, as opposed to gravitational phenomena. We can begin to get a richer view of modeling by examining the experiences of engineers.

Models in Engineering

Engineers use two very different kinds of model: *mathematical models* and *physical models*. A special type of physical model is known as a *scale model*. In mathematical models physical systems of interest are first given a theoretical description in terms of variables in equations (again, decisions are made as to which properties of the system under analysis are relevant to the behavior under study, and which are not). With the variables so interpreted, the equations can be manipulated to form expectations about the real system under analysis.

Engineers, like biologists, often have to model complex, interconnected systems—say a system of interconnected pipes to distribute water. In this context Petroski has observed:

> In a given piping loop in an ideal network, for example, we know the mass of water combined or divided at a junction where the ends of several pipes come together (known as a "node") is conserved, that is, the net inflow is equal to the net outflow. A continuity equation can thus be written for that node, and if a piping loop has a certain number of nodes, then that same number of continuity equations can be written for the loop. . . Additional equations can be written to express the fact that energy is balanced around a piping loop, taking into account the potential energy associated with the heads of reservoirs, losses through pipes due to friction and change in elevation, and the addition of energy, as through a pump in the loop. Friction losses can be expressed analytically as functions of the pipe characteristics of length, diameter and roughness, as well as the rate of flow through the pipe. . . These kinds of considerations lead to a system of algebraic equations equal in number to the number of pipes in the loop. Collectively, these equations are said

> to be a (mathematical) model for the piping loop, and a solution to
> the equations constitutes a prediction of how water will flow in a
> real system. [(Petroski 1996) p152]

As Petroski observes, the frictional losses in this system are nonlinear functions of the flow rate—so doubling the flow rate more than doubles the energy loss in the pipe. Because of this nonlinearity, it is typical to use approximation methods to solve the equations. Computers have been of enormous value in this regard, since they can manipulate large systems of equations—and engineers have come to rely on the resulting *computer models*.

The above example involving models of fluid dynamics has an interesting historical twist. This may be illustrated with an example from the history of economics. A. W. Phillips, famous for the *Phillips curve* in Keynesian economics (which showed a negative relationship between the rate of inflation and the rate of unemployment), also devised the *Phillips machine* (see figure 5.1). In the machine, colored water was pumped through a system of transparent pipes and tanks in such a way that the flows of water represented flows of income in the economy. The circular flow of income was treated as a literal flow of water. Backhouse [(Backhouse 2002) p293] refers to this as *hydraulic Keynesianism*. This is a nice example of the way in which the dynamical properties of one system, based on a given range of substrates, may simulate the dynamical behaviors of another system, based on very different substrates. We call such models *functional models*. They work because of dynamical similarities between the model and the system modeled.

Returning to the matters at hand, we will see later in this chapter, simple nonlinear equations (whose dependent variables are transcendental functions or variables raised to powers other than one) can have solutions describing extremely complex situations. Complex phenomena do not necessarily require the use of complex systems of equations for their description. But this is to look ahead. For the present, we note that mathematical models are fine so long as one remembers that in their construction, decisions have been made as to which features of the real system are relevant (and hence become the values of variables in model equations), and which are not. Errors here, even with the best mathematics and computers in the world, will lead to predictions that can go seriously astray.

The use of physical models, as the name suggests, involves the use of one or more physical systems to model the anticipated behaviors of another

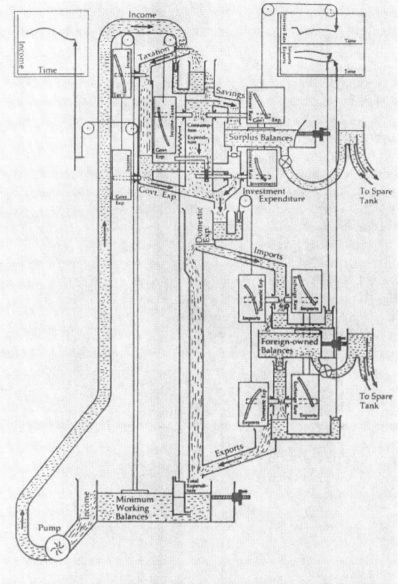

Diagram of the Phillips machine. Source: LSE Quarterly, Winter 1988, Nick Barr.

Figure 5.1. Phillips machine.

group of physical systems. In this context, one part of nature is manipulated in such a way as to enable us to form expectations about the behavior of other parts of nature. For example, engineers might build a *prototype* of a system they are interested in manufacturing in order to see if it behaves appropriately, and to iron out "bugs" before the system goes into the production phase. In a sense, a prototype simulates behavior in the system being modeled by actually being a *version* of that system (the system that goes into production may differ from the prototype in numerous ways that reflect solutions to problems encountered with the prototype.

A more common use of physical models in the context of engineering concerns scale models. Here, a physical model is constructed on one scale to simulate the behavior of another system on a different scale. The use of small-scale models of large systems (in wind tunnels and water tanks, for example) is commonplace, but it is not unknown for large-scale models of small systems to be constructed—especially in the context of education, where it is helpful to see in the large, what is hard to visualize in the small. Such large-scale models have also played a role in major discoveries (Watson and Crick built a physical scale model of DNA as part of their effort to unravel its structure [(Giere 1991) p18]).

There are many practical reasons why small-scale models are useful. While cost can be an issue, small systems are also often easier to manipulate than the large systems they model. Experimental tractability can be very important. A good example concerns the Wright brothers' use of scale models in their research on the possibilities of powered flight. As Cziko has recently observed of wind tunnel tests used by the Wright brothers:

> The Wright brothers' extensive use of such tests gave them an important advantage in the race to build the first airplane, since during the time it would take their French competitors to build and launch yet another complete prototype, the Americans built, tested, eliminated, and designed many different scale models and therefore made more rapid progress in their search for a successful design. [(Cziko 1997) p165]

The process of development here even bears a resemblance to a Darwinian evolutionary process whereby many variants are generated and tested for function, with a smaller number of variants surviving the trial process to serve as the subjects for further evolutionary modification. Designs can often be enhanced by producing variants on successful themes.

Indeed, before even building scale models it was historically possible to do "test runs on paper"—crude mathematical models to examine feasibility. With the advent of sophisticated computer models, it is now possible to test multiple variants on successful themes in real time. The technological analog of natural selection under discussion here, like natural selection itself, works by passing multiple variants through a selective filter. Cziko calls the use of computers to test multiple variants with a view to enhancing desired function in this way, a "technology of technology [(Cziko 1997) p166]."

Scale models, then, are valuable tools for scientist and engineer alike. But an issue that needs to be discussed here in connection with scale models is the issue of scaling itself—scaling up from the small-scale model to the larger system modeled, or scaling down. For some problems, this is fairly trivial. Recall Newton's Second Law, according to which $F = ma$. Consider a small mass m_1 and a larger mass m_2. If the masses are known, and it is known that a force F_1 is required to induce an acceleration a in m_1, then it is easy to scale up and deduce the size of the larger force F_2 required to induce the same acceleration a in m_2. There are numerous situations in physics, chemistry and biology where scaling such as this is possible and is indeed important.

However, uncritical use of scale models and scaling principles can lead the theorist astray from the real world. As Petroski has observed of engineered systems:

> Design errors attributable to overlooking scale effects in nature and artifacts have been especially persistent throughout history, and they continue to be so even in our age of high technology and computers. Although it had been known for more than two thousand years that pieces of timber grow weaker as they grow longer, Renaissance shipbuilders found inexplicable the fact that their large timber ships were breaking under their own weight. Galileo (1638) prefaced his seminal study on the strengths of materials by reciting the breakup of ships and other recurring failures of Renaissance engineering attributable to size, a problem, he noted, nature had well under control. [(Petroski 1994) p30-3]

The problem here is one of concern to us since it involves the business of extrapolation—the study of when phenomena observed in a scale model may legitimately be extended to the system being modeled.

The Renaissance shipbuilders got into trouble, possibly under the influence of ideas derived from the Greek philosopher Plato who extolled geometry and denigrated the study of matter, by scaling up from their scale models using pure geometry, and ignoring the properties of the materials they were working with, especially their internal heterogeneity. As Galileo himself put it:

> . . . you can plainly see the impossibility of increasing the size of structures to vast dimensions either in art or in nature; likewise the impossibility of building ships. . . of enormous size in such a way that their oars, yards, beams, iron bolts, and in short, all their other parts will hold together, nor can nature produce trees of extraordinary size because the branches would break down under their own weight; so also it would be impossible to build up the bony structures of men, horses and other animals so as to hold together and perform their normal functions if these animals were increased enormously in height; for this increase in height can be accomplished only by employing a material which is harder and stronger than usual, or by enlarging the size of the bones, thus changing their shape until the form and appearance of the animals suggest a monstrosity. [(Petroski 1994) p37]

As we have seen in our earlier discussion of Claude Bernard's reflections on the effects of toad's venom, it is not just a matter of animal bones.

In Bernard's discussion of the effects of toad venom on toads and smaller frogs, once differences in size had been allowed for, same toxic cause was to be followed, deterministically, by same pathological effect. The inner environments of animals were assumed to respond to toxic stimuli in essentially the same way, once size differences had been compensated for by allowing for differences in mass, for example. We will have more to say about scaling in the context of toxicology and pharmacology in later chapters. Suffice it to say here that evolved differences in the internal physiological environments of organisms can bring about marked differences in response to similar stimuli, even for organisms of the same mass or surface area. But this is to look ahead. Here we need to be clearer about the nature of models as they are used in science before such questions of scaling in toxicology can be meaningfully addressed in a principled manner.

A complicating factor in the use of nonhuman animals as models of human biomedical phenomena arises from the fact that humans and their nonhuman models (mainly rats and mice, but sometimes pigs, cats, sheep, monkeys and great apes) are all extremely complex, integrated systems. Before proceeding further, the issue of modeling complex systems requires some discussion.

Modeling Complex Systems

The study of complex systems has received a lot of ink lately and rightly so since it is a very important subject to many diverse fields (see Figures 5.2 and 5.3 for characteristics of complex systems). Thus Hiroaki Kitano has recently observed:

> To understand complex biological systems requires the integration of experimental and computational research - in other words a systems biology approach. Computational biology, through pragmatic modelling and theoretical exploration, provides a powerful foundation from which to address critical scientific questions head-on....
>
> It is often said that biological systems, such as cells, are 'complex systems'. A popular notion of complex systems is of very large numbers of simple and identical elements interacting to produce 'complex' behaviours. The reality of biological systems is somewhat different. Here large numbers of functionally diverse, and frequently multifunctional, sets of elements interact selectively and nonlinearly to produce coherent rather than complex behaviours.
>
> Unlike complex systems of simple elements, in which functions emerge from the properties of the networks they form rather than from any specific element, functions in biological systems rely on a combination of the network and the specific elements involved. For example, *p53* (a 393-amino-acid protein sometimes called 'the guardian of genome') acts as tumour suppressor because of its position within a network of transcription factors. However, *p53* is activated, inhibited and degraded by modifications such as phosphorylation, dephosphorylation and proteolytic degradation, while its targets are selected by the different modification patterns that exist; these are properties that reflect the complexity of the element itself. Neither *p53* nor the network functions as a tumour suppressor in isolation. In this way, biological systems might be better characterized as symbiotic systems.
>
> Molecular biology has uncovered a multitude of biological facts, such as genome sequences and protein properties, but this alone is not sufficient for interpreting biological systems. Cells, tissues, organs, organisms and ecological webs are systems of components whose specific interactions have been defined by evolution; thus a system-level understanding should be the prime goal of biology. Although advances in accurate, quantitative experimental approaches will doubtless continue,

insights into the functioning of biological systems will not result from purely intuitive assaults. This is because of the intrinsic complexity of biological systems. A combination of experimental and computational approaches is expected to resolve this problem. (Kitano 2002)

Others have likewise seen the need for a more systems-oriented approach to biological science. Thus, Ahn et al., have commented:

> The human genome contains 30,000 to 35,000 genes. Although this number is just five times the number of genes in a unicellular eukaryote (e.g., approximately 6,000 genes in Saccharomyces cerevisiae), the human genome encodes for nearly 100 trillion cells in the human body. The richness of information is derived not only in the genes themselves but also in the interaction between genes and between their respective products. The genes encode for messenger RNA, the messenger RNAs encode for proteins, and the proteins act as catalysts or secondary messengers, among other diverse functions. Between each hierarchical level, modifications (e.g., alternative splicing) are made, and at each hierarchical level (e.g., transcription), thousands of molecules interact with other molecules to create a complex regulatory network. What becomes evident from these molecular analyses is that phenotypic traits emerge from the collective action of multiple individual molecules. Therefore, the previous notion that a single genetic mutation is responsible for most phenotypic defects is overly simplistic. Complex diseases such as cancer, asthma, or atherosclerosis cannot generally be explained by a single genetic mutation. (Ahn et al. 2006)

In a similar vein, Van Regenmortel draws our attention to the dense pleiotropies that exist in real biological systems:

> Today, it is clear that the specificity of a complex biological activity does not arise from the specificity of the individual molecules that are involved, as these components frequently function in many different processes. For instance, genes that affect memory formation in the fruit fly encode proteins in the cyclic AMP (cAMP) signalling pathway that are not specific to memory. It is the particular cellular compartment and environment in which a second messenger, such as cAMP, is released that allow a gene product to have a unique effect. Biological specificity results from the way in which these components assemble and function together. Interactions between the parts, as well as influences from the environment, give rise to new features, such as network behavior, which are absent in the isolated components. . .

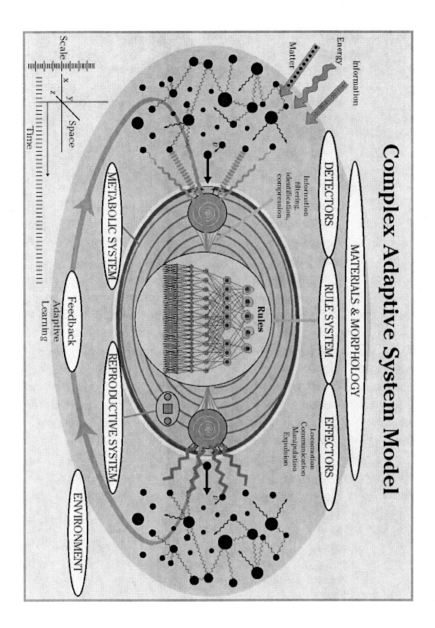

Figure 5.2. Complex adaptive system
Illustration copyright ® 2008 Marshall Clemens

> The constituents of a complex system interact in many ways, including negative feedback and feed-forward control, which lead to dynamic features that cannot be predicted satisfactorily by linear mathematical models that disregard cooperativity and non-additive effects. In view of the complexity of informational pathways and networks new types of mathematics are required for modelling these systems. (Van Regenmortel 2004)

How then, should we set about studying complex systems, including complex biological systems?

The study of complex phenomena as it concerns us here is a matter that can be illuminated by ideas drawn from dynamical systems theory. This is a branch of science that aims to cope with the nonlinearities and dynamical complexity generated by complex interactive systems, an account of whose dynamics involves networks with feedback loops of varying degrees of complexity. These are the very systems that exhibit internal heterogeneity and whose outputs are rarely linearly proportional to their inputs. Even in the context of deterministic dynamics, we will soon see that two complex systems may be qualitatively identical, and may begin in very similar, but slightly different states (initial conditions), and yet deliver output states (final conditions) that diverge exponentially. This phenomenon whereby there can be extreme sensitivity to initial conditions is inexplicable to theorists trained in the use of linear models where differences between initial conditions and final conditions grow at most as linear functions of increasing time.

In recent years, these issues about dynamics have become quite urgent as knowledge has grown about the nature of biochemical systems, for example. Thus Mohan has recently observed:

> The ubiquitous feature of biochemical kinetics is the presence of nonlinearity, particularly with respect to feedback or feedforward control loops that activate (sensitize) or deactivate (stabilize) the various processes. Developments in the theory of nonlinear dynamical systems and, in particular, that of deterministic chaos has helped to highlight, with the aid of computer simulations as well as in conjunction with experiments, the bewildering variety of dynamical behaviors that are possible in such systems. . . It is becoming increasingly evident that chaotic processes are more generic and widespread in biological systems than expected from the widely accepted concept of homeostasis in biology. . . In the complex circuitry of biochemical pathways, a variety of dynamical behaviors may

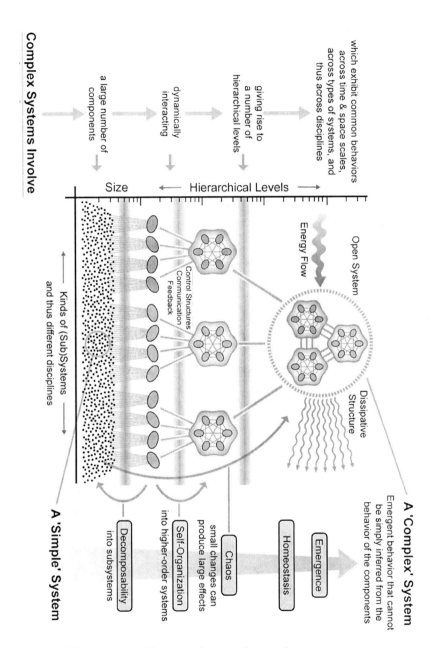

Figure 5.3. Characteristics of complex systems
Illustration copyright ® 2008 Marshall Clemens

be expected to arise because of ubiquitous nonlinear couplings and control loops which can have important roles in both normal and pathological functions of the systems. (Mohan 1998)

Mathematical Interlude: Dynamical Systems Theory

In this section we will illustrate some of these new ideas about modeling biological systems. The concepts are drawn from a branch of mathematical study known as dynamical systems theory. We begin our analysis with a discussion of mathematical models. (Readers with no interest in descriptions of mathematics may skip the details, but they are advised that some of the terminology introduced here will recur in the next chapter).

As with all models, so, too, here. We must take care not to confuse features of the mathematical model with the real system modeled. The real system generates time-series data, for example. The equations and concepts in the model are used to account for this data, and possibly to make predictions about future data. In this field it always makes sense to ask whether a real complex system (for example, an organism) is actually an instantiation or exemplar of some abstract complex system described in the mathematical model. Model properties cannot simply be assumed to be real world properties.

A deterministic mathematical model M of a real system S might be characterized as follows:

$$[1]\ M = <\Gamma, F>,$$

Here Γ is a set of possible system states γ_i and is known as a *state space*. F is an equation of motion describing how system states change with time. In the case of the accelerating rocket discussed in chapter 3, each state of the system was a point in the state space for the system, and consisted of an assignment of values for the variables of position and velocity. We saw there that for an initial assignment of a state $\gamma_o = (x_o, v_o)$ at initial time t_o for the rocket system, the equations of motion determined a unique state $\gamma_j = (x_k, v_j)$ for all other times.

If our real system was a chemical system in a reaction vessel (possibly a cell in an organism) then the dynamically relevant features of the system that we wish to model might be concentrations of reagents whose values change over time. If values of i-variables are needed for a complete characterization of the state of the chemical system under study, then the point γ representing a state of the system at any time will be a point in an *i*-dimensional state space. As the reaction takes place, and concentrations change, so the chemical state of the system changes.

In order to model the way in which our real system S changes with respect to time, our model M employs an equation of motion: a dynamical rule F whose role, for any initial assignment of state, is to determine a unique state at all other times. The easiest way to achieve this end is to introduce a function $F: \Gamma \times T \to \Gamma$, where $\Gamma \times T$ is the Cartesian product of Γ and a set T of times t, so that $F(\gamma_i, t_j) = \gamma_j$, for $\gamma_i, \gamma_j \varepsilon \Gamma$ and $t_j \varepsilon T$. Here, if γ_i is the initial state of the system, then γ_j will be the state of the system after elapsed time t_j. Clearly, where $t = 0$, $F(\gamma_i, t) = \gamma_i$. The model is deterministic in that for each initial state γ_i, and elapsed time t_j, there is a unique final state γ_j. Equivalently, we could write $\gamma(t_j) = F(\gamma(t_i))$, where $\gamma(t)$ is the state at time t. The dynamics may also be given in terms of infinitesimal time evolution as: $d\gamma/dt = F(\gamma(t))$. For discrete dynamics, $\gamma(t+1) = F(\gamma(t))$.

Given a deterministic dynamical rule F and an initial assignment of state γ_0, the model M will represent the time evolution of S by a succession of states in state space. The resulting curve is known as a *state space trajectory* (sometimes referred to as an orbit). If our system S is a chemical system in a reaction vessel, then the state space trajectory will convey information about the ways in which the concentrations of reagents characterizing the system change with respect to time, and so on.

We will be interested in the properties of F. It is possible that F is linear. Linearity means that the superposition principle holds, i.e., that for states $\gamma_i, \gamma_j \varepsilon \Gamma$, $F(\gamma_i + \gamma_j) = F(\gamma_i) + F(\gamma_j)$. A well known linear rule for the assignment of states to systems of interest is the time-dependent Schroedinger equation in quantum mechanics: $H\Psi = i(h/2\pi) \partial\Psi/\partial t$. Often, however, the dynamics of systems cannot be given by a linear equation of motion (typically where dependent variables are raised to powers other than 1 or where transcendental functions are involved). An example may help.

Consider the second order differential equation for an undamped pendulum, freely swinging in a plane:

$$d^2\theta/dt^2 + (g/L) \sin\theta = 0,$$

where θ is the angle between the vertical and the pendulum, g represents acceleration due to gravity, and L is the length of the pendulum. Here, if

θ_1 and θ_2 are solutions to the equation (i.e., points in state space), the linear sum $\theta_1 + \theta_2$ is not because $sin\ \theta_1 + sin\ \theta_2 \neq sin\ (\theta_1 + \theta_2)$. Failure of the superposition principle means we are cut off from using Fourier analysis that makes handling the dynamics of linear systems so tractable. In general, there are no systematic methods for the solution of nonlinear equations, and resort to numerical approximation techniques is thus often a necessity.

If we take the above equation for pendular motion and let $\alpha_1 = d\theta/dt$ and $\alpha_2 = \theta$, then the second order equation above can be written in terms of two coupled first order differential equations:

$$d\alpha_1/dt = -(g/L)\ sin\ \alpha_2,$$
$$d\alpha_2/dt = \alpha_1.$$

Here, values of L and g are assumed to remain fixed while the pendulum is swinging. They are known as *control parameters*.

Control parameters need to be considered by the mathematical modeler because as Eubank and Farmer have observed, the dynamics in the model M is typically only an approximation of the true dynamics of the system S. The true dynamics may differ from model dynamics because of interactions between the real system and other systems or because of factors internal to the system S. If these influences on the true dynamics of S are small, or occur slowly relative to the changes being modeled by the states γ in the model, then, at least in the short run, it is often possible to study the dynamics of the system by holding these factors constant. Eubank and Farmer observe:

> For a pendulum, for example, the natural parameters are the mass, length, and strength of the gravitational field. Assigning these constant values is an approximation; as the pendulum swings the local gravitational field varies from place to place, and the changes in stress cause the length of the pendulum to change. The "parameters" are aspects of the dynamics that change slowly enough that they can be approximated as constants, at least for a period of time. (Eubank and Farmer 1990)

They add, "The set of all possible values for the parameters is called the *parameter space*, or *control space* (Eubank and Farmer 1990)." If there are control parameters $p_1, \ldots, p_n$, the parameter space will be an n-

dimensional space. A point p in parameter space will then represent a specific assignment of values to these variables.

The dynamics of the pendulum can be explored for fixed values of L and g (the assumption of parameter stability), but we may also be interested in *structural changes* in the dynamics that occur as different values for L and g are explored. Sometimes small changes in the parameters bring about only small changes in the dynamics of the model. But sometimes small changes in parameter values can radically alter the entire dynamical situation, bringing about qualitative changes.

Many of the real systems of interest in dynamical systems theory are what are known as open-dissipative systems—systems that have exchanges with their environments, and whose complex internal dynamics is therefore affected by the external environment. In modeling such systems, it is important to be aware of the effects of small changes in environmental parameters. Moreover, even where the environmental parameters are held constant, two such complex systems may exhibit differences with respect to internal complexity. Hence, awareness of the effects of changes in the parameters internal to the system is also important.

An analogy may help. Consider a complex system such as an automobile. Some small changes have little or no effect (at least in the short term) on the behavior of the system (e.g., chipping a bit of paint off the fender). Other small changes (e.g., removing a connection to the battery) may bring about massive changes in the behavior of the system. (In organisms, some point mutations are silent, whereas other, equally small changes, are catastrophic).

This is a good place to introduce the concepts of *bifurcation* and *structural stability*. When a change in parameter values changes the quality of the dynamics in the model, a bifurcation has occurred. The parameter values where bifurcations occur define bifurcation boundaries. These boundaries divide the parameter space into disjoint volumes. Within a given volume, small changes in the parameters bring about small changes in the dynamical properties of the model. When a bifurcation boundary is crossed, however, dramatic qualitative changes in the dynamics in the model occur. In view of this, Kauffman has observed:

> The concept of *structural stability* concerns the idea that, typically, volumes in parameter space defined by bifurcation surfaces are like soap bubbles. The volumes are reasonably large relative to the bifurcation surfaces which divide them. Thus for *most changes* in parameters, the system remains within one volume in parameter space and the dynamical behavior does not change dramatically. Dynami-

> cal systems having this property are said to be structurally stable.
> [(Kauffman 1993) p181]

Structural stability is not a necessary property of model dynamics. Many models are known where structural instabilities exist, and virtually all small changes in parameters bring about bifurcations. Whether or not a model exhibits structural stability, bifurcations are a problem for the modeler since they mean that in some circumstances two systems that are very similar, differing only with respect to small values of the parameters, can nevertheless differ markedly in their behavior.

Consider the following example employed by Rueger and Sharp (Rueger and Sharp 1998). Suppose we have a law:

$$(1) \ f(x) = x^4 + ux^2 + vx,$$

where u and v are parameters. The concept of structural stability requires that if we add a perturbation to $f(x)$, the resulting function will not show qualitatively different behavior. A small perturbation of $f(x)$ is an added term $e(x)$ (which may have the precise form ex^2 or ex^3). Adding the perturbation results in variations of the parameters—$(u + e)$ or $(v + e)$—of $f(x)$. The qualitative behavior of $f(x)$ is evaluated in terms of the location and nature of the function's extreme or critical points (maxima and minima). It turns out that $f(x)$ characterized above is structurally stable. But consider now:

$$(2) \ g(x) = x^4 + ux^2.$$

Here the addition of a perturbation ex generates a function:

$$(3) \ h(x) = x^4 + ux^2 + ex.$$

For $u < 0$, this function has a minimum near $x = 0$, whereas $g(x)$ has a maximum near $x = 0$. Because the form of the original function $g(x)$ is different from the form of the perturbed function $h(x)$ we have an example of structural instability. Concerning this state of affairs, Rueger and Sharp observe:

> In physical applications in which $f(x)$ and $g(x)$ could be descriptions
> of a system's potential energy, the *extrema* of the functions corres-

pond to stable or unstable equilibrium states of the system. The number, location, and kind of such equilibrium points qualitatively characterize the system's behavior; a system described by $g(x)$ with a maximum (unstable equilibrium) at $x = 0$ behaves qualitatively differently from a system described by $h(x)$, which has a minimum (stable equilibrium) at $x = 0$. (Rueger and Sharp 1998)

In a deterministic model M of a system S, state space trajectories cannot merge or cross each other, though they may get very close. For a given initial state γ_0, a deterministic rule fixes a unique state γ_i for all other times. If trajectories crossed, the point γ of their intersection would no longer be unique to one of them. (In a deterministic model, systems cannot have more than one unique past history).

The state space trajectory for a system, given an initial state γ_0, may wander around aimlessly in state space, it may go off to infinity, but this is not necessarily the fate of such a trajectory. As noted by Smith [(Smith 1998) p8-9] the state space trajectories for open dissipative systems may either wander out to infinity, or they can converge, getting ever closer to, but never actually touching, mathematical objects called *attractors*. These are objects in the model that permit an explanation of the long-term behavior of a system, once initial perturbations called transients have been allowed to settle down.

To understand what an attractor is, it is helpful to understand the concept of invariant set. A set of states Ω is invariant under a dynamical rule F if $F\Omega(t) = \Omega$ for all times t. The simplest such set is a fixed point corresponding to a *steady state attractor*. In this regard Kauffman has commented:

> . . . two states that were initially quite far apart and on different trajectories may come to be arbitrarily close to each other, for their respective trajectories may *converge*. In particular, different trajectories may converge toward a single state which does not change in time—that is, a steady state—reaching it in the limit of infinite time. Then the steady state attractor is a zero-dimensional, or point, *attractor*, and the entire volume of states which lie on trajectories flowing to that attractor is its *basin of attraction*. [(Kauffman 1993)p176]

For a state space trajectory the fixed points are thus solutions to the equation: $dx/dt = F(x) = 0$. In the light of this observation Eubank and Farmer observe:

> Roughly speaking, an attractor is an invariant set that "attracts" nearby states. More formally, Ω is an attractor if there is an open neighborhood N about it such that $F^t(N) \to \Omega$ as $t \to \infty$, and Ω cannot be broken into pieces $\Omega_1, \Omega_2, \dots$ such that $F^t(\Omega_i) \cap F^t(\Omega_j) = \emptyset$ [i.e., the empty set]. The basin of an attractor is the set of points attracted to it; that is $\{x: \lim t \to \infty \, F^t(x) \subset \Omega\}$. A given dynamical system may have many attractors each with its own basin of attraction. (Eubank and Farmer 1990)

Quite generally, attractors are simply sets of points in the state space to which trajectories within some volume of the state space converge asymptotically with time [(Eubank and Farmer 1990) p177].

Commenting on the relationship between points in parameter space and points in state space, Kauffman has noted, "Any point in parameter space corresponds to a fixed set of parameters and thus to a fixed set of basins of attraction and attractors in the corresponding state space of the dynamical system. The set of basins of attraction is often called the *basin portrait* of the dynamical system [(Kauffman 1993) p180]." One effect of varying the parameter values (hence the point in parameter space in the model) is to change the basin portrait governing model dynamics. For a given point p in parameter space, which attractor in state space, of the set fixed by the choice of parameter values, models the long term behavior of the system modeled depends on choice of initial conditions (i.e., which basin of attraction in the basin portrait actually contains the initial state x).

Perturbations in a real system S may radically alter its dynamical behavior. In our model, such perturbations may have the effect of diverting a state space trajectory from one basin of attraction to another, thereby changing the character of the dynamics even for a fixed choice of parameters. As we saw above, some perturbations of the values of the parameters can also bring about significant changes in the dynamical behavior of the model system.

When a fixed point is an attractor, it is known as a *steady state* attractor. Some physical systems S, driven by energy flows, may oscillate (clocks, for example). In the model M for such a system, the oscillations are described by state space trajectories that orbit a closed loop of states called a *limit cycle attractor*. Shifts in parameter choices here are sometimes associated with bifurcations leading to period doubling. In a state space of 2-dimensions, where states x consist of values for pairs of variables, steady state attractors and limit cycle attractors are the only possibilities (this follows from the requirement that trajectories do not cross or merge). In 2-

dimensions, trajectories can diverge exponentially from each other, but only by going out to infinity.

In higher dimensional state spaces, other types of attractors are possible, and these can be used to model more complex behaviors in real systems. One such is the toroidal *quasi-periodic attractor*. Explaining this idea, Kauffman employs the following example:

> Suppose the dynamical system has a total of ten chemicals. Suppose four chemicals set up one oscillation and six set up another oscillation; then in the ten-dimensional state space, the representative point [characterizing the state of the system] moves on two hoops at once. This two-hoop movement can be can be represented by motion on a donut, with flow in one direction passing through the hole and flow in the other direction being around the donut but not passing through the hole. The simultaneous flow produces a spiraling trajectory which winds both through the hole and around the donut. [(Kauffman 1993) p178]

More complex still are attractors known as *strange attractors*.

To help explain what a strange attractor is, it is helpful to remember at this juncture that state space (an object in a mathematical model) is very different from physical space. Points in state space correspond to solutions to the differential equations describing systems of interest. The state space itself is simply the set of solutions to those equations for the system being modeled. Attractors, consisting of sets of points in state space, represent certain sets of solutions to those equations (they are not magically imposed on the state space).

To understand the more complex attractors that are possible, it is helpful to remember that the topology of the state space is important. To mathematicians, topology is the science of continuity. What we ordinarily think of as, say, 3-dimensional shapes, are simply sets of points to a topologist. Topology used to be known as "rubber sheet geometry" because early topological inquiries concerned geometric figures or surfaces that could be deformed *continuously* without tearing. These deformations result from topological mappings (squares can easily be visualized to deform into circles, and vice versa). For this reason, squares and circles are said to be homeomorphic under the mapping, i.e., topologically equivalent. Topological deformations involve stretchings, bendings and foldings—operations that can be described with mathematical precision.

Thinking now of a 3-dimensional state space, for example, it is possible to visualize the existence of highly complex topological structures where stretching of a 2-dimensional surface allows trajectories initially parallel to each other to diverge exponentially, but where careful folding of the surface back onto itself, exploiting the third dimension, can be done in such a way as to avoid the merging of trajectories. This, along with further, similar stretching and folding of the surface initially stretched and folded (a process iterated infinitely many times) results in an object in state space where trajectories that are initially close together diverge at an exponential rate but are nevertheless confined to a bounded region of state space. There are different ways in which these topological feats can be achieved, and textbook illustrations of such exotic objects of the Lorentz attractor and the Roessler attractor (see Figure 5.4 (Wikipedia Commons 2009)) give at least a visual appreciation for the nature of these complex structures.

Trajectories orbiting these structures—known as *strange attractors*—exhibit extreme sensitivity to initial conditions. Kauffman comments:

> The first novel feature found in strange attractors but not in steady states, limit cycles and so forth. . . is a *sensitivity to initial conditions*. Tiny differences in initial conditions make vast differences in the subsequent behavior of the system. In contrast, a system with a stable limit cycle squeezes all flows onto the same hoop of states, hence nearby initial states are still nearby later on. [(Kauffman 1993) p178]

Sensitivity to initial conditions means that microscopic errors in the determination of initial conditions can explode into enormous macroscopic errors about the expected behavior of the system modeled. (If $\delta x(0)$ is an infinitesimal change in the initial condition and $\delta x(t)$ is the corresponding change at time $t > 0$, then $|\delta x(t)| = e^{\lambda t} |\delta x(0)|$. Here γ is called the largest Lyapunov exponent. It is a measure of the rate of separation per unit time of trajectories passing near $x(0)$. If $\lambda > 0$, then trajectories diverge exponentially (Ruelle 1994)).

Nevertheless it is worth emphasizing that in dynamical systems theory, a *chaotic* system *is* a deterministic system in the sense that if the initial state is known with absolute precision, the states at other times can be calculated with absolute precision (at least in principle). But we seldom know the initial conditions with absolute precision, and therefore, the system may

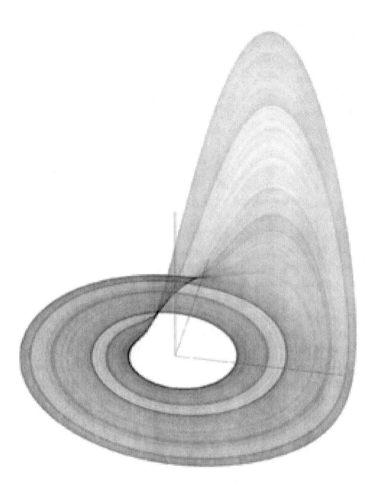

Figure 5.4. Roessler attractor.

appear to us to behave randomly because of our ignorance of the precise initial conditions (the *appearance* of randomness is thus fully consistent with an underlying deterministic reality). Thus Lam has pointed out:

> The *apparent* unpredictability of a chaotic, deterministic, *real* system (such as the weather) arises from the system's sensitive dependence on initial conditions *and* the fact that the system's initial conditions can be measured or determined only approximately in practice, due to the finite resolution of any measuring instrument. This difficulty precludes the long-term predictability of any chaotic real system. [(Lam 1997) p5]

So what looks like random behavior can contain hidden order, and the modeler must be prepared to try to distinguish signal from noise. Chaotic dynamics can be exhibited by systems whose states admit of as little as 3 degrees of freedom. Even simple systems may exhibit complex behavior. The good news is that complex behavior may be describable in terms of relatively simple equations.

Bringing out some of the potential biological implications of these newer approaches to the modeling of complex systems Mohan has pointed out:

> It is the paradigm now that biological systems, being nonlinear and far from equilibrium systems, achieve apparent homeostasis by maintaining themselves in a stable steady state... Multistability and multirhythmicity, where different initial conditions, for the same choice of parameter values, lead to different steady states or different periodic orbits, are also a feature of these nonlinear systems. Such behaviors have been found to be useful to explain the "switching" behaviors observed in biological systems. All this leads to the concept of "dynamic diseases" where pathological states of the system can be seen as due to an alteration with respect to basal dynamics which could be a stable steady state or periodic oscillations or even chaotic oscillations as, for example, in the understanding with respect to brain states. (Mohan 1998)

Examples of dynamic diseases include the onset of certain types of epileptic seizures, where a shift from chaotic to periodic dynamics with respect to brain waves seems to be indicated, the onset of ventricular fibrillation in mammals, and such phenomena as species differences with respect to the metabolism of xenobiotics. Shanks (Shanks 2001) has reviewed how a real chemical system (the Belousov-Zhabotinski reaction) and its mathematical model can be used to comprehend a variety of biological phenomena, including the onset of ventricular fibrillation.

The case of the Belousov-Zhabotinski (BZ) reaction is an interesting one, and may be used here to illustrate some of the basic ideas about mathematical modeling introduced above. Though based on very different substrates and products, the BZ reaction was originally proposed as an *in vitro* model of the Krebs cycle. Under appropriate conditions the reaction manifests a variety of non-equilibrium behaviors, including steady states and oscillations (depending on choices of parameter values).

Though the chemistry of the reaction is complex, it is often possible to model the dynamics of the system by focusing on a small set of time-dependent variables. Though this involves simplification and idealization, the resulting model is nevertheless to be judged on its ability to describe complex dynamical phenomena, and to suggest avenues of inquiry for relevantly similar dynamical systems (based on different substrates and products).

As observed by Babloyantz [(Babloyantz 1986) p175], the crucial time-dependent variables X and Y in the standard BZ reaction are concentrations of bromous acid ($HBrO_2$) and bromine ions (Br^-) respectively. Oscillations come into play when, as the concentration of X rises, it reacts with Y and where this has the effect of decreasing the concentration of X while increasing the production of Y, but then if too much Y favors the production of X, the system, maintained far from equilibrium, will find itself trapped in an oscillatory cycle (and this we can see in the lab).

A mathematical model of this situation, known as the Brusselator, has been proposed. The rate equations in the model are as follows:

(1) $dX/dt = A - (B + 1)X + X^2Y$;
(2) $dY/dt = BX - X^2Y$.

Here, A and B are parameters held constant during the reaction. In situations where $X = A$ and $Y = B/A$, a non-equilibrium steady state exists ($dX/dt = 0$; $dY/dt = 0$). However, in situations where $B > 1+A^2$, the system displays oscillatory behavior and its dynamics are described in terms of a limit cycle attractor.

Like the BZ reaction, oscillations have been observed in the behavior of the glycolytic pathway in yeast [(Babloyantz 1986) Ch. 11]. The reaction pathway is complex, but may be described as follows:

Glucose + 2 ADP + 2P_i + 2 NAD → 2(pyruvate) + 2ATP + 2NADH

As with the BZ reaction, is there a way of simplifying and idealizing a very complex chemical situation so as to describe its oscillatory dynamics? The PFK-mechanism has been the focus of dynamical inquiry (we use the following abbreviations: PFK = phosphofructokinase; F6P = fructose-6-phosphate; and FDP = fructose 1,6 diphosphate). The following step in the reaction pathway is crucial for the oscillatory dynamics:

F6P + ATP − PFK→ ADP +FDP

Babloyantz comments:

> In a given range of glucose concentrations, the enzyme is under the influence of two antagonistic factors. PFK is inhibited cooperatively by ATP and is activated by ADP . . . experience shows that the concentration ratio F6P/FDP does not influence the oscillatory behavior, whereas the ratio ATP/ADP is the determining factor. [(Babloyantz 1986) p259]

The focus of dynamical modeling may thus concern:

$$ATP - PFK \rightarrow ADP$$

As is now well-known, the situation can be described in terms of two coupled differential equations belonging to the same class of mathematical objects as the Brusselator equations. In this situation, a complex biological phenomenon can be described with the help of a physico-chemical system (the BZ reaction) and the mathematical model (the Brusselator) used to describe patterns of dynamical change.

The power of this kind of approach to modeling has been nicely illustrated by attempts by von Dassow et al. (2000) to consider the properties of a local developmental network (or developmental module) for the emergence of segment polarity in *Drosophila*. Their computer model, reflecting the greater complexity of the system under analysis than is illustrated by yeast glycolysis, consists of 136 coupled differential equations and some 48 free parameters whose values must be empirically determined. Von Dassow, et al. use their model to explore, among other things, the effects of changing parameter values and the robustness properties of the network (its stability in the face of parameter perturbations).

It is also interesting to note that Stuart Kauffman (1993) has exploited a class of discrete models known as *Boolean NK models* (which are amenable to computational analysis) to capture important features of the behavior of real biological systems. In these models, it is supposed that complex systems consist of N nodes, each of which has two possible states, 1 and 0 ("on" or "off"). Each node is supposed to receive K inputs from other nodes, with inputs determining node-states. Such models seem to be well-suited to the analysis of genomic regulatory networks, where the switching behavior of a given node reflects the switching behavior of other nodes in the network in which it is embedded. Commenting on the order and organization predicted in his models of complexity, Kauffman observes:

Complex systems, such as the genomic regulatory networks underlying ontogeny, exhibit powerful "self-organized" structural and
dynamical properties. The kind of order which arises spontaneously
in such systems is strikingly similar to the order found in organisms.
This raises the plausible possibility that the spontaneous order
found in such complex systems accounts for some or much of the
order found in organisms. [(Kauffman 1990) p67]

More generally, the basic structural issue in a network concerns the
linkages between nodes. Suppose you arrange some nodes representing
molecules, (substrates and products in a biochemical network, perhaps),
in two dimensions. These nodes can be linked together in various ways to
represent reactions and pathways. With this in mind, Bray has pointed
out:

> . . . you can connect them up in various ways—by linking nearest
> neighbors in a regular fashion, or by selecting them at random and
> joining them together. A third strategy—which is of great contem-
> porary interest because it seems to correspond to many naturally
> occurring networks—is to give a few nodes a very large number of
> connections and allow the rest to have relatively few. (Bray 2003)

These three linking strategies give rise to regular, random, and scale-free
networks respectively (see Figure 5.5 (Wikipedia Commons 2009)). In the
case of regular and random networks, the number of connections per
node will have a normal distribution with an average value that gives the
scale of the network. However, as Bray goes on to observe (Bray 2003), in
some networks the number of molecules (N) with a given number of
connections (k) may fall off as a power law $N(k) \sim k^{-\delta}$. Since $N(k)$ does
not have a characteristic peak value, it is said to be scale-free.

From these remarks it is clear that pursuit of a system-level understanding of organisms will involve at least 3 interrelated issues:

(a) *System structures.* Here interest will be focused on the origins, development and nature of structures in the system. These are the structures constitutive of the system at a given point in time.

(b) *System dynamics.* Here interest will be focused on patterns of interactions among subsystems that give rise to temporal patterns in the behavior of the system (steady states, oscillations, quasi-periodic behaviors,
chaos, and so on).

(c) *Control mechanisms.* Here the focus of attention is on regulatory mechanisms (e.g., positive and negative feedback loops), system-level redundancies (important both for evolution and the stability that complex organismal systems are capable of exhibiting in the face of environmental perturbations, and modularity of subsystems (important if local evolutionary changes in subsystems are not to necessitate global system-wide changes, and again for an account of the stability of biological systems in the face of local perturbations (Alon 2003)).

An early example of the systems approach to biological systems was metabolic control analysis (MCA) in biochemistry in the 1970s. Concerning the metabolic control analysis approach to biochemical complexity, Dipple et al. observe:

> A more recent application of MCA is the "top-down regulation analysis" which allows one to look at the elasticity of the system or how it adjusts to maintain homeostasis. . . Top-down regulation analysis approaches complexity in a manner strikingly similar to the model of a scale-free network with blocks of reactions represented by "modules" interacting through a few highly connected critical nodes. (Dipple, Phelan, and McCabe 2001)

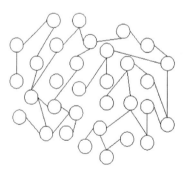

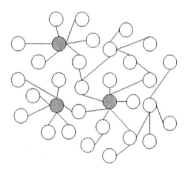

(a) Random network   (b) Scale-free network

Figure 5.5. Random and scale-free networks.

As noted above, scale-free network are networks that exhibit a high degree of inhomogeneity. Most nodes in such networks have one or two links to other nodes, but a small number of nodes act as hubs with many connections to other system nodes.

These models are useful in the context of molecular biology because, as pointed out by Dipple et al.:

> . . . biological systems show a remarkably robust tolerance for random failures of their components. This error tolerance is a fundamental characteristic of the scale-free design. The majority of nodes within the hub-and-spoke structure are located at the outer boundary of overall structure, and their loss does not affect any other nodes. This property, together with the high degree of connectivity within a scale-free network, means that even with as many as 5% of the nodes randomly removed, communication within the system will not be compromised. (Dipple, Phelan, and McCabe 2001)

But the good news is related to the bad news: failure of a hub-node can be catastrophic. This has some potentially interesting medical implications:

> Within molecular organization and communication of the cell, candidate sites for investigation of disease pathogenesis [the cellular events that occur in the development of a disease] will be those nodes with high connectivity. These highly connected nodes will include molecular "relay stations," represented by communication "bottlenecks" between modules, and "trunk lines," represented by major pathways between critical nodes. Cellular relay stations would include, for example, a metabolite transporter with high degrees of complexity on either side of the membrane, or a signaling molecule linking multiple input and output modules. Cellular trunk lines would include a critical metabolic pathway, or portion thereof, with limited alternative options or redundant paths through its course, or a transcriptional cascade linking key input variables with output responses (Dipple, Phelan, and McCabe 2001).

We will meet modularity again in our discussion of developmental regulatory networks in a later chapter where we discuss some basic concepts from evolutionary developmental biology—a context in which analogies with biochemical networks can be drawn quite freely. But for the present, is there any evidence to support the claim that biochemical networks are scale-free networks?

Jeong et al. recently performed a comparative, bioinformatic metabolic analysis of the metabolic networks in 43 organisms drawn from all three domains of life. Despite variation with respect to constituents and pathways, the metabolic networks were found to be scale-free networks. In these networks, substrates were represented as nodes with reactions being represented by connecting lines. Jeong et al. discovered:

> . . . only around 4% of all substrates that are found in all 43 organisms are present in all species. These substrates represent the most highly connected substrates found in any individual organism, indicating the generic utilization of the same substrates by each species. In contrast, species-specific differences among organisms emerge for less connected substrates. . . Thus the large-scale structure of the metabolic network is identical for all 43 species, being dominated by the same highly connected substrates, while less connected substrates preferentially serve as the educts or products of species-specific enzymatic activities. (Jeong et al. 2000)

It is too early to say what the properties are of other biological networks (for example, developmental networks), but this may be expected to be a matter to be illuminated as techniques of bioinformatics are extended to this and allied fields where networks are important. Interestingly, it has recently been shown, in the construction of a protein interaction map for *D. melanogaster*, that the network of interactions resemble those of a scale-free network (Giot et al. 2003). Similar observations have been made about the structure of the genetic interaction network in yeast (*S. cerevisiae*) (Tong et al. 2004).

Once again, mathematical modeling has a legitimate role to play in our understanding of biological systems of interest so long as it is clearly understood that key dynamical properties exhibited by the models must have analogs in the physical, biological systems modeled—correspondences which must be established through careful analysis of the real biological systems themselves, lest model and the system modeled drift apart!

In this chapter we have seen that models enter science in a variety of ways (and our presentation is by no means exhaustive). We have also seen that aspects of the behaviors of complex biological systems can be described with the help of physical models and mathematical models. But what of the use of animals in biomedical research, where members of one species are to serve as predictive models for complex biological phenomena anticipated in members of other species? To this matter we now turn. As a foretaste of things to come in the chapters that follow (and as we have just seen here), a major issue of concern will be the relationship between the animal model and the humans modeled, and the conceptual and empirical bases for the establishment of these relationships.

CHAPTER 6

◆

ANIMALS AS MODEL SYSTEMS

The price of employment of models is eternal vigilance.
—Richard Braithwaite (Braithwaite 1953)

Causal Analog Models

Causal analog models (CAMs) play an important role in various types of biomedical research, as well as other areas of scientific activity (LaFollette and Shanks 1995; LaFollette and Shanks 1996). Animal modelers such as Marilyn Carroll and Bruce Overmier, in their book, *Animal Research and Human Health*, state that animals are causal analogical models and thus can be used to study human disease and predict human response: "When the experimenter devises challenges to the animal and studies a causal chain that, through analogy, can be seen to parallel the challenges to humans, the experimenter is using an animal model (Overmier and Carroll 2001)."

The basic idea is simple enough. We study one physical system and the way it responds to perturbations in order to make predictions about how another physical system will respond in similar circumstances (once differences in size or mass have been compensated for). In biomedical research, typical *CAMs* (e.g., mice) and the systems they model (e.g., men) are complex, organized systems.

To understand how these models are supposed to work, it is helpful to begin with an abstract characterization. Causal analogical models are a proper subclass of *analogical models (AMs)*. An analogical model may be characterized as follows:

> (*AM*) *X* (the model) is similar to *Y* (the subject modeled) with respect to a range of properties $\{a, b, c, d, e\}$. *X* has the additional property f. It is likely that *Y* also has the additional property f.

In an analogical model, known similarities with respect to a range of properties lead to expectations about the existence of further similarities. These expectations need to be tested, and even if found to be correct, they at most establish further similarities between the model and the system modeled. Moreover, the discovery of f in the subject modeled tells us nothing, in and of itself, about deeper causal relationships between $\{a, b, c, d, e\}$ and f. For all we know the correlations and associations are purely accidental. A Ferrari and a Fiat both have 4 tires, gears, spark plugs, a driver's seat and so on. The Ferrari is red in color. Perhaps the Fiat is too. None of this turns Fiats into Ferraris, and none of this makes possession of wheels, etc, a cause of the color of the paintwork. Mere analogies, useful in some contexts, can be misleading in others.

In early work in atomic physics, questions arose concerning the relationship between electrons and the nucleus of the atom. Lord Rutherford suggested that the electron orbited the nucleus as a planet orbits the Sun. The analogy was a useful heuristic device, and prompted questions about the shapes, velocities and stabilities of the electron orbits. Subsequent inquiries, after the development of quantum mechanics, however, showed that electrons were nothing like planets whatsoever. Instead they were objects that had very different characteristics from macroscopic systems, and obeyed very different rules.

CAMs are *AMs* that satisfy certain commonsense constraints. These are as follows:

[1] The common properties $\{a, b, c, d, e\}$ must be causal properties— that is, they must be effects of various causes that are themselves capable of bringing about effects.

[2] The feature f to be projected from model to system modeled should stand in a causal relationship to the properties $\{a, b, c, d, e\}$ in the model. Ideally, it should be a cause or effect of $\{a, b, c, d, e\}$.

[3] There should be no relevant causal disanalogies between the model and the subject modeled.

Under these circumstances, all other things being equal, the presence of common properties $\{a, b, c, d, e\}$, and the causal relationship of f in the model to those common properties, gives some reason to suppose that f will be found in the system modeled, and found, moreover, to bear similar relationships to the properties $\{a, b, c, d, e\}$.

If the model and the system modeled are complex, organized systems such as organisms or economies, for example, there are various ways in which disanalogies can occur. First, the shared properties $\{a, b, c, d, e\}$ will typically be a proper subset of the actual set of properties of the systems under analysis. There will typically be other properties $\{s, t, u, v, w, \ldots\}$ and these may be *relevant* either to the occurrence of f (e.g., $f = 1$, or $f = 0$) or the magnitude of f (e.g., $f = 0.2$, or $f = 0.7$). Two systems may differ with respect to these other properties. Second, even if the shared properties $\{a, b, c, d, e\}$ are the only relevant properties, because of interactions among *these* properties, the occurrence of f may depend on the relationships that obtain between them (how they stand to each other). For some relations $R_i (a, b, c, d, e), f = 1$, for other relations $R_j (a, b, c, d, e), f = 0$, and for still other relations $R_k(a, b, c, d, e), f = 0.634$ (for example). Here, even if the model and the subject modeled agree on $\{a, b, c, d, e\}$, if $R_i \neq R_j$, all bets are off with respect to f. We will explore some consequences of this point about relations between components of interactive systems when we examine genetic networks in the context of evolutionary-developmental biology in a later chapter.

These points are all consistent with the principles of causal determinism: that all events have causes, and for qualitatively similar systems, same cause is followed by same effect. But it is here that dynamical systems theory, discussed in the mathematical interlude in the last chapter, becomes relevant, since in biology, for example, we are dealing with complex, interactive nonlinear systems. Our point here is that dynamical systems theory provides conceptual tools for analyzing aspects of the logic of causal analogical inference. At the very least, it supplies an important set of explanatory metaphors to help us conceptualize the basic issues as they pertain to modeling generally. (The analogy here is with the ways in which mechanical metaphors shaped the understanding of the early anatomists and physiologists in the renaissance).

In this context, we might think of *real* biological systems *as if* they were physical realizations of *abstract* mathematical models—in much the same way that the BZ reaction *in vitro* is a physical realization of the abstract Brusselator model. From our earlier discussion of dynamical systems theory it can be seen that there are several ways in which differences between model and the system modeled may be significant for our expectations about one system on the basis of observations of the other.

First, due to differences in parameter values (either internal to the system, or external to it), there may be qualitative differences in the dynamics exhibited by model and system modeled. In terms of their respective abstract mathematical models, their basin portraits may differ (attractors accessible to one system may be absent in the other, and so on). Second, even if the parameters are held constant in model and system modeled, the initial conditions may be such that the model can be considered, abstractly, to begin in one basin of attraction and the system modeled in another, so while the basin portraits are the same, long-term dynamical behaviors may differ because different attractors were accessed in model and system modeled. Third, if there is sensitivity to initial conditions, then even if model and system modeled have the same basin portrait and access the same (strange) attractor, minuscule differences beyond experimenter control may have the effect of generating radically divergent behaviors in model and system modeled.

If there are qualitative differences between systems—as there typically are when different species, with different evolutionary histories, play the role of model and system modeled—dynamical problems may be expected to accumulate. Indeed, because small differences between systems can be very important, intraspecific variation of the kind that typically exists in natural populations of organisms can be significant, raising questions about the use of one variant of a given species to model another. This is all by way of saying that when one complex system is used to model another, there are no straightforward expectations about similarity of stimulus being followed by similarity of response in both model and system modeled. (Compare this with the straightforward extrapolation, using $F = ma$, and given knowledge of how a known mass m_1 accelerates in a particular way to an applied force, we can make a larger system of known mass m_2 accelerate in the same way by applying a larger force.)

The study of the issues generated by the mathematical study of complex systems becomes relevant to the use of *CAMs* in biomedical research, because when one organism is used to make predictions about another, one highly complex, organized biological system is being used to make predictions about another that will typically be different in important respects (even two clones may differ with respect to initial dynamical condition, if only as a result of differential environmental exposures). There are good theoretical reasons, from both evolutionary biology and dynamical systems theory, for being cautious about non-human animal *CAMs* of human biomedical phenomena. But first we need to examine the properties of prediction and how the word is used.

Homology and Isomorphism

Theorists who discuss predictive models also discuss the relationship between the model and the subject modeled by considering whether the two are *homologous* or *isomorphic*. Since these terms are technical terms that have very different meanings in different theories, it will not go amiss to try to clarify their meanings in the context of the present discussion of modeling.

As we saw in the first chapter, biologists from different fields mean different things when they employ the word *homology*. Part of the reason for this is that it is a word with a curious history. *Homology* literally means the study of sameness (and homogeneous substances are the same throughout). But the word came to have special biological connotations prior to the advent of Darwin's theory of evolution in the works of the anatomist Robert Owen (circa 1847), where it came to refer to the study of *similar structures*. As Wake has recently commented:

> Owen envisioned a Platonic archetype. . . Two structures in different organisms, such as the seventh cervical vertebra of a mouse and of a monkey, are *special homologues* as two versions of the same structure. They are also each and collectively *general homologues* of a vertebra in the archetype that includes vertebra but not necessarily the seventh cervical vertebra. The seventh cervical vertebra was to Owen the *serial homologue* of the first thoracic vertebra, the eighth caudal, and so on in the same organism. (Wake 2003)

In this picture, there are abstract anatomical archetypes, and particular structures in particular physical organisms can be regarded as instantiations or manifestations of these archetypes. Structures are homologous if they are instantiations of the same archetype.

After Darwin's theory of evolution began shaping the intuitions of biologists, terms such as *homology* and *analogy* took on very different meanings, meanings that reflected the key evolutionary idea that organisms in different lineages descend historically from common ancestors and exhibit evolutionary modifications after divergence from their common ancestors. In these terms, two organisms have homologous characters if the characters have some recognizable similarity and arose from a character in a common ancestor. By contrast, characters are said to be analogous if they have some similarity, but arose independently as the modifications of characters inherited from different common ancestors. Sparrow's wings and penguin's flippers are homologous, even though the functions they serve are different (flying and swimming, respectively). The line leading to penguins and the line leading to sparrows diverged from the line of a winged bird ancestor. Bird wings and bat wings, by contrast are analogous because though they serve the same function of enabling flight, they arose independently from different wingless ancestors [(Lewontin 1995) p11].

In these terms, the fact that two characters are homologous does not imply that they serve similar functions (or indeed are instantiations of one and the same abstract anatomical archetype). For the evolutionary biologist, judgments of homology depend on careful historical analyses of the evolutionary origins of structures.

In contrast to both Owen's and Darwin's views on homology is the concept of homology as it is employed in molecular biology where it has been applied to the study of sequences of DNA nucleotides and sequences of amino acids. Two such sequences, if they are similar in sequence, are said to be homologous. Similarity of sequence is similarity of molecular structure. Homology in this sense does not necessarily refer to derivation from a single common ancestor. Chromosomes in a diploid individual are homologous if they have the same pattern of genes along the chromosome (the nature of the homologous genes may, of course, differ). Genes in different species may be homologous if there is a recognizable nucleotide sequence similarity. Homologues of human genes have been found in fruit flies, for example. Similarities are not necessarily absolute identities (for example, the amino acid sequence in human cytochrome c may differ from that in a horse or in baker's yeast with respect to the position of certain amino acids in the sequence), but the proteins are nevertheless homologous because they are recognizably similar in sequence structure. The explanation of these latter homologies is presumed to lie in descent from common ancestors with highly conservative modification in the respective organismal lineages. But a careful historical analysis of evolutionary origins is typically not undertaken. The judgment that the sequences are homologous depends only on recognizable sequence similarity. Nevertheless, function does not, of necessity, follow form or structure, and homologous proteins may perform very divergent functions in distinct organisms.

In contrast to homology, the word *isomorphism* is a word meaning literally *same form*. Today it is a theoretical term that finds precise meaning in mathematical contexts, and is used by analogy in other contexts. It will not go amiss to discuss the mathematical usage.

Suppose we wish to define the set of complex numbers, C. We begin with the set of real numbers, R, and construct the Euclidean plane, R^2, by taking the Cartesian product $R \times R$. The set R^2 is then the set of all ordered pairs (x, y) where x and y are real numbers. (The pair (x, y) then denotes a point in the Euclidean plane). A complex number is simply an element of R^2, i.e., an ordered pair of real numbers. Consider now the set of complex numbers of the form $(x, 0)$. This subset of the complex numbers is essentially a disguised version of the set of real numbers. It is said to be *isomorphic* to the set of real numbers. There is thus no point distinguishing between *(x, 0)* and *x* (Devlin 1981).

Animal models have been characterized using the terms *homology* and *isomorphism*. It is a good question as to what animal modelers are referring to when they speak of homology and isomorphism as properties of predictive animal models. Here is Nooneman and Woodruff:

> [T]he assumption is that to be useful a model must be completely *isomorphic* with that being modeled in all relevant relationships. Biomedical scientists are well aware that this is generally not the case with the models they use. Indeed there are many types of models, and their usefulness varies with the degree to which they are *isomorphic* with the human disease being modeled. [(Nonneman and Woodruff 1994) p7] (Emphasis added)

Nooneman and Woodruff continue, "If every aspect is fully isomorphic between the animal model and the human condition, including cause and mechanism, the model is *homologous*" [(Nonneman and Woodruff 1994) p9]. (Emphasis added.) They add:

> If a model is strictly *homologous* (complete *isomorphism* in all respects) then validity is guaranteed. But the converse is not true. A model that is not *homologous* with the human condition it represents may still have some validity. The key to validity rests in the choice of the right model for the right reason, not necessarily in *homology*. [(Nonneman and Woodruff 1994) p10]

(Homologous models are contrasted with *analogous* models, where there is only partial isomorphism).

The first thing worth noting here is that terms such as *homology* and *analogy* are *not* being used in the way that evolutionary biologists use these terms. Homology evidently means complete isomorphism (same causal pathways, same effects). In this sense, if rats, for example were said to be homologous to humans, they would in effect be humans writ small. Allow for differences in size, and same cause will be followed by same effect. This is very much in the physiological tradition of Claude Bernard. Analogous models are analogous not in the evolutionary sense, but rather because they are *partially* isomorphic (similar in some respects, but different in others). What could this mean? One thing it might mean is that humans and the animals used to model them could be conceived of as clusters of relatively independent modules. For a grossly simplified example, suppose rats consisted of two loosely coupled modules $M_1 = \{a, b, c\}$ and $M_2 = \{d, e, f\}$, and humans consisted of two loosely coupled modules $M_1 = \{a, b, c\}$ and $M_3 = \{x, y, z\}$ (for $a, b, c, d, e, f, x, y, z$, causally significant variables whose values at any time characterize the state of the modules they belong to). The organismal systems are partially isomorphic in that they have a module in common, and if this module is only loosely coupled to other modules in the respective organisms, one might hope to predict phenomena in the human M_1 module by studying the rodent M_1 module (making due allowance for perturbations from modules not common to both systems). We ignore here cautions about extreme sensitivity to initial conditions, as well as issues relating to structural stability—issues discussed at the end of the last chapter.

However, while it is true, as we shall see later, that from an evolutionary standpoint organisms can be thought of in modular terms—so changes in one module, in the course of *evolutionary time* (time over many successive generations), do not require corresponding changes in all the other modules to which it is loosely coupled—it does not follow from this that the modules constitutive of a particular organism, in the course of its life cycle, are only loosely coupled in *physiological time*. (In physiological time, brains, hearts, livers and kidneys conspire rather closely to maintain organismal homeostasis and viability.) Furthermore, it does not follow from the fact that evolution is modular, that humans and rodents have some modules in common that are isomorphic, at least in the sense discussed here.

Moreover, the fact that modules in two species that are homologous in the *evolutionary sense* does *not* guarantee that they will be homologous in the very different *animal modeler's sense*. Thus, Lewontin has observed in the context of an evolutionary analysis of animal modeling:

> When we study some characteristic in a model animal, all we really care about is that there really be a very close similarity between the traits at the level of anatomy, physiology and function that are of concern to us. The similarity at the appropriate level is not the same as [evolutionary] homology. A penguin's flipper has the same bones as a bird's wing, and when we study the internal anatomy of these two structures, we really do see that they are derived from the same ancestral bone arrangement. But if we want to study the dynamics of aerial flight, the penguin's flipper is a terrible model of the bird's wing. It has the wrong shape, and it has no feathers which are essential in bird flight. If all I knew about bird wings was what I

learned from studying penguin's flippers, I would conclude birds can't fly. (Lewontin 1995)

So when is a model the right model for the right reason? Nooneman and Woodruff tell us:

> In most cases models represent a compromise. A relatively simple experimental system is used to represent a more complex and less readily studied system. The model may represent only one aspect of a biological or behavioral system, but it is used because of the complexity of the natural system or because we believe we are modeling the most relevant component(s) of the problem, and the model allows a test of that belief. [(Nonneman and Woodruff 1994) p10]

An issue to be confronted later in this volume is whether complex, organized systems such as rodents can meaningfully be thought of as being simpler than human subjects (as opposed to being differently complex). Nevertheless, Nooneman and Woodruff discuss the issue of the relevance of phylogenetic closeness between the model organisms and the human subjects modeled:

> It is typically assumed that the more closely a species is related to humans the better the model it will provide. In some cases this is likely true. . . But presumed similarity of relationship to humans is not the only, and not necessarily the best, consideration in developing an animal model. Sometimes species that appear very dissimilar to humans in some respects share particular characteristics that suit them especially well to answer specific questions. [(Nonneman and Woodruff 1994) p11-12]

These observations by Nooneman and Woodruff seem right. Pragmatic considerations can thus sometimes trump the degree to which model and subject modeled are related in evolutionary terms. Again, Lewontin comments:

> Conversely, I may learn a great deal about a structure or function by studying another structure that is only analogous [in the evolutionary sense]. An octopus has a "camera" eye, as we do. That is, the eye has a focusing lens and a sensitive" photographic plate," the retina, on which images are formed and are translated into nerve impulses that are sent to the brain. The octopus eye originated in evolution completely independently of the vertebrate camera eye, yet

for some purposes of studying how lens defects may cause distortions on the retina, it would be a perfectly adequate model system. [(Lewontin 1995) p13]

But even in this case there are hidden problems. In the vertebrate eye, the blood vessels and nerves that serve the light sensitive cells are in front of these cells—so light has to travel through these nerves and blood vessels on the surface of the retina before it hits the rods and cones. The analogous eye of the octopus has the nerves and blood vessels *behind* the retina. These anatomical differences turn out to be medically relevant in the context of diabetic retinopathy, where bleeding from the vessels in front of the retina can lead to visual impairment—not a problem for the octopus whose vessels are behind the retina. It also needs to be noted that the phenomena involved in drug metabolism and its effects are often orders of magnitude more complex than the study of optical defects.

Hau also uses the terms *homology* and *isomorphism*. He remarks, "An animal model may be considered *homologous* if the symptoms shown by the animal and the course of the condition are identical to those of humans. Models fulfilling these requirements are relatively few, but an example is well-defined lesion syndromes in, for instance, neuroscience [(Hau 2003) p3]." Hau continues:

> An animal model is considered *isomorphic* if the animal symptoms are similar but the cause of the symptoms differs between human and model. However, most models are neither *homologous* nor *isomorphic* but may rather be termed "partial." These models do not mimic the entire human disease but may be used to study certain aspects or treatments of the human disease. [(Hau 2003) p3]

This use of isomorphism differs from that of Nooneman and Woodruff who appear to think in modular terms. Hau, by contrast, seems to be suggesting that models are isomorphic if the effects of underlying causes (the symptoms) are the same, while the underlying causal pathways leading to those effects may differ. The obvious problem here is that if the pathways are really different, then the fact that there are symptomatic similarities under one set of conditions by no means guarantees that those similarities will remain when the conditions (inputs into the different causal mechanisms) are varied. Models that are isomorphic in Hau's sense may be of value in the context of basic research—they may constitute *exploratory models*.

Hau's comments about lesion studies in the context of homologous models also deserves some discussion. Even for phylogenetically close relatives, say humans and primates such as chimpanzees or gorillas, the evolutionary phenomenon of recruitment or exaptation, whereby a structure that evolved to serve one function is co-opted or recruited in the course of evolutionary time to serve another function, can be a serious complicating factor. This is especially true in lesion syndromes studied in neuroscience. Lesion studies and electrical stimulation experiments reveal that roughly the same areas of the brain that are involved in the grunting and screeching vocalizations in chimpanzees are involved in complex speech in humans. As Lewontin has observed:

> When a region called Broca's area is damaged in humans, various speech disorders, aphasias, result that are not motor malfunctions, but have to do with making and comprehending sentences. . . When Broca's area is stimulated in lower primates, they move their tongues, lips, and mouths, but they do not make any vocalizations. So a region of the brain that is associated with speech production and comprehension, is nothing but a motor area in lower forms. . . What happened in the evolution of human beings is that certain regions of the primate brain were recruited from their simpler motor functions to create quite new functions, the functions of speech . . . There is no [evolutionary] homologue of speech in chimpanzees, certainly not grunting. [(Lewontin 1995) p16]

It is worth noting that the slender linguistic accomplishments of chimpanzees who have been taught American Sign Language (because they lack anatomical features needed for speech) are no better than those of dolphins, from whose lineage modern humans diverged some sixty million years ago [(Shanks 2002) chapter 11].

Hau's Taxonomy of Animal Models

The Handbook of Laboratory Animal Science is a good place to begin our examination of animal models according to the animal model community. There Jann Hau observed:

> What is generally understood by the term "animal model" is modeling of humans. It is not the image of the used animal that is the focus of research but the *analogy* of the physiological behavior of this animal to our own (or another) species. It would thus be more correct to refer to animals as "human models" in this context. Laboratory animal science, comparative medicine, and animal experiments are indeed much more about humans than about any other animal species. The significance and validity with respect to usefulness in

terms of "extrapolatability" of results generated in an animal model *depend on the selection of a suitable animal model.* [(Hau 2003) p2] (Emphasis added.)

How do we know, in advance of testing in humans, which animal model is the correct one for the phenomena we are studying? To this last question Hau has made the following suggestions:

> A good knowledge of comparative anatomy and physiology is an obvious advantage when developing an animal model. Animal models may be found throughout the animal kingdom, and knowledge about human physiology has been achieved in species far removed from the human in terms of evolutionary development. A good example is the importance of the fruit fly for the original studies of basic genetics. [(Hau 2003) p2]

We agree that as much knowledge as possible of the animal in question should be obtained before deciding upon its use as a model. As Hau is presumably aware, however, fruit flies play a rather different role in basic genetics, from that played by rats and mice, for example, in studies of the causes of, and cures for, cancer in humans; or of the relative safety of exposures to chemicals in the context of low-dose risk assessment.

Hau helpfully differentiates three concepts of *model* in the context of biological research. First there are *exploratory models*:

> A plethora of animal models has been used and is being used and developed for studies of biological structure and function in the human. The models may be *exploratory*, aiming to understand a biological mechanism, whether this is a mechanism operative in fundamental normal biology or a mechanism associated with an abnormal biological function. [(Hau 2003) p2]

Exploratory models are likely to play a crucial role in the context of basic biological research.

Secondly there are *explanatory models*. For the present we note that Hau characterizes these models as follows:

> Models may also be developed and applied as so-called *explanatory* models, aiming to understand a more or less complex biological problem. Explanatory models need not necessarily be reliant on the use of animals but may also be physical or mathematical model systems developed to unravel complex mechanisms. [(Hau 2003) p2]

Explanatory models will likely play roles in both basic and applied research. In the last chapter we distinguished between mathematical models and functional models, and gave an example of a chemical system that generated a functional model of biological phenomena—a physical model, moreover, with an accompanying mathematical model. Hau's concept of explanatory model covers models of both types.

Last, but by no means least, there are what Hau terms *predictive models*. Such models are characterized by Hau as follows:

> A third important group of animal models is employed as *predictive* models. These models are used with the aim of discovering and quantifying the impact of a treatment, whether this is to cure a disease or to assess toxicity of a chemical compound. [(Hau 2003) p2]

Predictive models play a crucial role in applied research. This is how animal models are, in fact, used in the context of drug testing and studying human disease. Animals in the case of predictive models are clearly used as substitutes for human subjects. This use of animal models is very much in the spirit of Claude Bernard's methodological prescriptions, which is why we think his work is far from irrelevant in the context of the current debate. Unless researchers believed that such predictive models were causally analogous to humans in relevant respects, there would be no rational basis for their use as predictive models. In the language we will use in the next section, animals used as predictive models are being used as *causal analog models* (*CAMs*). The critical arguments in the rest of this volume focus primarily on the use of animals as *CAMs* of human biomedical phenomena.

In light of the above, Hau continues his discussion by providing a valuable taxonomy of animal models. It will be helpful to briefly examine some of the details. The taxonomy is as follows:

1. Induced (experimental) disease models
2. Spontaneous (genetic) disease models
3. Transgenic disease models
4. Negative disease models
5. Orphan disease models. [(Hau 2003) p3]

Here we will examine the taxonomy and offer some comments on the different types of models in the light of our discussion of the properties of models in the last section.

We will start with a consideration of *induced (experimental) disease models*. In this regard, Hau comments:

> As the name implies, *induced models* are healthy animals in which the condition to be investigated is experimentally induced, for instance, the induction of diabetes mellitus with encephalomyocarditis virus, allergy against cow's milk through immunization with minute doses of protein, or partial hepatectomy to study liver regeneration. The induced-model group is the only category that theoretically allows a free choice of species. [(Hau 2003) p3]

Note that none of the models mentioned mimic humans in the etiology of the condition, which is very important. Induced models are prone to err in that the researchers may confuse the results of a disease—loss of pancreas function—with the cause of the disease. Simply cutting out a section of monkey brain may induce tremors but that hardly means the monkey has Parkinson's disease. Indeed, Hau himself concedes, "Few induced models completely mimic the etiology, course, and pathology of the target disease in the human [(Hau 2003) p3]."

The next category concerns *spontaneous (genetic) animal disease models*. Concerning such models, Hau observes:

> These models of human disease. . . use naturally occurring genetic variants (mutants). Many hundreds of strains and stocks with inherited disorders modeling similar conditions in humans have been characterized and conserved (see, e.g., http://www.jax.org). A famous example of a spontaneous mutant model is the nude mouse, which was a turning point in the study of heterotransplanted tumors and, for instance, enabled the first description of natural killer cells. [(Hau 2003) p4]

Spontaneous models are valuable research tools in the context of basic biological research in that they can serve as heuristic devices and prompt hypotheses about basic physiological mechanisms. It is less clear that such models can serve as *CAMs*, for as Hau himself points out:

> . . . if the object of a project is to study the genetic causes and etiology of a particular disease, then comparable genomic segments involved in the etiology of the disorder - *construct validity* - is normally a requirement. It should be remembered, however, that an impaired gene or sequence of genes very often results in activation of other genes and mobilization of compensating metabolic processes.

These compensatory mechanisms may of course differ between humans and the animal model species. [(Hau 2003) p4]

In this regard, much interest has focused on transgenic models. We will discuss transgenic animals later.

The next category concerns what are known as *negative disease models*, concerning which, Hau observes, "*Negative model* is the term used for species, strains, or breeds in which a certain disease does not develop, for instance, gonococcal infection in rabbits after an experimental treatment that induces the disease in other animals [(Hau 2003) p5]." The idea is, apparently, that if we could establish what it is that allows the rabbit to resist gonococcus, then we could exploit that knowledge to prevent the disease in humans. We find this suggestion problematic. AIDS is a good example. We studied chimpanzees for decades only to find out that they did not suffer from AIDS. Did we learn how to prevent or cure AIDS as a result of this knowledge? No. There are many evolved physiological differences between human and chimpanzees. Merely knowing what prevents members of one species from manifesting some pathological condition provides no rationale for supposing that organisms with different evolutionary histories will respond similarly to an analogous modification in their physiological makeup (members of different species may fail to succumb to a disease for different causal reasons, and so data from negative models will likely underdetermine which causal factors are relevant to the human biomedical condition).

The final category concerns what are known as *orphan disease models*. Hau explains the idea behind such models as follows:

> *Orphan disease model* is the term that has been used to describe a functional disorder that occurs naturally in a non-human species but has not yet been described in humans and that is recognized when a similar human disease is later identified. Examples include Marek's disease, Papillomatosis, bovine spongiform encephalopathy, Visna virus in sheep, and feline leukemia virus. When humans are discovered to suffer from a disease similar to one that has already been described in animals, the literature already generated in veterinary medicine may be very useful. [(Hau 2003) p5]

There can be little doubt that the veterinary literature in such cases can be a useful starting point, but relevance to the human condition will still have to be established through careful studies on humans.

Fidelity and Discriminating Power

Hau cautions that in selecting a predictive model one should not be misled by appearances, and to make this point he differentiates between the fidelity of the model and its discriminating power:

> The extent of resemblance of the biological structure in the animal with the corresponding structure in humans has been termed *fidelity*. A high-fidelity model with close resemblance to the human case may seem an obvious advantage when developing certain models. What is often more important, however, is the discriminating ability of the models, in particular the predictive models. When using models, for instance to assess the carcinogenicity of a substance, it is of the essence that at least one of the model species chosen responds in a manner that is predictive of the human response to this substance. [(Hau 2003) p2-3]

This is an important distinction. High fidelity has been alleged for both rodents and chimpanzees. In each case, claims about the relative phylogenetic closeness of these species to humans has been used to justify claims about high fidelity. As we will see in later chapters, phylogenetic closeness, even between humans and primates, guarantees little when it comes to extrapolation from one species to another, and hence the importance of the issue of discriminating power—the ability of the model to actually predict the human response. In this connection, however, we find Hau's claims about carcinogenicity testing to be interesting.

One can, *in retrospect*, usually find an animal that mimics the human condition. But what is often left unsaid in stating this observation is how many species were tested and failed to predict the human response. If one cherry picks the data, one can usually find an animal that mimics the human condition. The purpose of testing drugs and other substances, however, is to *predict* what will occur in humans, not to obtain 10 different results from 20 different species, one result of which will turn out to be correct. Which one of these divergent results is predictively correct can only be known in retrospect, after human trials or exposures. The retrospective cherry picking of animal models is a bit like saying that one picked 15 teams to win the World Series and one of them did, so therefore one predicted the World Series winner.

An example of this would be drug testing such as toxicity testing, carcinogenicity, or teratogenicity testing. There are roughly 1600 known chemicals that cause cancer in mice and other rodents, but only approximately 15 of these cause cancer in humans (Coulston 1980). Put another way, if the animal modeler said that all 19 drugs that cause birth defects in humans did so in at least some animals, that would be correct. But it would be a mistake on the basis of this to say that all human teratogens were *predicted* by animals. That would be an example of *sensitivity* but a complete lack of *specificity*.

The concepts of sensitivity and specificity are worthy of a brief discussion here. Suppose we have a new model by means of which we hope to study carcinogenicity. The first thing we need is a *gold standard*. Here we have to identify a group of substances that are known to have the property of interest (e.g., the property of being a carcinogen), relative to an accepted reference test. This test is known as the gold standard. Then, as noted by Hoffmann in his discussion of carcinogenicity testing using short-term *in vitro* mutagenicity tests:

> Sensitivity refers to the proportion of carcinogens that are positive in the test, whereas specificity is the proportion of noncarcinogens that are negative in the test. Both sensitivity and specificity contribute to the concordance of short-term tests with carcinogenicity, because concordance is simply the percentage of qualitative agreement (+ or -) between the results of different assays. [(Hoffman 1993) p216]

So sensitivity is the *True Positive Rate* of test. This is the probability that a carcinogen will have a positive test result:

> [1] Pr (T+ | C+) = the probability of a positive test result given that the substance is a carcinogen = number of carcinogens with positive test/ all carcinogens (identified by the gold standard). (Equations 1-6 are modified from (Knapp and Miller 1992).)

But as Hoffmann goes on to point out:

> Before arguing that sensitivity is more important than specificity, because false negatives can pose a health risk, whereas false positives cannot, one must acknowledge that the sensitivity of a test that is always positive is 100%, and such a test would be useless. To be useful a test must have a reasonable predictive value for both carcinogens and noncarcinogens. [(Hoffman 1993) p216]

Specificity is the *True Negative Rate*. This is the probability that a non-carcinogen will have a negative result:

[2] $Pr(T- \mid C-)$ = the probability of T- given C- = non-carcinogens with negative test/all non-carcinogens (again, identified by the gold standard).

In these terms we can define the *False Negative Rate*:
[3] $Pr(T- \mid C+)$ = carcinogens with negative test/all carcinogens.

The *False Positive Rate*:
[4] $Pr(T+ \mid C-)$ = non-carcinogens with positive test/ all non-carcinogens.

The *Positive Predictive Value* of the test:
[5] $Pr(C+ \mid T+)$ = carcinogens with positive test/ all substances with positive test.

The *Negative Predictive Value* of the test:
[6] $Pr(C- \mid T-)$ = non-carcinogens with negative test/all substances with negative test. (See Figure 6.1.)

Concordance, it will be recalled, is simply the percentage of qualitative agreement (+ or -) between the results of different assays. A new model can be evaluated in terms of its concordance with the model constituting the gold standard. In these terms we might hope to evaluate the adequacy of our *in vitro* mutagenicity test by comparing it to a suitable gold standard. The concordance of short-term mutagenicity tests with carcinogenicity has been evaluated by comparing the results of these tests with long-term tests in rodents. The trouble here is that rodents constitute a very shaky gold standard, for rodent assays often disagree with each other:

> Correspondence between mouse and rat, the two most commonly used species in carcinogenicity tests, is not especially high. For 73 compounds evaluated by Tennant et al. the concordance between mouse and rat was 67%. Moreover, in an evaluative study by Griesemer and Cueto, only 44 of 98 agents that were carcinogenic in either rats or mice were carcinogenic in both species. Besides interspecific variation, factors that can complicate the interpretation of carcinogenesis assays include the high spontaneous incidence of some tumors in rodents, the lack of replicate experiments in expensive bioassays, equivocal results owing to lack of statistical power, and shortcomings in dosage regimens or other aspects of assay design. (Hoffman 1993)

		Gold Standard	
		S+	S-
Test	T+	TP	FP
	T-	FN	TN

T+ = test positive
T- = test negative
T = True
F = False
P = Positive
N = Negative
S+ = standard positive
S- = standard negative

Sensitivity = TP/TP+FN

Specificity = TN/FP+TN

Positive Predictive Value = TP/TP+FP

Negative Predictive Value = TN/FN+TN

Figure 6.1. 2x2 Table for Prediction

Methodological Issues

Even with this elaborate characterization of the different types of animal model, methodological questions remain concerning the quality of the scientific results gained in experiments involving animal subjects. For example, Pound et al. have commented:

> Although randomization and blinding are accepted as standard in clinical trials, no such standards exist for animal studies. Berbarta et al. found that animal studies that did not report randomization and blinding were more likely to report a treatment effect than studies that used these methods. (Pound et al. 2004)

Pound et al. went on to highlight some of the methodological flaws observed in their examination of animal experiments (Pound et al. 2004). These were as follows:

- Disparate animal species and strains, with a variety of metabolic pathways and drug metabolites, leading to variation in efficacy and toxicity.
- Different models for inducing illness or injury with varying similarity to the human condition.
- Variations in drug dosing schedules and regimen that are of uncertain relevance to the human condition.
- Variability in the way animals are selected for study, methods of randomization, choice of comparison therapy (none, placebo, vehicle), and reporting of loss to follow up.
- Small experimental groups with inadequate power, simplistic statistical analysis that does not account for potential confounding, and failure to follow intention to treat principles.
- Nuances in laboratory technique that may influence results may be neither recognized nor reported—e.g., methods for blinding investigators.
- Selection of a variety of outcome measures, which may be disease surrogates or precursors and which are of uncertain relevance to the human clinical condition.
- Length of follow up before determination of disease outcome varies and may not correspond to disease latency in humans.

These methodological issues may account for some of the variable results reported in the animal experimentation literature. But even if they were properly dealt with, it would not follow that animal experiments would yield data relevant to humans. The remainder of this book explores reasons why this is so.

Horrobin poses the challenge this way, and it remains to be seen whether it can be sustained:

> *Does the use of animal models of disease take us any closer to understanding human disease? With rare exceptions, the answer to this question is likely to be negative. The reasoning is simple. An animal model of disease can be said to be congruent with the human disease only when three conditions have been met: we fully understand the animal model, we fully understand the human disease and we have examined the two cases and found them to be substantially congruent in all important respects.*

These conditions have not been fully fulfilled for any human disease, although perhaps the closest examples come from research into infections and endocrine-deficiency diseases. Even in infectious-disease research, the animal model is often very different from the supposed human disease because of differences in the immune response. We have largely forgotten that when we apply the term 'guinea pigs', rather than 'rats' or 'mice', to experimental human subjects, we do so because in the early days of animal research the guinea pig was the only common animal which reacted to some infections in a near-human way.

All the other animal models — including those of inflammation, vascular disease, nervous system diseases and so on — represent nothing more than an extraordinary, and in most cases irrational, leap of faith. We have a human disease, and we have an animal model which in some vague and almost certainly superficial way reflects the human disease. We operate on the unjustified assumption that the two are congruent, and then we spend vast amounts of money trying to investigate the animal model, often without bothering to test our assumptions by constantly referring back to the original disease in humans. (Horrobin 2003) (Emphasis added.) (See Appendix 5 for full Horrobin article.)

These are serious charges. They must be subjected to critical scrutiny from both a theoretical and an empirical perspective. We begin this process by looking at the theory of evolution.

PART III

♦

ANIMAL MODELS IN THE LIGHT OF EVOLUTION

CHAPTER 7

◆

EVOLUTION AND GENETICS

*A new scientific truth does not triumph by convincing
its opponents and making them see the light, but rather
because its opponents eventually die, and a new
generation grows up that is familiar with it.*

—Max Planck

In this chapter we will consider some of the important ways in which the science of biology has been transformed by the emergence of evolutionary theory and genetics in the 19th century, and the subsequent fusing of these two disciplines in the 20th century. These theoretical developments are important for the present purposes because they culminate in biological perspectives concerning the similarities and differences between species that are different in fundamental respects from those found in the physiological tradition in biomedical research established in the 19th century. The physiologists, working in the light of Bernardian methods and ways of conceptualizing biological systems, confronted problems that were somewhat different from those confronting the naturalists who established and developed evolutionary biology.

One major consequence of this is that there emerged a tension between the methods, theories and focus of biological inquiries (in and out of the laboratory) in these two traditions of biological research. These tensions will be seen to have an important bearing on the issue of expectations surrounding the use of animals as predictive or *causal analog models* (*CAMs*) of human biomedical phenomena. Needless to say, where there are differential expectations about some matter, the issue of evidence comes to the fore to settle the matter in question, or at the very least to open new avenues of inquiry into the puzzles in question.

In fact, Darwin's theory constituted a revolution in science precisely because it challenged an entire way of thinking that had guided the practice of science since the seventeenth century. Indeed, as Ernst Mayr has observed:

> Physiology lost its position as the exclusive paradigm of biology in
> 1859 when Darwin established evolutionary biology. When beha-
> vioral biology, ecology, population biology, and other branches of
> modern biology developed, it became even more evident how un-
> suitable mechanics was as the paradigm of biological science . . .
> [(Mayr 1988) p12]

In other words, the appearance of evolutionary biology challenged a way
of doing biology that had emerged from the rise of science itself during
the Renaissance. It did so by challenging the central metaphor of the
biomedical approach to biology, that of the *organism-as-machine*. The question we must briefly address now is this: if Darwin challenges the metaphor of nature as a machine, then how does his theory stand with respect
to the tradition of mechanistic physiology?

As will emerge in this chapter, these differences are so fundamental as
to constitute a basis for the claim that the two traditions in biological research embody incompatible views about the nature of biological systems
common to their respective research interests. Though we point to the
existence of this conflict in this chapter, and analyze some of its roots, the
analysis of the nature of this conflict, and its consequences for biomedical
research, will be developed in the chapters that follow, where these matters are examined in the light of genomics, and evolutionary-
developmental biology. For this reason, this chapter should be read, like
those that preceded it, as setting the background for the discussion and
analysis of a serious problem.

Evolution, Adaptation and the Origin of Species

For many scientists prior to Darwin, the observed complexity and organization of organisms, along with their adaptedness, i.e., their possession of
suites of characteristics suiting them to the occupation of their respective
places in nature, was seen by many to be so great as to demand a theological explanation. No less a figure than Newton himself saw in biology evidence of the hand of God, having occasion to remark:

> Opposite to godliness is atheism in profession and idolatry in prac-
> tice. Atheism is so senseless and odious to mankind that it never
> had many professors. Can it be by accident that all birds, beasts and
> men have their right side and left side alike shaped (except in their
> bowels); and just two eyes, and no more, on either side of the face
> . . . and either two forelegs or two wings or two arms on the
> shoulders, and two legs on the hips, and no more? Whence arises

this uniformity in all their outward shapes but from the counsel and contrivance of an Author? [(Thayer 2005) p65]

For Newton, morphological similarities were evidence of deliberate intelligent design. Atheism was odious because it could offer no good account of the similarities, save that they were, perhaps, fortuitous accidents. But Newton does not rest his case simply with the observation of morphological similarities. There is also evidence of adapted complexity:

> Whence is it that the eyes of all sorts of living creatures are transparent to the very bottom, and the only transparent members in the body, having on the outside a hard transparent skin and within transparent humors, with a crystalline lens in the middle and a pupil before the lens, all of them so finely shaped and fitted for vision that no artist can mend them? Did blind chance know that there was light and what was its refraction, and fit the eyes of all creatures after the most curious manner to make use of it? These and such-like considerations always have and ever will prevail with mankind to believe that there is a Being who made all things and has all things in his power, and who is therefore to be feared. [(Thayer 2005) p65-66]

For Newton, such adapted complexity had two possible explanations. First, that it was the result of intelligent design, or second, that it all came about by chance and happenstance. Newton is inclined to the former, as the latter is, and everyone will admit this, so implausible as to be silly and beyond belief. Part of Darwin's achievement, as we shall see, is to offer a third possibility, one that Newton never considered, to explain the same appearances: the morphological similarities and the existence of adapted complexity.

Darwin was not the first person to consider the possibility of biological evolution. His importance lies not in seeing evidence for evolution for the very first time in human history. His importance lies rather in the fruitful ways in which he conceptualized and explained what he, and others, had seen in nature. Like Bernard, Darwin is important for the science he conducted. Darwin's scientific interests lay not in physiology, but in geology, ecology and natural history. But also like Bernard, Darwin is important because he introduced methods, concepts and ways of thinking that have shaped the inquiries of subsequent investigators—investigators, who, in the long course of the 20th century, have entered scientific territory, especially genetics, physiology and biochemistry, of which Darwin knew little or nothing. But this is the very territory already occupied by investigators of Bernardian ancestry—creatures very different from the Darwinian invaders. For this reason alone, much is at stake in this debate from internalist and externalist perspectives in the history of science; and from the perspective of those interested in evaluating the conduct of scientists, and the rationales given for their research activities.

Charles Darwin published the *Origin of Species* in 1859. His achievement reflects his intellectual inheritance. Darwin is an heir to what William Whewell called *the method of gradation*, according to which, when confronted with two objects of inquiry that appear to be absolutely and categorically different, the hypothesis that they are indeed so distinct can be tested by seeking intermediate cases connecting them. Faraday, for example, undermined the claim that conductors were categorically different from non-conductors by showing that there existed a range of semiconductors of varying degrees of conductivity linking the two groups of objects.

In terms of the method of gradation, species differences would come to be seen by evolutionary biologists not as absolute, categorical discontinuities in nature, but as endpoints on a spectrum linked by a range of intermediate cases. Two closely related species will be similar to each other in some respects, and different in others. Consider humans and chimpanzees, for example. The line that leads to modern humans diverged from the line that leads to modern chimpanzees several million years ago. Evolution at the species level is thus a branching process. These two lineages diverge from the lineage of the common ancestor of humans and chimpanzees. Chimpanzees thus did not evolve into humans, rather humans and chimpanzees descend with modification from a common ancestor in the distant past that was neither human nor chimpanzee.

The similarities between humans and chimpanzees reflect this common evolutionary ancestry from a now extinct parental species. The differences between humans and chimpanzees reflect evolutionary modifications that occurred after the respective lineages diverged. Humans and chimpanzees, like Faraday's conductors and non-conductors, are connected by a series of intermediate cases. In the line leading to modern humans, for example, intermediate cases of interest concern our distant *Australopithecine* ancestors, *Homo habilis*, *Homo erectus*, and so on. This is known as the *principle of phylogenetic continuity*. If Darwin is right, then the taxonomic order seen in nature reflects the historical facts resulting from the operation of evolutionary mechanisms driving biological changes in populations over many successive generations. The taxonomic order in no way, then, reflects God's plans for critters.

In fact, Darwin's crucial insight may be found in his decision to consider the problem of evolution from the standpoint of *populations*. First of all, individuals come and go, but populations typically exist for many generations. Individuals live and die, reproducing if they are lucky, but they do not evolve. *Populations of individuals evolve over time.* Evolution thus occurs across generations and its pace is governed in part by generational time, which in humans is about twenty years, but in a microorganism like *Staphylococcus aureus*, may be as little as twenty minutes. One effect of evolution is to gradually change the way in which a population of organisms is structured—particularly with respect to the statistical frequencies of characteristics that are found in individuals making up the population. But what mechanism could bring about such effects in populations over many successive generations?

Darwin observed that members of natural populations of organisms typically show *variation* with respect to *heritable* traits. Since the dawn of agriculture, animal breeders had long exploited naturally occurring *intraspecific* (within species) variation to make new varieties: only animals with desirable traits (woolliness of coat, milk yield, domesticity, etc.) were allowed to reproduce and pass these traits on to the next generation, where the process would be repeated. Over time, animal breeders were able to change the way in which domestic populations of animals were structured. But natural varieties—Darwin called them *incipient species* [(Darwin 1970 [1859]) p39]—pervade nature and not just the farmer's yard.

What natural mechanisms might be at work to this end? Darwin's answer reflects his acquaintance with some ideas originally explained by Thomas Malthus (1766-1834) in his *An Essay on the Principle of Population* (1798). According to Malthus, much human misery arises from the tendency of populations to grow faster than they can increase food supply to support their numbers. Starvation, conflict and disease are the consequences of this process, and they are consequences whose effects trim expanding populations back, changing their structure in the process.

As applied to natural populations generally, this suggested to Darwin that a *struggle for existence* arises naturally from the fact that organisms tend to produce more offspring than can be supported by the environment:

> Every being, which during its natural lifetime produces several eggs
> or seeds, must suffer destruction during some period of its life, and
> during some season or occasional year, otherwise, on the principle

of geometrical increase, its numbers would quickly become so inor-
dinately great that no country could support the product. [(Darwin
1970 [1859]) p41]

It is here, in the context of the superabundance of organisms, that heritable variation plays its crucial role.

In this ongoing struggle for existence, some organisms—variants—
will have characteristics that hamper their ability to survive and reproduce, other variants will have characteristics that enhance these same abil-
ities. Such traits aiding survival and reproduction are said to confer *fitness
advantages*:

> Owing to this struggle, variations, however slight and from whatev-
> er cause proceeding, if they be in any degree profitable to the indi-
> viduals of a species, in their infinitely complex relations to other or-
> ganic beings and to the physical conditions of life, will tend to the
> preservation of such individuals, and will generally be inherited by
> the offspring. [(Darwin 1970 [1859]) p32]

For Darwin, this mechanism is the primary engine of evolution:

> This preservation of favorable individual differences and variations,
> and the destruction of those which are injurious, I have called *Natu-
> ral Selection*, or the *survival of the fittest*. Variations neither useful nor
> injurious would not be affected by natural selection, and would be
> left either a fluctuating element, as perhaps we see in certain poly-
> morphic species, or would ultimately become fixed, owing to the
> nature of the organism and the nature of the conditions. [(Darwin
> 1970 [1859]) p44]

Natural selection thus works on heritable variation found in *populations*
of organisms. In the environment in which the struggle for existence
takes place, the traits favored by selection increase in frequency over successive generations, and they come to represent adaptations to the environment in which the struggle for existence occurs. *Adaptations* are those
features of organisms that are the quintessential fruits of the operation of
natural selection.

Importantly for our purposes, it was Darwin's contention that the
same selective mechanisms that bring about adaptations *within* populations of organisms will also, as an unintended by-product, gradually bring
about and amplify differences *between* populations great enough to consti-

tute their designation as separate species. In this way, Darwin's understanding of species differences reflects the method of gradation. What appear to be absolute categorical differences turn out instead to be extreme differences that arose gradually by degrees through the accentuation of differences between varieties of a given species. As Darwin put it:

> On the view that species are only strongly marked and permanent varieties, and that each species first existed as a variety, we can see why it is that no line of demarcation can be drawn between species, commonly supposed to have been produced by special acts of creation, and varieties which are acknowledged to have been produced by secondary laws. On this same view we can understand how it is that in a region where many species of a genus have been produced, and where they now flourish, these same species should present many varieties; for where the manufactory of species has been active, we might expect, as a general rule, to find it still in action; and this is the case if varieties are incipient species. [(Darwin 1970 [1859]) p108]

In this way, the varieties that result from microevolutionary processes within populations, are driven still further apart by the continued action of the same mechanisms so as to constitute new species in their own right. In this way, microevolutionary changes, continued long enough, give rise to macroevolutionary phenomena: the origin of new species.

Thus, the processes driving the origin and accentuation of varieties within a species, given enough time, will turn varieties into good and true species in their own right. Over long periods, these processes result in increasing biodiversity:

> As each species tends by its geometrical rate of reproduction to increase inordinately in number; and as the modified descendants of each species will be enabled to increase by as much as they become more diversified in habits and structure, so as to be able to seize on many and widely different places in the economy of nature, there will be a constant tendency of natural selection to preserve the most divergent offspring of any one species. Hence, during a long continued course of modification, the slight differences characteristic of varieties of the same species, tend to be augmented into the greater differences characteristic of species of the same genus. [(Darwin 1970 [1859]) p108]

This process of *adaptive radiation* explains what happens when animals from an ancestral species move into a multiplicity of *ecological niches*, each niche being characterized by a particular complex of features that affect an animal's way of making a living: nature and availability of food, type and number of predators, pathogens and parasites, climates, and so on. Thus, as small differences between populations accumulate through adaptive specialization over many successive generations, the *invisible hand* of natural selection will accentuate differences between these populations until they are so distinct as to be recognized as different species.

Darwin's argument can be neatly summarized as follows. There are 3 basic Darwinian principles:

[1] *Principle of variation.* Organismal populations naturally display variation with respect to their characteristics.

[2] *Principle of differential reproductive success.* In virtue of differences with respect to characteristics, different organisms in such populations leave different numbers of offspring behind.

[3] *Principle of heritability.* Characteristics leading to differential reproductive success will probably be inherited by the offspring of the organisms bearing those characteristics.

If these 3 principles are true of an organismal population, then over successive generations, the statistical frequencies with which characteristics are found in those populations will undergo changes and *adaptive evolution* will occur.

Adaptations are the characteristics of organisms that result from adaptive evolution. Sober gives the following definition:

Characteristic c is an adaptation for doing task t in a population if and only if members of the population now have c because ancestrally there was selection for having c and c conferred a fitness advantage because it performed task t. [(Sober 1993) p84]

This definition is helpful because it focuses our attention on populations and their history as part of the causal explanation of why organisms have certain characteristics. It is not enough to look just at the forces impinging on an object now, or even on events in Bernard's *milieu intérieur*, for an organism's very internal environment is a unique product of historical causes that may differ in important respects from those found in other organisms in the population, and more radically from those of organisms of different species in different populations. If Darwin is right, the prime effect of evolution is to accentuate differences between varieties within populations, and ultimately between populations themselves.

Classical Genetics and the Evolution of Population Genetics

Neither Darwin nor Bernard knew much about the mechanisms of inheritance or the principles governing inheritance. Darwin's argument relied on the fact—known to farmers for millennia—that certain characteristics can be inherited. These characteristics—adaptations—were functional properties of complex biological systems, and they can often be understood reasonably well even in the face of considerable ignorance of the precise means of their generation. Sober and Wilson have pointed out that an important feature of adaptive explanations is that they:

> . . . can be employed with minimal knowledge of the physiological, biochemical and genetic processes that make up the organisms under examination. For example, imagine studying the evolutionary effects of predation on snails, seeds, and beetles. Suppose you discover that for all three groups, species exposed to heavy predation have harder, thicker exteriors than species not so exposed. The property "hard exterior" can be predicted from the knowledge of the selection pressures operating on the populations. Since the exteriors of snails, beetles and seeds are made of completely different materials, there is a sense in which these materials are irrelevant to the prediction. That is why Darwin was able to achieve his fundamental insights in almost total ignorance of the mechanistic processes that make up organisms. Adaptationist explanations have the power to unify phenomena that are physiologically, biochemically and genetically quite different. [(Wilson and Sober 1994) p588]

An important point to bear in mind here is this: the unifying power of adaptive explanations in terms of functional properties arises because there are often multiple causal pathways to given functional properties (e.g., "hard exterior"). Where this is the case, we shall say that functional properties lack *substrate specificity* (i.e., they can be brought about by a multiplicity of lower level causal mechanisms employing different substrates. Complex systems may thus be similar to each other in some respects (some of their functional properties), but different in other respects (substrates, mechanisms and pathways, and possibly, indeed, other functional properties).

Modern genetics, however, can trace its ancestry back to the work of Gregor Mendel in the 1860s—work whose significance would not become apparent until after 1900 when it was rediscovered by Hugo DeVries and others. Here we will use this introduction to classical genetics as an opportunity to introduce modern terminology and ideas. Mendel was working with small groups of plants (*Pisum sativum*). His inquiries resulted in the formulation of principles of inheritance governing observable patterns of events. These principles are known as Mendel's laws. Mendel hypothesized the existence of *factors*—the so-called *particles of inheritance*. Today the particles of inheritance are called *alleles*. Alleles are variant forms of genes that exist with varying frequencies in organismal populations.

The *Human Genome Project* has revealed that the human genome contains fewer than 30,000 genes. Recently reported low estimates are at 24,500, of which 3,000 might be functionless pseudogenes. The mouse genome is about the same size as the human genome (Boguski 2002). This may be compared to approximately 13,600 for the fruit fly *D. melanogaster*, 18,400 for the worm *C. elegans* and about 25,500 for the plant *Arabidopsis* (Szathmáry, Jordan, and Pal 2001). Differences between *M. musculus* and *H. sapiens* will not be found simply by counting genes.

In diploid organisms—organisms with two sets of chromosomes, one from each parent (we are diploid organisms)—the matched pairs of chromosomes are called *homologous chromosomes*. In humans, barring chromosomal abnormalities, each cell contains 46 chromosomes (22 matched pairs, and one pair of sex chromosomes, with females having XX-pairs and males having XY-pairs). The number of matched pairs is called the *chromosome number*, n. In humans, $n = 23$.

The *locus* of a gene is its position on a chromosome. For a given locus, a population of organisms may contain two or more variant forms of the gene associated with that locus. These variant forms of a gene are called *alleles*. In diploid organisms (e.g., mammals), there are two alleles of any gene, one from each parent, which occupy the same relative position on homologous chromosomes. (Counterparts of genes found in humans have been identified in other organisms. Such genes are referred to as *orthologs*.)

Mendel's laws of inheritance for diploid organisms may be stated as follows:

[1] *The Law of Segregation*. Each hereditary characteristic is controlled by 2 alleles which segregate and pass into different germ cells in meiosis.

[2] *The Law of Independent Assortment.* Pairs of alleles segregate independently of each other during meiosis.

In Mendel's original work it was assumed that alleles determined organismal characters. To explain his observations in various crosses among his pea plants, Mendel classed alleles as *dominant* and *recessive* according to which was actually manifested in the organism. The dominant allele influences the particular characteristic that will appear in the ontogeny of an organism. It has since become apparent that matters are much more complex. It is possible for both alleles to be fully expressed; this is called *co-dominance.* There are also cases where neither allele is fully expressed, and a characteristic results from the partial expression of each. This is called *incomplete dominance.*

If we restrict attention to a particular locus and suppose there are two alleles A_1 and A_2 of a gene A, there are three possible allelic combinations: $A_1 A_1$, $A_1 A_2$, and $A_2 A_2$. These are then the three possible genotypes which an organism may exhibit.

The distinction between an organism and its genes underlies one of the most basic distinctions in genetics, that between *phenotype* and *genotype*:

> The "phenotype" of an organism is the class of which it is a member based upon the observable physical qualities of the organism, including its morphology, physiology, and behavior at all levels of description. The "genotype" of an organism is the class of which it is a member based upon the postulated state of its internal hereditary factors, the genes. [(Lewontin 2000) p138]

Corresponding to this distinction is that between *genome* and *phenome*:

> The actual physical set of inherited genes, both in the nucleus and in various cytoplasmic particles such as mitochondria and chloroplasts, make up the *genome* of an individual, and it is the description of this genome that determines the genotype of which the individual is a token. In like manner there is a physical *phenome*, the actual manifestation of the organism, including its morphology, physiology and behavior. (Lewontin 2000)

The phenome of an organism typically reflects the alleles present, the nature of the external environment (abiotic as well as biotic—e.g., predators, prey, pathogens and parasites), and the internal environment in which developmental interactions occur (Lewontin 2000). Genome does not determine or fix phenome.

Indeed, for a given genome G instantiating genotype $\mathbf{G}$, a function f: $E \rightarrow P$ (from environments encountered to phenomes) can be constructed. This function is known as the *norm of reaction*. One and the same genotype can give rise to a range of phenotypes. Phenotypic variability is thus a feature even of inbred or clonal lineages. The study of norms of reaction has been the context for much work on the phenomenon of *phenotypic plasticity*. The phenome of an organism can then be thought of in terms of the equation $P = G + E$, where G is the organism's genome and E is the environment actually encountered in the course of developmental time. The equation simply records the observation that phenome reflects contributions from the particles of inheritance, the alleles present, and the environment encountered.

In this context of gene-environment interactions, it is worth giving some consideration to the Baldwin effect and genetic assimilation of acquired characters. In the early 1950s, Waddington studied the effects of heat shock on the pupal development in fruit flies. Heat shock induced "crossveinless" phenotypes. After fourteen generations of selective breeding, some individuals became crossveinless without heat shock. As Gottlieb has recently observed, this is *not* a case of Lamarckian inheritance of acquired characteristics:

> . . . all phenotypes are a consequence of gene-environment interaction. Obviously in Waddington's experiment the genes were there from the start in the organisms that responded with the crossveinless phenotype. Breeding among themselves could have heightened their temperature sensitivity so that, after fourteen generations, some crossveinless individuals did not require the original heat shock but required only exposure to the inevitable small cyclic increases in temperature fluctuations that occur during normal incubation. (Gottlieb 2006)

Recently West-Eberhard (West-Eberhard 2003) has gone on to develop the importance of genetic assimilation effects in the context of an extended study of the evolutionary relevance of phenotypic plasticity.

Polygenic characteristics of organisms are generated by the contributions of many alleles (these characteristics are the subject of the study of *quantitative genetics*). Consider height in a population of humans. Typically there will be a mean height and a range of sizes above and below the mean. There will thus be phenomic variance $V_p = V_g + V_e$, where V_g is genomic variance, and V_e is variance in environments encountered. In quantitative genetics, *heritability in the broad sense* is defined as $V_g / V_p = V_g / (V_g + V_e)$. By contrast, *heritability in the narrow sense*, h^2, can be understood by realizing that $V_g = V_a + V_d + V_i$ where V_a represents variation among individuals due to the additive effects of genes, V_d represents variation among individuals due to dominance effects, and V_i represents epistatic interactions among genes at different loci. Only additive effects contribute to heredity in the narrow sense, even though dominance effects and epistatic interactions probably make a significant contribution to phenotypic variation among individuals. Thus $h^2 = V_a / V_p$, where $V_p = (V_a + V_d + V_e + V_i)$. Heritability so defined is the proportion of phenomic variance owing to genomic variance. It is thus a population statistic. In a clonal population, $V_g = 0$, and hence $h^2 = 0$. In such a case, though, V_p may be far from negligible if $V_e \neq 0$. In many natural populations $V_g \neq 0$ and $V_e \neq 0$. Concerning the environmental component of phenotypic variance, Hallgrimsson has recently observed:

> A useful distinction in evolutionary contexts is between environmental factors that contribute to differences among individuals and those that contribute to variation within individuals. This distinction defines the boundary between canalization and developmental stability. Canalization refers to the reduction of variation among individuals, whereas developmental stability refers to the minimization of variation within individuals. [(Hallgrimsson 2006) p369]

While it is tempting to leave the discussion of the components of variation here, a further contribution to phenotypic variation needs to be included. This concerns the effects of developmental noise—the consequences of random events within cells at the level of molecular interactions [(Lewontin 2000) p36]. Developmental noise is associated with the phenomenon of *fluctuating asymmetry*. In *Drosophila*, for example, the *average* fly is symmetrical with respect to sensory bristles on the left and right hand sides. *Individual* flies typically differ with respect to bristle number. These fluctuations reflect developmental noise. The long reach of developmental noise may be far from negligible, for, as Lewontin has recently noted:

> A leading current theory of the development of the brain, the selective theory, is that neurons form random connections by random growth during development. Those connections that are reinforced from external inputs during neural development are stabilized,

while the others decay and disappear. But the connections must be randomly formed before they can be stabilized by experience. Such a process of neural development could give rise to differences in cognitive function that were biological and anatomically innate, yet neither genetic nor environmental. [(Lewontin 2000) p38]

This and related phenomena will be discussed later under the heading of *delegated complexity*.

The quantity h^2 is important for studies of natural selection of polygenic characteristics because once we have h^2 and the selection differential S (the difference between the mean value of the characteristic of interest in the individuals that reproduce and the mean value for the characteristic for the entire population), we can calculate the response to selection R (i.e., the difference between the mean value found in the offspring and that of the population in the previous generation) as follows: $R = h^2 S$.

Fredrik Nijhout has recently observed that though Mendel thought that each allele coded for a single trait, the vast majority of traits are in fact affected by many genes (Nijhout 2003). To use his example, many genes are involved in pigment biosynthesis in flowers:

> Some of these genes code for enzymes that transform colorless precursors, such as amino acids and sugars, into variously colored pigments. These biosynthetic pathways can include more than a dozen steps, each regulated by a different enzyme. Other genes code for proteins that regulate enzyme synthesis and activity; these regulators affect the time and place where pigments are produced. Yet other proteins control the stability and subcellular localization of the pigments. The genes that code for these regulatory proteins are, in turn, regulated by another set of proteins called transcription factors, which are each encoded by a different gene. . . This kind of interminable regression of regulation and interaction among genes . . . appears to be the norm even for the simplest of traits. (Nijhout 2003)

How then, can a single allele appear to control the manifestation of a given trait? Nijhout's answer is to consider a biosynthetic pathway. A single allele can code for an enzyme that is a *rate limiting step*—the point in the pathway that most impedes the production of product to serve as substrate for the next step in the pathway. If this is the case, then variation in this gene will appear to control variation in the trait of interest (Nijhout 2003). The lesson here is that gene interactions are crucial even for simple traits. Single genes do not have special "powers" to generate phenotypic features, and it can be grossly misleading to implicate single genes as the sole causes of traits of interest. We will return to this point in a later chapter on transgenic modeling, where issues pertaining to oligogenic phenomena will become important.

Nijhout emphasizes the importance of context—not just in the production of traits, but also in the effects of mutations. These too turn out not to be intrinsic properties of the mutated gene. Nijhout observes:

> The importance of context is also illustrated by studies of the effects of "knockouts" of specific genes in mice, a method that completely eliminates the function of a gene's product. For example, knockout of a retinoblastoma-related gene causes severe abnormalities and embryonic death in one strain of mice, but the same mutation in another strain has no effect: The mutant mice are viable and become fertile adults . . . (Nijhout 2003)

This observation is likely to be relevant for expectations concerning the results of splicing human genes into the mouse genome, or extrapolating to humans the effects of knockouts of orthologous genes in mice. It will be explored later in this book.

Alleles, as noted above, are typically found with various frequencies in populations of interest. The transition from Mendelian genetics to the study of population genetics arose from arguments about the ways in which these allele frequencies might change from one generation to the next. The modern theory of evolution emerges from the study of the dynamics of allele frequencies in organismal populations under the effects of a wide range of processes and mechanisms.

Consider a single locus and suppose there are only two alleles A_1 and A_2. Let p and q be the respective relative frequencies (or proportions) with which these alleles are found in a population of interest. Since these are the only alleles then $p + q = 1$. In 1908, the mathematician G. H. Hardy assumed (a) a large population; (b) random mating; (c) sexes with equal allele proportions; and (d) uniform fertility. Since diploid organisms have two alleles at each locus, one from each parent, then the relative allele frequencies in the next generation, given values of p and q in the parental generation can be calculated by multiplying the relative frequencies with which these alleles are found in the sperm pool of the population under study by the relative frequencies with which they are found in the ova pool of the population. That is to say, $(p + q)^2 = p^2 + 2pq + q^2$. In this case, p^2 represents the relative frequency with which the $A_1 A_1$ genotype is instantiated in the next generation, q^2 represents the relative frequency with which the A_2A_2 genotype is instantiated, and similarly $2pq$ represents the relative frequency with which the A_1A_2 genotype is instantiated. *All other things being equal*, these relative frequencies will remain the same over successive generations. This result is known as the *Hardy-Weinberg Law*. But as with all *cateris paribus* clauses in scientific arguments, all other things are seldom equal. Modern evolutionary biology grows in large measure out of an understanding of the ways in which all other things can fail to be equal [(Price 1996) p316-18].

To the modern evolutionary biologist, evolution occurs when the frequencies with which alleles are found in populations change, *for whatever reason*. Natural disasters or epidemics can radically reduce population size resulting in a genetic bottleneck. Genotypes common on one side of the bottleneck may be rare or absent on the other. In a very small population, the loss or gain of a single individual can make a marked difference to relative allele frequencies. These relative frequencies can also change if there is assortative mating on the basis of phenotypes instantiated (though found in nature, assortative mating plays a prominent role in animal breeding, and what is known as *artificial selection*). Gene flow within and between populations (and its cessation) can have major effects on relative allele frequencies. Linkage of alleles on the same chromosome, along with genetic hitchhiking, sex-linkage, and interactions between genes at different loci can cause relative allele frequencies to change. Last but not least, there is natural selection as a mechanism bringing about change.

Natural selection operates on variation occurring in natural populations. Variation arises in several ways. *Recombination* is the process whereby genes are shuffled during *meiosis*—the formation of reproductive cells (sperm or egg)—and results in offspring having a combination of characteristics different from either of their parents. By contrast, *germ-line mutation* is the process that results in genetic changes in an organism's reproductive cells, and hence results in heritable changes in an organism's genetic constitution. Both these processes add to variation in populations. (There are other mutations called *somatic mutations* which result in genetic changes in cells other than the reproductive cells, and which are thus not heritable. These latter mutations may have adverse effects for the organisms possessing them, such as cancer, a type of disease given further consideration in this book under the heading of delegated complexity).

If a base-pair changes in an allele, this is called a *point mutation*. When structural alleles are expressed, they make proteins, which serve many functions and roles in the lives of organisms. A point mutation in a critical location on a gene can change the nature of a protein, for good or ill, because of its implications for survival or reproduction (and hence natural selection). Because the genetic code contains redundancies, point mutations may have no effect whatsoever, and so are sometimes said to be *neutral*. But when such changes occur, often more than one base-pair is affected. Common changes also include the deletion of extant base-pairs, or the insertion of additional base-pairs.

Important for our present purposes are genetic changes known as *duplications*. Entire genes can be duplicated, and when this happens, the resulting genome has two copies of a gene where before it had one. Duplication events are very important for evolutionary biologists. First because with two copies of a functional gene, one can continue its old job, while the new copy can undergo mutation and acquire new functions that can participate in the life of an organism in novel ways. This may have important implications for natural selection, by contributing to reproductive success. The process by which a gene acquires new functions in this manner is known as *exaptation*. Second, duplication is the way in which organisms acquire new genes. They do not appear by magic, they appear as the result of duplication. Duplications can also occur at the level of chromosomes, and can cause serious problems. Down's syndrome is a well-known result of chromosomal duplication. But entire genomes can be duplicated, with some very interesting consequences, as we will shortly see. These large-scale genomic duplications are discussed under the heading of *polyploidy*.

Mutations are sources of variation, and hence they help provide the grist for the evolutionary mill, for selective mechanisms can only operate on the organismal consequences of variation among the alleles present in a population. Mutations are neither good nor bad in and of themselves. Instead, one must always look at the consequences of the mutational change for the life of the organism that contains it. This will often mean an examination of the way an organism is trying to make a living in its ecological context, and the challenges it faces. However, it is worth noting at this point that some genes are *conserved*. This means they have stayed the same in many lineages. What this usually means is that these genes perform essential roles in enabling the basic functions needed for life. They are the same in many lineages because mutational variants are lethal or debilitating, and have been weeded out by selection. Other genes, especially duplicates, are much more tolerant of mutation, and therefore can play a positive role in evolution.

Important for evolution, then, is the existence of multiple alleles in populations of organisms. A given allele may be found with a given relative frequency in a population. Evolution occurs in a population when the relative frequency with which alleles are found in that population changes (for whatever reason) from one generation to the next. An important part of the *new synthesis*—the fusion of evolutionary theory with genetics—was the development of sophisticated techniques to analyze relative allele frequencies in populations. The resulting theory is thoroughly *allele-centered*. By this it is meant that what gets replicated are alleles, and it is alleles that travel down the generations—alleles, barring mutations, that are identical by descent. Underlying the heritable variation in developmental, morphological, physiological and behavioral characteristics observed in populations is variation with respect to alleles. Parents pass on alleles to their offspring who receive 50% of their alleles from each parent.

Natural selection changes relative allele frequencies in populations because alleles that contribute positively to reproductive success are more likely to find themselves in the next generation, in higher frequencies, than alleles that do not. Such alleles are said to confer *fitness* advantages. Members of a population of organisms typically differ from each other with respect to their *fitness*. Differences in fitness are defined in terms of differential reproductive success. Thus the effect of natural selection is to change the frequencies with which alleles are found in populations over time. As Ewald [(Ewald 1994) p4] has noted, natural selection favors characteristics of organisms that increase the passing on of the genes (alleles) that code for those particular characteristics.

Evolutionary biologists argue that biodiversity results from speciation: the complex array of processes that give rise to new species. But what exactly is a species? In pre-evolutionary views of species there are absolute discontinuities between species. Associated with these views of species were various *morphological* species concepts according to which species membership could be determined by reference to *shape* (especially the shapes of anatomical features) construed as a measure of form. This idea fell into disrepute first through the observation of *polytypic* species, in which individuals of a given species display a great deal of variation with respect to characteristics, especially morphological characteristics. Second there was the observation of *sibling species*. That is, good and distinct species that were sometimes so similar as to show no obvious morphological differences, implying that speciation can occur without change of form. A good example here is a type of frog that used to be known as *Rana pipiens* (we now speak of the *R. pipiens* complex). This was a standard frog-model in physiological research. But laboratories started getting anomalous results, and careful studies revealed that what had been thought of as one species was in fact at least fifteen similar species [(Berlocher 1998) p8].

What is needed is a way of thinking about biological species that reflects the facts of evolution. Any good textbook in evolutionary biology (Futuyma 1998) will provide you with an introduction to modern thinking about species and speciation, but the following observations will be helpful here. From the standpoint of modern evolutionary biology, species are "individuals" that exist in space and time. They come into existence with speciation events; while they exist, they have geographic (spatial) distributions, and they go out of existence with extinction events. So what are they? Evolutionary biologists interested in mammals and birds (organisms that reproduce sexually) formulated the *Biological Species Concept* (*BSC*) as a first attempt to deal with this issue.

The *BSC* was one of the early fruits of the new synthesis that gave rise to modern evolutionary biology. As formulated by Ernst Mayr, who is one of the architects of modern evolutionary biology:

> A species . . . is a group of interbreeding natural populations that is reproductively (genetically) isolated from other such groups because of physiological or behavioral barriers. . . Why are there species? Why do we not find in nature simply an unbroken continuum of similar or more widely diverging individuals, all in principle able to mate with one another? The study of hybrids provides the answer. If the parents are not in the same species (as in the case of horses and asses, for example), their offspring, mules, will consist of hybrids that are usually more or less sterile and have reduced viability, at least in the second generation. Therefore there is a selective advantage to any mechanism that will favor the mating of individuals that are closely related (called conspecifics) and prevent mating among more distantly related individuals. This is achieved by the reproductive isolating mechanisms of species. A biological species is thus an institution for the protection of well-balanced, harmonious genotypes. [(Mayr 1998) p129]

In these terms, morphologically indistinguishable sibling species, along with species whose members display a great deal of morphological variation, count as distinct species because they are reproductively isolated from other such groups of interbreeding natural populations.

From the standpoint of the *BSC*, it is necessary to think of species in terms of populations. A species may consist of a single population or several geographically distributed populations. The integrity of a species is thought of as being maintained by gene flow, that is, the exchange of genes within and between populations constitutive of the species. Consequently, processes and mechanisms that result in cessation of gene flow between populations are capable of driving the speciation process.

The central idea here is that with the cessation of gene flow between populations constitutive of a given species, genetic differences between those populations can accumulate to the point at which they become so different as to be reproductively isolated from each other (either physiologically or behaviorally). For example, with the cessation of gene flow between two populations adapting to new environments, mutations (contributing to variation among the alleles circulating in those populations) and natural selection (favoring some alleles at the expense of others) will drive genetic divergence between populations by bringing about changes in the frequencies with which alleles are found in those respective populations. Eventually these genetic divergences become so great that populations, once capable of interbreeding can no longer do so. At this point, speciation has occurred.

Many mechanisms capable of driving the occurrence of speciation events can be devised and tested in the laboratory [(Rosenzweig 1995) ch. 5]. Typical experiments might involve short-lived organisms, such as fruit flies, which can be subjected to various forms of selection and tracked in real time for 50 or more generations. Disruptive selection often plays a role in these experiments by favoring individuals with extreme traits at the expense of individuals with average values for those traits. (For example, if the trait was height, disruptive selection might work in favor of very short and very tall individuals—they would reproduce—while individuals of average height would face a reproductive penalty. The result of such selection, over many generations would be two populations, one made up of tall individuals, the other made up of short individuals.) In this regard Rosenzweig has recounted the following anecdote:

> Bruce Wallace once showed me a new species of *Drosophila* [a fruit fly] he and his graduate students produced in his laboratory at Cornell. It fed exclusively on human urine, a previously unexploited ecological opportunity for flies. They forced the speciation with artificial disruptive selection. Unfortunately the species is now extinct. The demigods at Cornell tired of the novelty and the fly lost its niche. [(Rosenzweig 1995) p105-106]

Which of the possible mechanisms (derived from theory and laboratory experiments) actually play roles in driving the speciation process in nature is a current matter of scientific inquiry, one requiring careful field observations.

Of particular interest in connection with the issue of actually observing the occurrence of speciation is the possibility of speciation through polyploidy (or genome duplication). As noted above, genome duplication is a mutational event. When it happens, the organism with the duplicated genome is reproductively isolated from its ancestors because it has twice the number of chromosomes. Speciation happening this way occurs in a single generation, and has been observed to do so. It is conservatively estimated that at least 30% of speciation in plants has involved polyploidy. Some plants can, of course, fertilize themselves, so being cut off from their ancestors, and their ancestors *other* descendants, is not so important as it would be for mammals, and their ability to hybridize more viably than animals is also believed to be important [(Li 1997) p395-6] [(Smith et al. 1997) p207-09]. Speciation occurring this way has been observed, and hence macroevolution, as well as microevolution, has been observed.

While the *BSC* is helpful in the study of species and speciation, it has known limitations, and these become clear as one moves away from mammals and birds. Some clearly recognizable species consist of organisms that reproduce asexually (examples can be found among bacteria, where, even though different species may share genetic material, they do so in ways decoupled from sexual reproduction), whereas other species (for example, many plant species) have members that hybridize readily and viably with members of other clear and distinct species. In the case of these hybridizing species, gene flow between species can be an important source of genetic variation for evolution within the species. For these cases, the *BSC* is not helpful at all. As Price has recently observed, "Many species do not have enough sex: they are parthenogenetic, self-fertilizing, cloning or otherwise do not meet the criterion of biparental sexual reproduction. . . . Many other species have too much sex: they are promiscuous beyond the bounds of species identity, forming genetically open systems [(Price 1996) p69]." How can we cope with this situation?

Either the *BSC* is not a species concept with general applicability, or we have been mistaken about what is to count as a species. Perhaps bacteria and hybridizing plants are not, contrary to appearances, good and true species after all. This is not a conclusion that many biologists find to be satisfactory. There is a now a growing consensus among evolutionary biologists that the *BSC* provides an incomplete understanding of the nature of species, and recent developments in evolutionary biology have taken this into account (Pigliucci 2003). Notice that the strategy adopted by those who champion the *BSC* is to take causal processes that create and sustain some good and distinct examples of species (in this case, processes inhibiting gene flow between sexual populations) and then to formulate a species concept in terms of an important result of these processes (reproductive isolation). In order to get beyond the *BSC* we need to give due consideration to the causal processes that create and sustain asexual species, hybridizing species, and so on. Moreover, we need to characterize what species are in such a way that it does not simply reflect an end result (say, reproductive isolation) of just one of these causal processes.

To accomplish this end, we need to see if the processes that create sexual non-hybridizing species, sexual hybridizing species, asexual species, and so on, though different in mechanism, nevertheless share some functional similarities. It may then be possible to formulate a general species concept in terms of one or more of these functional similarities so that the different mechanisms can be seen as distinct causal pathways to a common functional end. This idea has recently been discussed under the heading of the cohesion species concept or *CSC*. Alan Templeton, who first formulated the *CSC*, characterizes a biological species as, "the most inclusive population of individuals having the potential for phenotypic cohesion through intrinsic cohesion mechanisms [(Templeton 1989) p12]." What does this mean? The strategy is to adopt a general concept of what a species is, while giving fair consideration to the plurality of mechanisms—*intrinsic cohesion mechanisms*—by means of which they are brought about and sustained. This way of proceeding allows us to talk of biological species by focusing on species in functional terms as maximally cohesive units, while simultaneously refusing to reduce our conception of biological species to the consequences of a particular causal mechanism to this end (for example, reproductive isolation).

Intrinsic cohesion mechanisms include (a) gene flow; (b) stabilizing selection (where individuals whose phenotypes diverge too far from the norm for the population are penalized through natural selection); (c) developmental constraints (while the phenotype of an organism reflects complex interactions between the genotype and the environment, so that one and the same genotype might give rise to distinct phenotypes if the environments encountered are sufficiently different, it is nevertheless true that many phenotypes are not accessible from a given genotypic starting point because there is no developmental pathway leading in that direction); and (d) reproductive isolation.

In any given species, one or more of these cohesion mechanisms may be at work, but it may also be the case that mechanisms at work in one species may not be at work in another. Stabilizing selection, for example, might maintain the cohesion of an asexual bacterial population, while gene flow and developmental constraints might be at work in a sexual population. Some sexual populations are reproductively isolated from other such populations, while others hybridize. As Price [(Price 1996) p69] has noted, even hybridizing species usually retain distinctive species characteristics, with the hybrid zones where the hybrids flourish typically being narrow.

Evolutionary biologists have thus come to realize that the natural discontinuities that constitute species differences are the results of complex dynamical processes involving a multiplicity of mechanisms. How, then, do species differ? Where do new forms or morphologies come from? These questions will be deferred to a later chapter, where evolutionary-developmental biology will be discussed. There is an issue that must be mentioned here, however, because of the way in which we have presented evolutionary biology from the standpoint of Mendelian genetics and population genetics. There is a growing consensus that evolutionary biology, as it emerged from the study of population genetics, is fundamentally incomplete. Currently, efforts are underway to incorporate into evolutionary biology insights derived for developmental genetics. In particular, as Gilbert and Burian have recently noted:

> . . . the population genetics approach to evolution focuses primarily on genes that affect adults and their impact on competition for reproductive success, whereas the developmental genetics approach to evolution focuses on genes expressed during development, their interactions, and their impact on the ontogeny of the organism. [(Gilbert and Burian 2006) p71]

The incorporation of developmental genetics into evolutionary biology has resulted in considerable enrichment of an already fruitful branch of science. We explore some relevant details in the next chapter.

For the present, however, we note that the relationship between structural genes and the proteins they make has turned out to be somewhat complicated. The old idea of "one gene, one protein" no longer holds true. Alternative splicing of the primary RNA transcript has shown that different proteins can arise from the same DNA segment. Post-transcriptional modification of RNA transcripts (RNA editing, e. g., C to U conversions in an RNA transcript) also influence coding specificity [(Li

1997) p308]. It has recently been estimated by Szathmáry et al. (Szathmáry, Jordan, and Pal 2001) that in humans at least 35% of the gene transcripts undergo alternative splicing (see Figure 7.1). Theoretically, alternative splicing and RNA editing could generate 1,032,192 mRNA transcripts (each encoding a slightly different protein) from the *Drosophila para* gene (encoding a sodium channel) (Szathmáry, Jordan, and Pal 2001). Various spliceome projects are underway to study this phenomenon.

In a similar vein, according to Ledford:

> Genome-wide surveys of gene expression in 15 different tissues and cell lines have revealed that up to 94% of human genes generate more than one product . . . Only about 6% of human genes are made from a single, linear piece of DNA. Most genes are made from sections of DNA found at different locations along a strand. The data encoded in these fragments are joined together into a functional messenger RNA (mRNA) molecule that can be used as a template to generate proteins. (Ledford 2008)

From the standpoint of human evolution, the role of alternative splicing in the evolution of the human brain has been of much interest. As observed by Bray:

> . . . humans have just three genes encoding neuronal cell surface proteins called neuropins. However, through alternative promoters and alternative splicing, these three genes yield thousands of different isoforms expressed in distinct combinations on the surfaces of different neuronal cell types. Taken together with all other sources of molecular variation, it seems likely that each cell in an organism—even the estimated 10^{11} nerve cells of the human nervous system—is chemically unique. (Bray 2003)

More recently, Lu et al. (Lu, Peng, and Su 2007) have shown that alternative splicing accounts for the existence of a human-specific, type II form of neuropsin, a protein involved in learning, memory and other aspects of human cognition.

Goldstein and Cavalleri (Goldstein and Cavalleri 2005) estimate that the human genome has around 10 million polymorphisms. Genes can vary because of numerous factors (see Figure 7.2). A gene can be deleted or inserted, can be inverted or a segment can be duplicated, the number of copies of a gene can vary. Genes can have single nucleotide polymorphisms (SNPs) where the usual nucleotide is substituted (adenine for cytosine, for example). These changes have consequences.

Redon et al. (Redon et al. 2006) found a surprising number of copy number variants (CNVs) among humans. A copy number variant is exactly what it sounds like: a different number of copies of a given gene. For example you might have one copy of a gene that metabolizes a drug while your neighbor might have 10 copies. Depending on the nature of the copy number variant, your neighbor may likely metabolize the drug more rapidly then you will. Redon et al. discovered that at least 10 percent of genes in the human population can vary in the number of copies of DNA sequences they contain. These variations can greatly influence enzyme activity and a human's response to drugs and disease. For example, Gonzalez et al. (Gonzalez et al. 2005) studied genetic variation among African, European, Asian and American populations. They found that extra copies of the gene responsible for making CCL3L1, helped protect people against HIV-1 and discovered that if people with extra copies became infected with HIV-1, they progressed more slowly to AIDS.

As reported by Check, different patterns of copy number variation exist even between very closely related species and may indeed be medically significant:

> And studies comparing us with our chimp cousins have already linked structural variation to our divergence from the apes. Last year, scientists from the University of Colorado in Denver and Stanford University found 1,005 genes that differed in copy number among humans and four other primates. This month, Eichler's group reported 651 likely structural rearrangements between chimps and humans. The group counted 245 genes contained in these variants, including some genes involved in reproduction and drug metabolism. Eichler's group has also found that segmental duplications have created much more of our genomic differences from chimps than single base-pair differences. There are 177 genes contained within the human-specific duplications. As such duplications are hotspots for evolution, those 177 genes could be partly responsible for creating the traits that make us human.

> These genetic differences could also be useful. Scherer's lab has just released a targeted analysis of inversions between the chimp and human genomes. The group found 1,576 probable inversions, and confirmed 23; three of these differed among human individuals. Not only does this shed some light on primate evolution, but as inversions can often predispose DNA to harmful mutations, these inversions might be involved with human disease. (Check 2005)

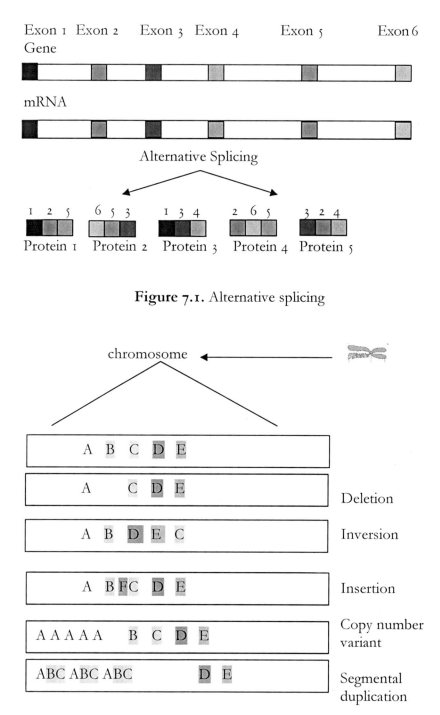

Figure 7.1. Alternative splicing

Figure 7.2. Gene variations

We will have more to say about copy number variation when we focus our attention on matters generated by the metabolism of xenobiotics both from the standpoint of intraspecific variation and interspecific variation. It will turn out to be a complicating factor for the use of nonhuman species as *CAMs* of human biomedical phenomena.

CHAPTER 8

◆

EVOLUTION AND DEVELOPMENT

What gets us into trouble is not what we don't know, it's what we know for sure that just ain't so.

—Mark Twain, 1835-1910

The *New Synthesis*—the fusion of evolutionary biology with population genetics—was arguably one of the great scientific achievements of the 20th century. While evolutionary biology became a research specialty in its own right, its implications have gradually come to pervade most branches of biological inquiry. As we will see in this chapter and those that follow, evolutionary thinking has gradually begun to make serious inroads into the field of medicine, both from the standpoint of studies of disease and from that of xenobiotic metabolism (see also (Shanks and Pyles 2007)). In these fields of biomedical inquiry, the mechanistic study of the proximate causes of biomedical phenomena has been usefully illuminated by evolutionary inquiries that are concerned with more distant causes of the same phenomena. Nevertheless, until comparatively recently, developmental biology—a branch of biology that has been traditionally concerned with the proximate causes of ontogenetic phenomena—has only had distant, and often ill-defined connections, with evolutionary inquiries into phylogenetic phenomena. West-Eberhard even refers to the interface between developmental molecular biology and Darwinian thought as the "Wild West of evolutionary biology [(West-Eberhard 2003) p335]."

Traditional accounts of the diversity we see in nature rooted the explanation in supernatural creation, which, though still attractive in some quarters, can hardly commend themselves to a scientist since claims about the supernatural are opaque to rational, evidential scrutiny. Darwin, knowing little of the mechanistic details of the processes he so ably described in functional terms, was able to offer an alternative naturalistic, evolutionary account of the nature and origins of biological diversity.

This was an enormous step forward, but it did not take us as far as we need to go, for we also need a mechanistic understanding of the origin and causes of nature's diversity and novelty. Important steps in this direction have already been taken. Thus Gerhart and Kirschner have pointed out:

> For the most part, evolutionary biologists have relied on theories of selection and population genetics as explanations for evolutionary change. . . However, selection only provides a filter on the possible [organic] forms. It screens the forms presented to it by development and the activities of cell biology. It is the processes of embryonic development that are responsible for all the morphological details, and selection only affects those that significantly influence fitness. [(Gerhart and Kirschner 1997) p1]

Elsewhere these same authors observe:

> Novelty in the organism's physiology, anatomy, or behavior arises mostly by the use of conserved processes in new combinations, at different times, and in different places and amounts, rather than by the invention of new processes . . . The surprisingly small number of genes for humans and other complex animal forms reflects the anatomical and physiological complexity that can be achieved by the reuse of gene products. The conserved processes are fundamentally cellular processes; they operate on many levels in the development and functioning of the organism. They are the core processes of the organism. The core processes reveal conservation and economy. On the conservation side is evidence that the genes encoding the RNAs and proteins of these processes are highly conserved across diverse animals, from jellyfish to humans We have seen that the core processes were established in several great waves of innovation, and since then they have remained basically unchanged. On the economy side is recognition of just how few genes a complex organism has to work with—only 22,500 in humans, about one and a half times those of a fruit fly.
>
> These two facts, conservation and economy, suggest that complexity must arise through the multiple use of a relatively few conserved elements. Complexity arises when different parts of the adaptive range are selected. It also arises when different combinations of conserved elements are chosen. Conservation can be compatible with economy, as long as the elements of the core processes have properties that allow them to make diverse combinations with differing consequences. . . In evolution, these preexisting combinations of cell be-

haviors and expressed genes must be altered to give new combinations of behaviors and genes . . . [(Kirschner and Gerhardt 2006) p109-110]

Evolutionary biology leads us to expect that there will be similarities as well as differences between organisms belonging to distinct species. Where do these similarities and differences originate? Ptashne and Gann point out:

> . . . it is generally believed that mammals—humans and mice, for example—contain to a large extent the same genes; it is the differences in how these genes are expressed that account for the distinctive features of the animals...changes in patterns of gene expression (rather than evolution of new genes) have had an important, perhaps even determinative, role in generating much of that diversity (that occurred during the Cambrian explosion)...a relatively small number of genes and signals have generated an astounding panoply of organisms. Thus, the regulatory machinery must be such that it readily throws up variations—new patterns of gene expression—for selection to work on. [(Ptashne and Gann 2002) p136]

Put this way, it is very natural to expect that a fruitful conceptual mutualism should emerge from a much closer association between developmental biologists and evolutionary biologists. In the last twenty years a new field of inquiry known as *evolutionary developmental biology* has gradually begun to emerge from the conceptual background. This fusion of ideas from developmental biology and evolutionary biology has already effected a radical transformation of the ways in which traditional problems in both fields are being thought about and analyzed. This chapter is concerned with some of the implications of this *Newer Synthesis* for questions about the nature of biological complexity and the nature of differences between species.

The phenome of a multicellular organism is a dynamical phenomenon. It is something that manifests itself and changes in the course of ontogenetic time. The changing phenome reflects a complex dynamical interaction between genes and causal factors in their immediate environments. The relevant proximal environment for genes is the Bernardian *milieu intérieur* of the cell that contains them. This is an environment in turn whose states change in complex ways in response to changes from without. After all, cells belong to tissues, tissues form organs, and these in turn are integrated into the organism—the creature that must negotiate and respond to causal factors, both biotic and abiotic—that lie beyond, and impinge upon, and sometimes penetrate, its organismal boundaries. It is the organism that reproduces or fails to reproduce in this broader environmental context. Organisms belong to populations. They differ from each other in terms of their inherited constitutions. Because these differing constitutions are responsible for differential reproductive success of their organismal bearers, the frequencies with which the alleles constitutive of those successful unions of genome and phenome will change over successive generations.

A lot of work in evolutionary developmental biology has focused on the regulation (Figures 8.1 and 8.2) of the expression of genes (Figure 8.3), on the mechanisms governing communication between cells and the coordination of their behaviors. Concerning this, Minelli has commented:

> All these behaviors, mechanisms and genes are not there to *ensure* the deployment of the wonderfully complex shapes of living beings. Much more modestly, they are simply there and consequently affect other cellular behaviors, mechanisms or genes and set in place those forms of self-regulation that are the key to avoid developmental bankruptcy. From this perspective, development is deprived of the mysterious finalistic overtones which have thus far constrained our ability to understand it. On the other hand development becomes an even more pervasive dimension of biology than we are accustomed to accept. Everything important in the biology of multicellular organisms belongs to development . . . organisms are not just adults—they are life cycles and life consists of a succession of life cycles. Development is thus a key aspect of the unending continuity of life [(Minelli 2006) p5].

Too narrow a focus on genes might be one sin, and too narrow a focus on adult organisms might be another. How then is development to be understood? How are we to understand the evolution of the differences we see between, say, mice and men? To understand this we must tell a tale of two paradoxes.

The Cuvier Paradox

Georges Cuvier (1769-1832) did not think evolution was possible. His argument is fairly straightforward and can be restated in modern terminology as follows. Every organic being is an exquisitely adapted functional unit. Each subsystem of an organism is co-adapted to fit in and function with every other subsystem of the organism. Evolutionary change in one part would require corresponding (and very unlikely) simultaneous changes

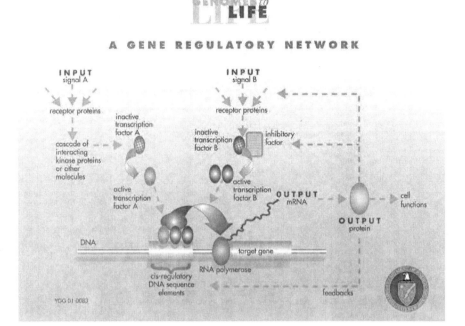

Figure 8.1. Gene Regulatory Network

GRNs are remarkably diverse in their structure, but several basic properties are illustrated in this figure. In this example, two different signals impinge on a single target gene where the cis-regulatory elements provide for an integrated output in response to the two inputs. Signal molecule A triggers the conversion of inactive transcription factor A (green oval) into an active form that binds directly to the target gene's cis-regulatory sequence. The process for signal B is more complex. Signal B triggers the separation of inactive B (red oval) from an inhibitory factor (yellow rectangle). B is then free to form an active complex that binds to the active A transcription factor on the cis-regulatory sequence. The net output is expression of the target gene at a level determined by the action of factors A and B. In this way, cis-regulatory DNA sequences, together with the proteins that assemble on them, integrate information from multiple signaling inputs to produce an appropriately regulated readout. A more realistic network might contain multiple target genes regulated by signal A alone, others by signal B alone, and still others by the pair of A and B. (Text and figure from U.S. Department of Energy Genomes to Life Program http://www.ornl.gov/hgmis.)

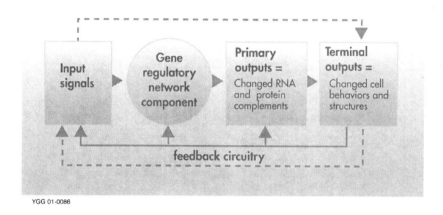

YGG 01-0086

Figure 8.2. Gene Regulatory Network II

Gross anatomy of a minimal gene regulatory network (GRN) embedded in a regulatory network. A regulatory network can be viewed as a cellular input-output device. At minimum, a gene regulatory network typically contains the following components: (1) an input signal reception and transduction system that mediates intra and extracellular cues (left box; often, more than one signal impinges on a given target gene); (2) a "core component" complex composed of transacting regulatory proteins and cognate cis-acting DNA sequences (circle; functionally similar components may be associated with multiple target genes, resulting in similar gene-expression patterns); and (3) primary molecular outputs from target genes, which are RNA and protein (box to right of circle). The net effects are changes in cell phenotype and function (right box). Direct and indirect feedbacks typically are important. More realistic networks often feature multiple tiers of regulation, with first-tier gene products regulating expression of another group of genes, and so on. Beyond GRN boundaries are signaling responses and feedbacks, such as those that drive bacterial chemotaxis, which do not involve regulation of gene expression but instead act directly on proteins and protein machine assemblies (dashed arrows). Some regulatory networks have no embedded GRN component. (Text and figure from U.S. Department of Energy Genomes to Life Program, http://doegenomestolife.org.)

in all other parts of the organism, viewed as a tightly integrated functional system. Minelli states the paradox this way:

> If changes in every part of an organism had a non-trivial developmental effect on every other part of the body, evolution could never have occurred. Selection acting on the smallest part would have produced simultaneous selective pressure on all the other body parts. Consequently the organism would have to be totally rebuilt in response to the slightest selective change. [(Minelli 2006) p234].

The Cuvier paradox provides a good jumping off point for a discussion of the evolving conceptual mutualism between evolutionary biology and developmental biology.

For the present, it is worth noting that in 1954 De Beer used the term *mosaic evolution* to refer to the capacity of different parts of an animal body to be modified in separate and discrete fashion during evolution. This concept of mosaic evolution was a conceptual precursor of the more recent idea of *modularity*. Organisms consist of a number of modules whose coupling, contrary to Cuvier, is not tight. Thus Minelli observes:

> A *module* produces an integrated character complex and is thus both a developmental and an evolutionary unit. Characters within such a complex evolve together in a coordinated fashion because they are genetically correlated. Modules may arise by differential integration of previously independent characters serving a common functional role, or by parcellation of an originally larger character complex, by selective elimination of pleiotropic effects among characters. Complex organs are made by adding new modules, each controlled by a few genes. [(Minelli 2006) p234]

These modules may be characterized by anatomists in terms of organs and tissues, or by geneticists in terms of underlying modular genetic networks.

Carroll et al. [(Carroll, Grenier, and Weatherbee 2004) p10] identify 4 kinds of morphological change that are prevalent in animals organized in terms of clusters of modules:

[1] Changes in the number of *repeated* parts (*meristic* changes).
E.g., changes in segment number or vertebral number.
[2] Diversification of *serially homologous* parts.
A series of repeated parts are serially homologous. Arthropod appendages

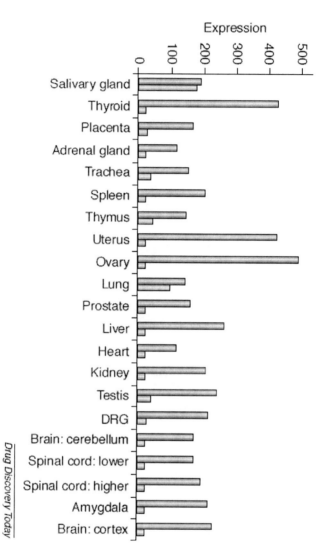

Figure 8.3. Gene expression profile.

Here is an example of a gene expression profile for relative expression of the cystic fibrosis transmembrane conductance regulator gene (CFTR) in 21 tissues for humans (top bars) and mice (bottom bars).

are serially homologous. Ancestrally similar appendages have diversified into antennae, mouthparts, legs and genital structures.

[3] Diversification of *homologous* parts.

Homologous parts share a common evolutionary history, though their functions may differ across lineages. All tetrapod forelimbs are homologous. Bird wings, bat wings and human forelimbs have all conserved the basic structure of the tetrapod forelimb.

[4] Evolution of *novelties*.

New characters may arise from pre-existing structures, or arise *de novo* and become adapted to a new function e.g., feathers, fur, teeth.

Carroll et al. comment:

> Considering that modularly organized animals are among the most diverse groups (in terms of both the number and morphology of species) could there be a correlation between body design and evolutionary diversity? One possible explanation for this relationship is that modular organization allows one part of the animal to change without necessarily affecting other parts. The evolution of genetic mechanisms that control the individualization of parts would allow for the uncoupling of developmental processes in one part of the body from developmental processes in another part of the body. In this fashion, for example, vertebrate forelimbs can evolve into wings while hindlimbs remain walking legs. Dissociation of the forelimb and hindlimb developmental programs allows further modifications to occur selectively in either structure, such as the development of feathers in the forelimb of birds and scales in the hindlimb. [(Carroll, Grenier, and Weatherbee 2004) p10]

This solution to the Cuvier paradox, however, only works if we can find a way to understand modular genetic systems and their relationships to modular developmental processes. Wilkins has recently developed an approach to pathway analysis to help address this problem.

Wilkins' [(Wilkins 2001) p9-10] suggested methodology involves a distinction between *developmental pathways* and *genetic pathways*. In this context a *developmental pathway* is a conceived of as a sequence of causal events that propels a particular developmental process from beginning to end. Underlying a developmental pathway is *a genetic pathway* consisting of the sequence of key gene activities that underlie the developmental pathway. Following Wilkins' exposition, we can introduce two pathway schemas (i.e., formal skeletons) as follows:

Developmental Pathway: *State 1* → α → *State 2* → β → *State 3* → γ → *End State*

Here the states are (phenotypic) developmental states that might be characterized in terms of anatomical features. The pathway could involve a sequence of cellular states, or other developmental sequences (depending on how you choose to put the flesh on the bones in the schema). The letters α, β, and γ denote events that mediate or trigger the transition from one developmental state to another.

A corresponding genetic pathway schema represents a sequence of genetic activities underlying the developmental sequence. Such a pathway might be represented as follows:

Genetic Pathway: *Genes A1,A2* → *B1* → *C1,C2,C3* —| *D2* → *End State*

(In such diagrams, → are used to indicate activation relationships in the sequence of changes, whereas symbols such as "—|" represent relationships of inhibition in the sequence. After all, the absence of something can be as important in one context as its presence is in another. In the pathway schema above, gene *A1*, for example, is said to be upstream of *C1*, whereas *D2* is downstream from *C1*).

In order to flesh out the schema we note here that *cis*-regulatory elements are discrete regions of DNA that affect the transcription of a gene. By contrast, *transcription factors* are proteins that regulate gene transcription by binding to *cis*-regulatory elements. These factors never work alone, and a *cis*-element typically needs several different transcription factors of various types for its regulation. This means that outputs from *cis*-elements typically result from integration of diverse, multiple inputs. Changes in the expression of a gene may thus result from changes in its *cis*-regulatory DNA or from changes in transcription factors upstream (reflecting *cis*-element changes in upstream genes).

It is also worth noting that one and the same transcription factor can control different target genes in different tissues at different stages of development. Furthermore, changes in *cis*-regulatory elements enable changes in gene expression to occur in one structure independently of another. Local changes within a module don't require changes in signaling and regulation that affect multiple modules (the answer to Cuvier's paradox). When discussing heritable variation in the context of evolutionary biology, it is thus important to bear in mind that organisms display variation not just with respect to the coding sequences of structural genes whose proteins form an organism's infrastructure, but also with respect to the sequences coding for transcription factors and *cis*-regulatory elements.

Genetic pathways can be thought of as basic dynamical units out of which more complex functional units can be constructed. That is, they can be assembled into more complex interactive functional units such as genetic circuits, and these in turn may be linked to form genetic networks. Regulatory hierarchies can also exist in which the products of higher circuits can modulate the behavior of circuits lower in the hierarchy. Circuits and networks can form functional modules whose dynamics may depend in complex ways on inputs from other such functional modules. Importantly, two functional modules may behave independently of each other, or they may be only loosely coupled—so that the course of evolutionary change in one module is relatively uncorrelated with that in the other (as required by the proposed resolution to the Cuvier paradox). These complex, dynamical genetic systems are the subject matter of developmental genetics. Wilkins has commented that:

> The idea of the genetic pathway, however, is more than a convenient abstraction. Extensive work in countless laboratories has shown that many developmental processes can be dissected into sequences of events dependent on sequences of gene action. Although neither pathway concept is free of complexities or ambiguity, the two concepts provide a foundation for thinking about development and its evolution. [(Wilkins 2001) p9]

The relationship between developmental pathways and genetic pathways will be examined more closely later in this chapter.

The two pathway schemas we have just introduced (developmental and genetic, respectively) will often represent fragments of more complex structures. The following observations by Wilkins are helpful in this regard:

> The realization that genetic pathways in development are usually part of more complex networks is part of a conceptual evolution that parallels the earlier history of biochemistry. Simple linearity in the biochemical pathways of the 1930s and 1940s gave way to far more complex patterns of metabolic conversion, involving cycles, feedback loops, and dichotomous outcomes at individual steps during the 1950s and 1960s. Correspondingly, it has become abundantly

clear that biological development involves circuitry of at least comparable complexity. Like metabolic pathways, a genetic or developmental pathway can have multiple inputs and complex outputs at early, middle or late points in its sequence.

> Where the inputs can influence the *type* of outcome, selecting between, for instance, two alternatives, then a potential bifurcation is introduced. If several such bifurcations exist, then the flow path will correspondingly take on the look of a branching tree . . . [(Wilkins 2001) p118] (See also figure 8.4 below.)

As in biochemistry, so, too, in the study of development: positive feedback can give rise to multistability (the existence of several steady states), and negative feedback to oscillations and homeostasis. The reader is referred back to the mathematical interlude in chapter 5 for a more detailed discussion of issues linked to network structures and models.

The Hox Paradox

The distinction between developmental pathways and genetic pathways allows us to formulate a solution to another paradox, the so-called *Hox* paradox. The solution to this paradox takes us to the heart of modern thinking about the evolution of development and the nature of biological complexity—issues that need to be discussed if we are to understand the nature of species differences in the light of evolution. First, however, some terminology needs to be introduced.

Of particular importance to developmental geneticists are *homeotic* genes. These are genes that regulate the identity of body regions. Mutations in these genes can bring about *homeosis*, that is, the transformation of one part of an organism into that resembling another part normally found at another location. A classic example is the *aristapedia* mutation, which for homozygous *Drosophila*, causes the tip of the insect's antenna to develop leg-like structures. Another interesting example is discussed by Wilkins as follows:

> The mutant sex determination genes identified in *Drosophila*. . . are "switch," or regulatory, genes. Thus, for instance, mutant individuals. . . with the chromosomal constitution of females that develop as males have not lost a fundamental capacity for building female structures. Instead, they have simply turned on the male development "program." The comparable interpretation holds for the opposite sexual transformation. Sex determination mutants are a class of homeotic mutants, in which the sex-specific characters of the "wrong" sex (in terms of chromosomal constitution) appear instead of the normal ones due to the incorrect switching on of the opposite sex's genes. [(Wilkins 2001) p112]

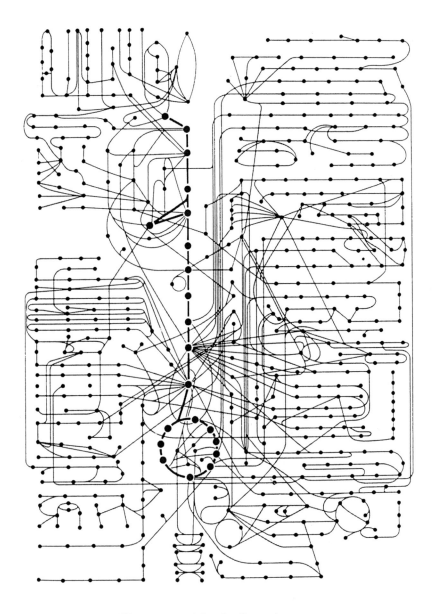

Figure 8.4. Metabolic pathways
(Alberts et al. 2002)

This is the classic metabolic network and interconnection diagram for a typical cell displaying among other things glycolysis and the Krebs cycle.

The now-famous *Hox* genes are examples of conserved developmental genes. More precisely, they are homeotic, homeobox-containing genes found in linked clusters in all bilaterians. *Hox* genes, first observed in *Drosophila* have since been found in a wide range of other animals. Essentially, the *Hox* paradox is a biological form of an ancient puzzle in western metaphysics: the problem of *unity* in *diversity*. Humans, mice, zebra fish, fruit flies, and sea urchins, for example, are very different organisms. Wilkins has commented on this apparent diversity as follows:

> If these visible differences are a faithful reflection of the underlying range of genetic architectures, then few generalizations will be possible, and the task of understanding this genetic diversity will be correspondingly large. It is possible, however, that the visible diversity of morphology and development is misleading as to what lies beneath. Might there not be some significant, but hidden, genetic identities that exist between these seemingly highly different forms? [(Wilkins 2001) p128]

This question could not be answered until the molecular revolution had taken place, and biologists had PCR (polymerase chain reaction) machines to clone genes from many different animal species. The issues here may be illustrated with some examples. These examples are relevant to matters concerning the nature of species differences.

Consider for a moment humans and chimpanzees. Hochachka and Somero ask, "How much new or different genetic machinery is required to make a new species . . . ?" They comment:

> The problem (and in some senses the paradox) is that protein and gene sequences in the common chimpanzee and in the human are remarkably similar. In fact, human and chimpanzee proteins appear to be 99% identical at the amino acid level, and it is widely assumed that the same percentage similarity prevails at the DNA level. Yet no one would mistake the two species as one. [(Hochachka and Somero 2002) p17]

Price observes:

> No one would call a chimpanzee a sibling species with man; no one would ever confuse the two taxonomically. Nearly every bone in the body differs in a chimp-human comparison, and differences in posture, locomotion, feeding, communication, and general ecology are easily observed. Chimpanzees and humans are so different phe-

notypically that they are placed in different genera, *Pan* and *Homo*, and even different families, Pongidae and Hominidae. [(Price 1996) p269]

So even if we are so similar to chimps at the genetic level, we are also clearly different too. How could this be explained?

The focus of much current research is on gene regulation. Genes do not work in isolation, but work together in complex, interconnected networks—in fact, the study of this phenomenon belongs to a new branch of biology known as *genomics* [(Carroll, Grenier, and Weatherbee 2004) p167-8]. In such interconnected genetic networks a single mutation in a regulator gene could have very large, nonlinear effects, bringing about changes in large patterns of gene expression [(Kauffman 1993) p412]. Two organisms may be said to be similar not simply because they have roughly similar numbers of similar genes (as do humans, chimpanzees and mice), but because in addition to these similarities, they have the same types and numbers of interactions between them—and differences here, with respect to interactions, not genome size, may be what differentiates humans from chimps and mice. But if this is right, then when comparing such interactive systems across species lines, there will be no simple, linear, quantitative rules relating degree of genetic (base-pair) similarity in organisms of distinct species, to degree of phenotypic similarity.

In this context, what matters is not enormous similarity with respect to structural genes, but rather a few changes in gene regulation that affect patterns of activity in other genes. In a classic paper King and Wilson put it this way:

> Small differences in the timing of activation or in the level of activity of a single gene could in principle influence considerably the systems controlling embryonic development. The organismal differences between chimpanzees and humans would then result chiefly from genetic changes in a few regulatory systems. . . (King and Wilson 1975)

If this is right, important parts of the biology of species differences arise because of differences with respect to gene regulation. King and Wilson contended that regulatory mutations could have large phenotypic effects, while manifesting only small changes at the genetic level (they recognized two kinds of regulatory mutation: point mutations in a promoter or operator gene, with consequences for amino acid production, but not amino acid sequence; and chromosomal inversions, translocations, additions, deletions, fissions and fusions, with consequences for patterns of gene expression [(Price 1996) p269-70].

This point has recently been made forcibly by West-Eberhard who warns of what she terms, *the informational fallacy*:

> . . . the belief in a necessary, quantitative correspondence between informational (genomic) and phenotypic evolution, is inherent in the now refuted expectation that "genetic distance," or the degree of genetic difference (e.g., measured in base pairs, or allele frequency changes) between populations would correspond to phenotypic distance. This expectation was strikingly contradicted by the very small genetic difference between humans and chimpanzees, estimates to be only about 1.1% of their total base pairs. [(West-Eberhard 2003) p335]

The study of the major phenotypic differences between humans and chimpanzees, on the one hand, and the very small differences at the genetic level of description (a real case of near unity in the face of diversity) should serve as a caution to those who claim to infer physiological similarities between humans and chimpanzees on the basis of major genetic similarities.

Another way to make this point is to consider not humans and chimpanzees, but humans and insects. Over the last ten years many genes (including the so-called *Hox* genes) have been found to regulate similar developmental roles in animals as distantly related as mammals and insects. The *eyeless* (*ey*) mutation in *Drosophila* results in phenotypes in which the organism is missing one or both compound eyes, or in which they appear in reduced form. But as Wilkins has pointed out:

> When the *ey* gene was cloned and sequenced, a search of sequence databases revealed it to be the fruit fly orthologue of the murine *Pax6* gene. The mouse gene, however, had already been identified as *Small eye* (*Sey*), the gene responsible for a dominant heterozygous mutant condition described by its name. Furthermore *Pax6* in humans had been previously identified as *Aniridia*, the gene responsible for a dominant condition of that name, a genetic deficiency of iris development. The shared phenotype of eye defects resulting from loss-of-function mutations in *Pax6* in fruit flies, mice and humans provided early evidence for yet another widely conserved developmental function in animals, one that plays an essential role in eye development. [(Wilkins 2001) p148]

What is initially surprising here is not so much that humans and mice (or, indeed, chimpanzees) are similar from a genetic standpoint, but that humans and insects are as well, in certain fundamental respects. Largely because of this, developmental biologists have been confronted with a puzzle known as the *Hox* paradox:

> How can bodies as different as those of an insect and a mammal be patterned by the same developmental regulatory genes? Very few anatomical structures in arthropods and chordates can be traced back to a common ancestor with any confidence. Yet to a rough approximation, we humans share most of our developmental regulatory genes not only with flies, but also with such humble creatures as nematodes and such decidedly peculiar ones as sea urchins. (Wray 2001)

One approach to this paradox was to simply deny that distantly related animals were that different after all. Indeed, never mind mammals, perhaps all bilaterians are the same animal dressed up differently. Davidson states the argument this way:

> . . . the same regulatory genes are needed for development of insect and vertebrate brains, insect and vertebrate hearts, appendages eyes and so forth. . . an almost automatic response has been that though they may look different these body parts are actually homologous. . . Conservation must be the reason the same genes are used in the development of each part in diverse bilaterians . . . The conclusion that developmental processes are conserved features seems . . . to be required by the observation that orthologous genes are used in the development of the same body part. [(Davidson 2001) p189]

But as Davidson goes on to point out:

> . . . if we sum all the assertions of this sort we produce an impossible and illogical image of the bilaterian ancestor. It would have been equipped with brain, seeing eyes, moving appendages, beating heart, etc. Something is very wrong with this picture because the way these body parts develop in diverse branches of the Bilaterian actually share little in the details of their respective pattern formation processes. If the common ancestor had appendages, for example, it could not easily be ancestral to insect and mouse appendages both, because the structures and processes through which these develop

are completely different. The pattern development processes for appendage development cannot depend to any large extent on a broadly conserved regulatory network. So there must be another explanation for the fact that the same genes are used to make body parts that we call by the same name in very different kinds of animal; and indeed there is. [(Davidson 2001) p189]

It has become clear, in fact, that developmental regulatory genes have acquired new roles in both insect and vertebrate lineages since divergence from a common ancestor. A consequence of this, as Wilkins has observed is that:

A bat cannot be regarded as a slightly different form of fruit fly, nor is an earthworm just a small snake. To reduce the reality of these different animals to the common denominator of their conserved regulatory genes is to reduce them to genetic abstractions; they differ dramatically from each other not only in their developmental processes and adult morphologies, but also in their physiology, ecological roles and suites of behaviors. How can one reconcile the existence of shared regulatory genes of apparently similar functions with the undeniable differences in development these animals display? [(Wilkins 2001) p8]

The alternative (and preferred) approach to the *Hox* paradox is one in which it is recognized that though developmental regulatory genes have been conserved (so that similar genes are found in distantly related organisms) their *interactions* are not. The reconciliation of conserved regulatory genes with diverse developmental outcomes is effected through the idea that many of the changes we see in animal evolution are the result of rewiring developmental gene networks. Using the conceptual framework of pathway analysis introduced above, Wilkins observes:

A general conclusion that emerges . . . is that most of the highly conserved patterning genes act as *intermediate steps in genetic pathways*. From that conclusion, it follows that the evolution of developmental pathways within large groups, such as the bilaterian Metazoa, is, to a large extent, a matter of *evolved differences in pathway components surrounding those key, conserved regulators*. These changes involve components that act either before, or "upstream" of, these regulators or after, "downstream" of, them. The evolution of highly divergent pathways employing the comparable regulatory gene will often, of course, involve multiple changes, both upstream and downstream.

In general, genetic changes upstream of the conserved regulatory genes can alter either their timing or their spatial domains of expression (or both). In contrast, alterations downstream of the conserved regulators can affect the precise sets of "target" genes—themselves often other regulatory genes—that are turned on or off. [(Wilkins 2001) p10-11.]

Suppose organisms X and Y descend from a common ancestor and have genes in common. They might differ as follows with respect to genetic pathways:

X: UR_1, UR_8 → H_1 → DT_1, DT_2, DT_4, DT_8,

Y: UR_1, UR_3, UR_7 → H_1 → DT_2, DT_5, DT_8, DT_9.

Here the same *Hox* gene (H_1)—a conserved regulator—is affected by different upstream regulators (UR) with differential consequences for downstream targets (DT).

Some analogies might be helpful. Consider a piano. When a key is pressed on the keyboard, the hammer to which it is attached strikes a piano wire and a note is produced. Now consider two pianos with identical (conserved) keyboards. The *phenotypic tunes* they produce can differ because *upstream fingers* are pressing groups of keys in different orders, with different timing. The tunes may also differ if the hammers strike different *downstream piano wires* (so whereas the keys on the left hand side of the keyboard usually sound lower notes than the keys on the right, this may no longer be true on one of the pianos). In either or both ways, the existence of a great diversity of tunes is quite consistent with a conserved keyboard.

For another analogy, West-Eberhard, criticizing the claim that evolutionary innovation cannot occur without expansion of the genome, draws our attention to what she calls *combinatorial evolution* at the molecular level:

> By analogy with human language, biological evolution proceeds not only by making many words (molecules) from a single set of letters (the genetic code). By taking a limited vocabulary of words, it can rearrange those into an even more enormous number of distinctive sentences, and then can continually and simultaneously reorganize paragraphs and pages and chapters and books (the phenotype at different levels of organization) without increasing the volume of the basic vocabulary at all. [(West-Eberhard 2003) p334]

Students of genomics are interested in the interactive complexity of genetic switching networks, their implications for systems elsewhere is the biological hierarchy, and the influence of these systems in turn for the behavior of the genetic switching networks. Important aspects of the biological significance of species differences between organisms arise because of differences with respect to this particular kind of organized complexity. Thus, commenting on the evolution of novelties, Carroll et al. observe:

> The recurring theme . . . is the creative role played by evolutionary changes in gene regulation. The evolution of new regulatory linkages—between signaling pathways and target genes, transcriptional regulators and structural genes, and so on—has created new regulatory circuits that have shaped the development of myriad functionally important structures. These regulatory circuits also serve as the foundation of further diversification. [(Carroll, Grenier, and Weatherbee 2004) p167-8]

There is a good sense, then, in which evolutionary developmental biology is showing that diversity of body forms is in the details of the genetic interactions.

Genes and Development

The human genome has less than 30,000 genes. The mouse genome is about the same size as the human genome. That of *Drosophila* has about 14,000 genes. These genes perform a variety of roles in organisms. We can divide these genes into 3 broad groupings:

(a) *housekeeping genes*. These genes encode proteins that do essential biochemical work in the body such as metabolism and biosynthesis of macromolecules.

(b) *specialist genes*. These do specialized work such as oxygen transport, or immune defense.

(c) *toolkit genes*. These genes make the products governing the construction of the body (body plan, number and identity of parts).

(The handy metaphor of *the toolkit* is derived from Carroll et al. (Carroll, Grenier, and Weatherbee 2004).)

This division should not be thought of as a ranking in descending order of importance (toolkit genes are not luxury "add ons" relative to housekeeping genes, moreover, normal organismal development requires the activities of genes of all types. Nevertheless, developmental geneticists have a special interest in the toolkit genes.

It will not go amiss to summarize briefly what is known about the genes in the developmental toolkit (derived from Carroll et al. (Carroll, Grenier, and Weatherbee 2004)):

[1] Only a small fraction of the genes in an animal genome appear to be devoted to development of body plan and body parts.

[2] Toolkit genes produce two classes of gene product of particular importance:

(a) transcription factors, i.e., proteins that regulate the expression of genes during development.

(b) proteins in signaling pathways that mediate interactions between cells.

[3] Toolkit genes are generally conserved among different animal phyla.

The developmental toolkit is very similar across morphologically different animal lineages. It is the unity underlying morphological diversity.

So what does this tell us about the differences between, say, humans and other mammals? At the genetic level of description there will certainly be some differences in the coding content of structural genes. But much more importantly, a common developmental genetic *keyboard* has evidently been used to play distinct developmental (phenotypic) tunes. As pointed out by Carroll et al.:

Although the expansion of the toolkit may be related to morphological *complexity*, no correlation appears to exist between toolkit expansion and animal *diversity*. The morphological diversity of vertebrates, from humans to hummingbirds, or from whales to snakes, evolved around a common set of developmental genes. For example mammals, birds and amphibians share the same set of 39 *Hox* genes. [(Carroll, Grenier, and Weatherbee 2004) p113]

A growing body of evidence now exists to support the claim that morphological diversity results from changes in gene regulation rather than (a) the acquisition of new genes for novel features, or (b) the evolution of functional changes in protein-coding sequences of structural genes. Humans and chimpanzees or mice are not the same animal dressed up differently, rather they are complex biological systems whose manifest differences reflect underlying regulatory differences with respect to the ties to a common set of developmental genes.

While the role of developmental evolution in processes giving rise to speciation itself is controversial (Wilkins 2001), morphological diversity is a demonstrated consequence of speciation. Accompanying the morphological changes that result from changes in the regulation of developmental genes—changes that, in part, enable organisms to insinuate themselves into a wide variety of ecological niches—are physiological changes, that while perhaps less visible to the naked eye, are of equal importance, and are also reflective of underlying regulatory changes. As Hochachka and Somero, commenting on the consequences of organizational complexity of organisms, observe:

> Here we emphasized that the requirements for intercellular and interorgan coordination in complex metazoans create regulatory challenges not faced by unicellular species. These requirements for complex integration and coordination of biochemical adaptations appear to have played major selective roles in the proliferation of genes encoding proteins that play regulatory roles, for instance, protein kinases and phosphatases. The question, "Why do complex organisms contain so much DNA?" may be answered in part by the complexity entailed in regulating the adaptive responses found in metazoans. [(Hochachka and Somero 2002) p18]

In the course of this chapter, we have explored, using a wide variety of contemporary sources, the issue of similarities and differences between organisms in the context of our current understanding of evolutionary developmental biology. The similarities reflect profound evidence of descent from common ancestors, whereas the morphological and even biochemical differences between organisms reflect, in no small measure, regulatory modifications brought about by a variety of evolutionary mechanisms that affect the evolutionary trajectories taken by divergent lineages.

Hochachka and Somero draw our attention to the curious phenomenon (also seen in our discussion of morphological evolution) whereby differences between organisms with respect to adaptive biochemical modification are evident notwithstanding the existence of highly conserved biochemical elements (something normally discussed under the heading of the *unity of biochemistry*—a matter to be touched on in the next chapter). In the present context, differential biochemical adaptations seen in different organisms reflect regulatory differences that ensure the activation of appropriate genes to effect the adaptive changes in question (Hochachka and Somero, 2002:18). Another case, in fact, of genetic unity in the face of adaptive diversity.

What are we to make of this observation? Are the similarities so great and pervasive, and the differences so minor and superficial, to support biomedical inferences from test results in members of one species to the formation of reliable expectations about the effects of similar causes as applied to members of another species? In other words, do we have any good reason to suppose that animal models (whatever else they may be useful for in science) are predictive of human responses? We are not yet ready to discuss this, but the present chapter raises issues that prompt these questions.

It is an enduring theme in contemporary evolutionary biology that the evolution of metazoans proceeds in large measure through the re-use and modification of existing parts. Intuitively, these modifications, reflected in differential patterns of regulatory action in distinct lineages, ought to be relevant to a discussion of the prediction question with which we are concerned in this book. However, since many things deemed intuitively plausible turn out to be illusory, it behooves us to examine in more detail the nature of inter-individual differences as it applies to both intraspecific and interspecific variation. Consistent with our interest in events occurring in the lifecycles of organisms, our inquiries begin with a discussion of delegated complexity and its implications for biological individuality. We then turn to issues raised in the context of the metabolism of xenobiotics and their implications for the issue of prediction in the biomedical sciences.

CHAPTER 9

◆

ORGANISMS AND DELEGATED COMPLEXITY

If rats are experimented on they will develop cancer.

—Morton's law

Multicellular organisms, for example, mice and men, are the products of more than 3 billion years of evolution. *The cumulative result of these continuous and unbroken chains of evolutionary causes and effects, from the origin of life itself to extant biodiversity, are organismal systems that exhibit hierarchical organizational complexity.* The biological hierarchy of organization reflects the fact that there are various levels of description that can be used to describe and conceptualize an organism. Organisms can be described in terms of their constitution and dynamics not just as members of populations or species, which are themselves components of ecologies, but also in terms of the constitution and interactive dynamics exhibited by their organs, the tissues out of which their organs form, the cells out of which those tissues are made, the intracellular structures within those cells, and so on down to the macromolecules of interest to biochemists.

These levels of description reflect the nature of organisms themselves. Thus Jacob observed three decades ago:

> At each level, units of relatively well-defined size and almost identical structure associate to form a unit of the level above. Each of these units formed by the integration of sub-units may be given the name *integron*. An integron is formed by assembling integrons of the level below it; it takes part in the construction of the integron of the level above. (Mayr 1998)

There is, thus, complexity at each level of organizational integration, and complexity in the interactions between the levels themselves. Molecules are organized into intracellular structures, which are organized into cells, and so on. Organisms are integrated systems, so while you could not have an organism without its DNA, for example, DNA will not prosper without a viable organism whose cells provide a suitable *milieu intérieur* to buffer the DNA from perturbing, degrading influences in the external environment.

At the biochemical level in the hierarchy of organization, there is mutual causal interdependence among the subsystems constitutive of the biochemical system of an organism. As Cairns-Smith has observed:

> ... proteins are needed to make catalysts, yet catalysts are needed to make proteins. Nucleic acids are needed to make proteins, yet proteins are needed to make nucleic acids. . . The manufacturing procedures for key small molecules are highly interdependent: again and again this has to be made before that can be made—but that had to be there already. The whole is presupposed by all the parts. The interlocking is tight and critical. [(Cairns-Smith 1986) p39]

At the cellular level, liver cells depend on kidney cells, yet kidney cells depend on liver cells. At higher levels still, lungs depend on hearts and hearts depend on lungs, and so on. Sometimes the dependency is tight and critical. Remove a heart from a mammal and the whole system collapses, for example. Sometimes the dependence is not so critical, a kidney can be removed, but the redundancy of the system (possession of two kidneys) means that such disruptions are not catastrophic. Sometimes complete removal of a system impairs function of the organism, but not catastrophically. Many humans manage to function in the absence of their tonsils or spleens.

These remarks apply at the biochemical level where there are also redundancies, and where metabolic disorders (arising from defective or absent genes) are sometimes fatal, but sometimes merely inconvenient. Shanks and Joplin (Shanks and Joplin 1999) have argued that the existence of biochemical redundancies is a footprint of evolution in action, and is also essential for an account of the stability shown by biochemical systems in the face of what would otherwise be catastrophic systemic disruption.

According to Shanks and Joplin, the existence of biochemical redundancies adds a dimension to our understanding of the nature of biological complexity in the form of *redundant complexity*. For an example, consider the glycolysis pathway, and in particular the first step of that pathway which involves the transformation of glucose into glucose-6-phosphate—a transformation catalyzed by the enzyme hexokinase. In this step there are multiple redundancies. If glucose is unavailable, the pathway can use other hexose sugars. Due to gene duplication, vertebrate tissue contains multiple isoforms of hexokinase. If all these isoforms were removed, there are alternative pathways that can provide the needed products, such as the pentose phosphate pathway (Martini and Ursini 1996). Shanks and Joplin observe:

> It is a hallmark characteristic of evolved biochemical systems that there are typically multiple causal routes to a given functional end, and where one route fails another can take over. The existence of multiple isoforms of a given enzyme are evolutionary legacies— legacies by means of which one and the same enzyme can be co-opted to serve specialized functions in specialized tissues. (Shanks and Joplin 1999)

Representations of biochemical networks that do not depict redundancy are inevitable simplifications of a much more complex reality. For example, when a gene is absent or ceases to be functional, its product is not available to the pathways in which it would typically have participated. If there is redundancy in the biochemical network, the products of other genes may be able to nevertheless be able to take over for the missing product and preserve at least some degree of function.

A good example can be drawn from the study of blood clotting. Plasminogen deficient mice (Plg -/-) have been produced. This is significant because the activated form of plasminogen—plasmin—is needed for clot degradation (it degrades fibrin). However, as Bugge et al. have recently observed:

> Plasmin is probably one member of a team of carefully regulated and specialized matrix-degrading enzymes, including serine-, metallo-, and other classes of proteases, which together serve in matrix remodeling and cellular reorganization of wound fields. . . However, despite slow progress in wound repair, wounds in Plg-/- mice eventually resolve with an outcome that is generally comparable to control mice. Thus an interesting and unresolved question is what protease(s) contributes to fibrin clearance in the absence of Plg? (Bugge et al. 1996)

A complex system such as a mouse can evidently achieve the same function—clot degradation—by different means. Moreover, as Minelli has recently pointed out, "Living beings do not care much which gene is encoding a particular function. Within one developing organism, there often seems to be many equivalent ways of doing the same thing—a circumstance that is usually described as an instance of redundancy (Minelli 2006)."

An example will be helpful. Researchers can now target a specific gene in mice and "knock it out." Such *knockout mice* have been touted as valuable models for human diseases in gene function experiments, especially where a specific gene is suspected as the causal agent. However, such mice do not always give the expected result—they do not exhibit the predicted functional deficits—and when this happens, they serve as examples of the type of redundant complexity we have been discussing.

One example concerns the gene *p53*, originally identified as a tumor suppression gene, but which has subsequently been found to be involved in a number of fundamental cellular processes. For example, it plays a role in gene transcription, the cell cycle, programmed cell death (*apoptosis*), DNA replication and DNA repair processes (Elledge and Lee 1995). Looking at this case from a naive, mechanistic standpoint, one would naturally predict that the removal of this gene, involved as it is in all of these vital processes, would lead to catastrophic collapse of the developmental process—a bit like removing the distributor cap in an automobile. Such is not the case, since *p53* knockouts in mice yield viable, fertile offspring, although they are susceptible to the early appearance of spontaneous tumors (Donehower et al. 1992). This suggests the following dilemma: either *p53* is not required for embryonic development after all or there are redundant ways in which the function of the missing component is compensated for (Elledge and Lee 1995). The evidence at hand supports redundant complexity, since there are at least 400 proteins associated with the proper control of the cell cycle alone, (Murray and Hunt 1993) and it would appear that some of these other proteins pick up the slack created by the missing *p53*.

Developmental biologists are very interested in characterizing developmental pathways by which an organism gets from one developmental state to another. In such a pathway, we might have a gene-mediated step as follows:

[1] $A - (a) \rightarrow B$,

where A and B are successive developmental states and a is the gene whose biochemical product mediates the step. But we can also have redundancy:

[2] $A - (a + b) \rightarrow B$.

Of this case, Wilkins has recently observed:

> If, however, two gene products contribute to the same step, and
> their activities are similar and additive at this step, then mutational
> inactivation of one gene will often be masked by the continued ac-
> tivity of the other. . . . The consequence is that mutational inactiva-
> tion of either gene is frequently insufficient to block the sequence,
> and correspondingly, activity of both genes must be eliminated to
> prevent step B from occurring. *In general, pathway steps with dual, or*
> *multiple, inputs of this kind will be missed in conventional mutant hunts, since,*
> *in general, only a single gene of the pathway is affected in each mutant line.*
> [(Wilkins 2001)P114]

So *mutant hunts* on carefully inbred strains of research animals are complicated by the existence of redundancy. Wilkins provides some examples
[(Wilkins 2001) p114-116].

It is also worth pointing out that absences of functional genes are not
always compensated for at the biochemical level. Humans lack a functional gene for L-gulano-gamma-lactone oxidase. The lack of this gene product disrupts the synthetic pathway that makes vitamin C in many other
mammals (in fact a pseudogene is present in humans). We compensate for
this feature of our biochemistry at the organismal level, by having a diet
containing vitamin C. This points to a need to discuss hierarchical dependency (Shanks and Joplin 1999). Hierarchical dependency occurs when
events at lower levels in the hierarchy of organization become dependent
on events, structures, and processes at higher levels. Cells are the primary
arena for biochemical interactions, but they are far from being passive
receptacles in which chemical events occur. The intracellular structures (at
the next level up from the biochemical level of the hierarchy) are of great
importance:

> . . . certain glycolytic enzymes form specific non-covalent complex-
> es with structural components of the cell, which may serve to or-
> ganize reaction sequences and assure efficient transfer of interme-
> diates between cellular compartments. Certain glycolytic enzymes
> bind to microtubules or to actin microfilaments, bringing those en-
> zymes into close association and holding them in a specific region
> of the cytoplasm. [(Lehninger, Nelson, and Cox 1993) p415]

For these reasons, it is a mistake to think of organisms as bags of chemicals or buckets of genes. Biomolecules and genes are important, but only
because of the organismal context in which they are found.

In previous chapters we saw that it was not possible to reduce organisms to their genes: the phenome reflected the interaction of genome and environment. The point here is slightly different. It is that organisms are integrated complex systems that need to be understood from the standpoint of *systems biology* (this is a matter we shall return to in our concluding remarks at the end of the book). Nearly 40 years ago Florkin and Schoffeneils made the point this way:

> If animals are generally endowed with motility (function of organism), this is related to the fact that certain cells biosynthesize macromolecules of contractile proteins (molecular structures related to function considered at the level of the organism). In fact, molecular biology, in spite of the concentration of its studies at the molecular level, brings us back to the organismic viewpoint too often neglected in biochemical studies. [(Florkin and Schoffeneils 1970) p162]

These comments were made at the dawn of the molecular revolution in biology. Systems biology has been overshadowed by this revolution. Yet it is now becoming clear that there is an urgent need to consider the characteristics of integrated biological systems.

Evolution and Delegated Complexity

Various proposals have been put forward to provide measures of biological complexity in the light of the new sequencing information flowing from the various genome projects. Merely counting genes does not seem to provide a good way to measure biological complexity—the gene count for *Arabidopsis* (25,500) is at the low end for estimates of the human genome count. Proposals have been put forward to consider complexity in terms of networks of transcription factors (the *transcriptome*) and the genes they regulate with a view to constructing mathematical measures of network connectivity (Szathmáry, Jordan, and Pal 2001).

But it seems that there are dimensions of organismal complexity, especially in vertebrates, that such approaches ignore. In particular, these approaches give no good account of an important aspect of evolutionary developmental complexity, a phenomenon known as *delegated complexity*. This idea has been explained as follows:

> With a limited number of genes, vertebrates manage to code for two highly complex subsystems that are specialized for information accumulation, storage and retrieval: namely the immune system and the nervous system. Both systems operate on a generative basis, that

is, they can store huge amounts of information based on a fixed set of rules. These rules reside in variation-generating mechanisms (such as the reshuffling of immunoglobulin genes) and internal selective filters. In the case of the vertebrate immune system, reshuffling of immunoglobulin genes produces an enormous variety of antibodies. An internal selective filter then recognizes cells producing antibodies against self antigens, weeds them out and destroys them. Although less well characterized, the vertebrate nervous system contains similar Darwinian elements. During development, a large surplus of nerve cells and their myriad connections are produced, from which only those that best innervate a given territory are retained. . . . vertebrates have delegated a large part of their complexity to their immune and nervous systems. . . Delegated complexity, achieved by genetically encoded information-processing systems such as the nervous systems and immune systems of vertebrates, adds another dimension to biological complexity. (Szathmáry, Jordan, and Pal 2001)

In fact, there is a now growing realization that Darwinian principles generating adaptive responses to environmental challenges operate within organisms in the course of their life cycles. Thus Gerhart and Kirschner have recently remarked:

> Physiological systems based on variation and selection are much more prevalent in biology than has been appreciated. The power of Darwinian selection as a cellular mechanism in the short term (rather than a genetic selection mechanism used only in the long term) has recently become clearer. In many biological systems, several, and often a large number, of alternative responses to external stimuli are in fact produced, and one is selected. [(Gerhart and Kirschner 1997) p147]

Since the study of *ontogenetic* phenomena involves the study of the development and life cycles of individual organisms, Shanks (Shanks 2004) has referred to these extensions of Darwinian principles to events occurring within an individual organism in the course of its life cycle as *ontogenetic Darwinism*. This may be contrasted with traditional or *phylogenetic* Darwinism, whose concern is with evolutionary phenomena in populations of organisms over many successive generations.

As noted above, the immune system affords a good example of ontogenetic Darwinism in action in each of us. The example shows how Darwinism can be used to explain some important processes whereby individual organisms themselves become adapted to short-term changes in their environments and which cannot be directly encoded and foreseen in the genome they inherit.

The immune response is the reaction of the body (self) to invasion by foreign substances (non self) known as *antigens*. B-lymphocytes produce circulating antibodies and are responsible for *humoral immunity*. The response involves the production, by specialized white blood cells, of proteins known as *antibodies*. Antibody molecules are coded for by *immunoglobulin genes*. Antibodies react with antigens to flag them for further immunological action that (with luck) renders them harmless. B-lymphocytes play an important role in *adaptive immunity*.

We focus here on *B* cells and the antibodies they produce. The population of antibodies available to attack a given antigen will vary with respect to their ability to bind to that antigen. Some antibodies won't bind at all (or rarely), others will bind more frequently, and some will bind virtually every time they encounter the antigen. Antibodies are said to have *specificity* for the antigens to which they bind, and one of the things we will be concerned to discover is how this specificity is improved upon during the course of an infection. This will enable us to see how Darwinian mechanisms can tune an adaptive response *within an individual in the face of novel threats from outside that might otherwise disrupt immunological homeostasis.*

To be effective, the immune system must produce an enormous range of antibodies. There are up to 10 billion *B* lymphocyte cells, and the system is (very conservatively) capable of recognizing more than 100 million antigen shapes. What vertebrates inherit from their parents are immunoglobulin genes. As inherited, these are said to be in *germ-line configuration*. But what is inherited does not code for the immense diversity of antibody molecules. There is not enough information in the vertebrate genome. In 1976, Susumu Tonegawa discovered that immunoglobulin genes are not inherited complete, but rather as fragments that are shuffled together to form a complete immunoglobulin gene that specifies the structure of a given antibody. This process is known as *somatic recombination*, since it occurs in body cells that are not germ-line (reproductive) cells. As these fragments are combined to form a complete immunoglobulin gene, new DNA sequences are added at random to the ends of the fragments, ensuring still more antibody diversity.

This random reshuffling of immunoglobulin genes, together with the random insertion of DNA sequences during the somatic recombination process, results in a high probability that at least one antibody, though perhaps not binding perfectly, will fit at least one of the many *determinants* (molecular handles) presented by a new antigen. Once an antibody is selected by binding to antigen (along with appropriate *T* lymphocyte chemical signals), it stimulates the *B* lymphocyte to which it is attached to make exact copies—*clones*—of itself. Some of these clones remain as circulating *B* lymphocytes, serving as the immune system's memory. Increased numbers of these cells provide for a faster immune response to subsequent infections and to establish the immunity that follows some infections and vaccinations. Other clones stop dividing, grow larger and turn into plasma cells whose sole function is to produce large numbers of free antibodies to fight the current infection. What about the observation that antibodies produced in the later stages of an infection are more effective at binding than the antibodies initially produced?

The fine-tuning of the antibody response is accomplished by another Darwinian mechanism that changes the genetic makeup of the immunoglobulin genes through mutation. This random mutation of the immunoglobulin genes is known as *somatic hypermutation*. By randomly producing variations on a successful theme, some antibody variants will be better than the original clones at binding to a given antigen, and specificity will be enhanced [(Parham 2000) p21]. We have in the *B* lymphocyte population the random production of a wide range of variants, with differential reproduction—cloning—of selected variants, depending on the specific challenges to an individual's immune system. The clones of the selected variants inherit the genetic properties that made their progenitor cells successful. The immune system's adaptive response to novel antigen presentation is based on the same evolutionary principles that shaped the organism itself, and adapted it to its external environment. Each of us has a unique immune system, whose current features reflect the historical contingency of fast evolution occurring during our life cycles.

In this way our best theory of the adaptive immune system, with enormous implications for the way we think about infection, is thoroughly Darwinian. Thus commenting on the history of modern immunology, Peter Parham observes:

> At some point this century the experimental biologists, in an echo of Henry Ford, divorced themselves from the evolutionary biologists. This artificial and regrettable separation remains with us today. For the immunologists it was always a sham for the very foundations of their subject are built upon stimulation, selection and adaptive change. Now we see clearly the immune system for what it is, a vast laboratory for high speed evolution. By recombination, mutation, insertion and deletion, gene fragments are packaged by lymphocytes, forming populations of receptors that compete to grab

hold of antigen. Those that succeed get to reproduce and their progeny, if antibodies, submit to further rounds of mutation and selection. There is no going back and the destiny of each and every immune system is to become unique, the product of its encounters with antigen and the order in which they happen. This all happens in somatic tissues in a time frame of weeks . . . . (Parham 1994)

This is but one example of the way in which delegated complexity contributes to the uniqueness (in this case, immunological uniqueness) of individuals of a given species. Other examples can be found in developmental biology where the production of a superabundance of cells, with differential retention of a smaller number, plays a crucial role, for example, in developmental sculpting of such structures as fingers and toes. Other important examples can be found in cell biology, neurobiology (Gerhart and Kirschner 1997), and in oncology (Greaves 2000). The delegated complexity manifested in oncological phenomena will be explored below, since it points to the existence of a dimension of complexity that is apt to be ignored in non-evolutionary biomedical investigations.

Epigenetics and monozygotic twins

Another way to approach the issue of the medical uniqueness of individuals is to consider the hard case: monozygotic twins. Here the concept of *epigenetics* is important. Jeneen Interlandi explains the meaning of *epigenetics*, as well as its potential medical importance, as follows:

> The term "epigenetics" refers to the study of changes in gene expression that do not involve changes in the genetic code; they include an expanding roster of subtle molecular modifications that tell cells which genes to activate (or transcribe) and which to suppress. In cancer cells, these small-molecule regulators can act like broken dimmer switches, turning genes that promote cell growth all the way up and those that suppress tumors all the way down . . . . (Interlandi 2007)

It is important to note that the term epigenetics has several meanings. The one referred to here is a meaning drawn for genetics and molecular biology. Other, different, meanings can be found in the context of quantitative genetics.

Barring somatic mutations, just about all cells of different types that are constitutive of the human body, for example, are genetically identical (exceptions include certain specialized cells in the immune system). Cells belonging to different types have the same genes but are differentiated on the basis of gene-activation states (genes active in cells of one type may be inactive in cells of another type). Differential states of gene activation are brought about by a number of regulatory mechanisms, but DNA methylation has been well-studied in connection with gene silencing. These different gene activation states propagate through mitosis in the different cell lineages (kidney cells of a given type give rise to kidney cells of the same type, and so on), and are important for the organization of cells into tissues, for example.

The process whereby a cell "remembers" its progenitor cell type and differentiates accordingly, is known as *cell memory*. (Again, this is something undergirded by a variety of regulatory mechanisms.) Humans are diploid organisms which means we get two sets of genes, one from each parent. Sometimes gene expression depends on which parent the gene is derived from—something known as *genomic imprinting*. The mechanism of DNA methylation is important here, and has medical implications insofar as loss of imprinting via defective methylation plays a role in several genetic diseases. Mueller and Olsson observe:

> Loss of imprinting is also an important factor in tumor formation, possibly because the effect of a mutation on an imprinted allele will not be rescued by the other, nonmutated allele. The study of epigenetic gene regulation has become an important field of biomedicine including cancer and stem cell research, somatic gene therapy, transgenic technologies, cloning, teratogenesis and so forth. [(Mueller and Olsson 2003) p120]

The study of monozygotic twins—nearly identical with respect to inherited DNA—allows us an opportunity to see the effect of epigenetic changes in the course of development: environmentally-induced changes in regulation and expression. Fraga, et al. (Fraga et al. 2005) found more epigenetic changes, arising from DNA methylation and histone acetylation, in older monozygotic twins than younger. They also found that epigenetic changes were proportional to differences in lifestyles.

The notion that monozygotic twins were subject to different diseases and reacted differently to drugs is not new. Thus, Machin has observed:

> The use of the adjective "identical" rather than monozygotic leads to misunderstandings about the biology of monozygotic twinning. Most monozygotic twin pairs are not identical; there may be major discordance for birth weight, genetic disease, and congenital anoma-

lies. These indicate that postzygotic events may lead to the formation of two or more cell clones in the inner cell mass and early embryo that actually stimulate the monozygotic twinning event. There is
also evidence that there may be unequal allocation of numbers of
cells to the monozygotic twins; this may have widespread implications for the cascade of developmental events during embryogenesis, formation, and vascularization of the placenta. Large-scale zy-
gosity testing at birth could be the template for analysis of twin
outcomes and their biologic causes. (Machin 1996)

These medically significant cascade effects of what might be called, for
want of a better term, *developmental noise* are also the reason why monozygotic twins do not have the same fingerprints. The lesson here, as we saw
in the last chapter, is that developmental factors must be taken seriously.
It is a mistake to simply look at adults, on the one hand, and their genes
on the other. As noted earlier, the phenome is dynamical and subject to
change, and the causal factors shaping phenomic change are many and
varied.

Discussions of monozygotic twins are frequently used as a platform
for the emphasis of similarity among individuals, rather than differences,
especially medically significant differences. Yet as Horrobin points out,
even here the differences are important:

> For example, I am puzzled as to why most people other than pro-
> fessional geneticists seem uninterested in the high levels of non-
> concordance among identical twins for common diseases. And even
> the geneticists are more interested in the concordance than the non-
> concordance. For almost every common disease, whether it be in-
> flammatory, malignant, degenerative, psychiatric or any other type,
> when one identical twin is affected the other twin is not affected
> from 40– 90% of the time. Therefore some factor or combination
> of factors in the environment must have prevented or switched off
> the disease process in the non-affected twin. Whatever factors are
> involved, they have ocurred during a normal life and so do not re-
> quire high-tech genetic or pharmaceutical manipulation. The corol-
> lary is that it cannot be that difficult to switch off genetic disease
> processes if only we can understand them better. Is it really too
> much to think that a direct assault on human disease by studying
> humans might be at least as productive as the massive investment in
> investigation of unvalidated animal and in vitro models? At least
> what we find in humans will be both real and relevant. (Horrobin
> 2003) (See Appendix 5 for full Horrobin article.)

Biological individuality, relevant for medical extrapolation between monozygotic twins, plays a more dramatic role in connection with the dynamics of diseases such as cancer, where ontogenetic Darwinism plays a grim role. To this we now turn.

Cancer as a Disease of Delegated Complexity

Cancer is a major cause of death in humans (and other animals). It is a phenomenon where rapid growth in knowledge of causes has been largely (though not entirely) decoupled from the practical ability to develop effective treatments. A major reason for this is that cancer is a Darwinian disease of delegated complexity. One consequence of this is that cancer exhibits a high degree of individual variability—what is true of the phenomenon in one patient may not be in another. Another consequence is that within a given patient, the operation of the principles of ontogenetic Darwinism gives the cells generating the disease the ability to adapt to therapeutic measures in the course of the (often) diminished lifespans of patients.

There are not simply differences between individual human patients, there are evidently differences between species with respect to oncogenic phenomena. Hahn and Weinberg observe that, "Primary rodent cells are efficiently converted into tumorigenic cells by the coexpression of cooperating oncogenes. However, similar experiments with human cells have consistently failed to yield tumorigenic transformations, indicating a fundamental difference in the biology of human and rodent cells (Hahn and Weinberg 2002)." It is here, once again, that the importance of genetic context rears its ugly head. For as Nijhout has recently pointed out:

> Whether or not a particular oncogene actually causes cancer often depends on the genetic background of the individual, as well as particular environmental variables such as vitamin deficiency or smoking habits. When an oncogene is introduced into a mouse by genetic engineering, it typically induces cancer in only a few tissues, even when the gene is expressed broadly. This suggests that only some tissues provide the conditions that are necessary for the defective gene to have its deleterious effect. (Nijhout 2003)

As we will see, this is hardly surprising, given the variability that cancer shows within humans themselves. This variability is no accident.

Comparisons have often been made between the bodies of multicellular organisms such as ourselves, and ecological communities. A liver, for example, might be thought of as a population of cells that respects territorial limits. Liver cells have a niche in the economy of the body, and tend to confine their activities to that niche. No organ is an island, however, and there appear to be complex, mutually beneficial relationships between cells comprising different populations in the various organs constitutive of the body. But there are some peculiarities and disananlogies with ecological communities. All cells in the body (with exceptions such as red blood cells and specialized cells involved in humoral immunity) are genetically identical, with differences between cell types, e.g., liver cells and kidney cells, being reflective of different patterns of gene activation, a matter discussed in the previous section.

Somatic cells carry the same genes as germ cells. Somatic cells cooperate and die so that the genes in the germ cells can get through the reproductive bottleneck into the next generation. Insofar as somatic cooperation occurs to get genes *identical by descent* into the next generation, it is a phenomenon reflective of a type of natural selection known as *kin selection*. Legend has it that J.B.S. Haldane was once asked under what circumstances he would jump into a lake to save a drowning man. He is reputed to have replied that he would do it for two brothers or eight first cousins. This contains the basic insight behind kin selection. What matters in evolutionary genetic calculus is the getting of genes identical by descent to *your* genes into the next generation. You can do this by reproducing, or by helping relatives to reproduce (who carry a proportion of genes identical to yours by descent).

In this way, the cooperation among somatic cells makes evolutionary sense, given the genetic relatedness between somatic cells and germ cells. Thus Alberts et al. observe:

> As a result, each cell behaves in a socially responsible manner, resting, dividing, differentiating, or dying, as needed for the good of the organism. Molecular disturbances that upset this harmony mean trouble for a multicellular society. . . Most dangerously, a mutation may give one cell a selective advantage, allowing it to divide more vigorously than its neighbors and to become a founder of a growing mutant clone. A mutation that gives rise to such selfish behavior by individual members of the cooperative can jeopardize the future of the whole enterprise. Repeated rounds of mutation, competition, and natural selection operating within the population of somatic cells cause matters to go from bad to worse. These are the basic in-

gredients of cancer: it is a disease in which individual mutant clones of cells begin by prospering at the expense of their neighbors, but in the end destroy the whole cellular society. [(Alberts et al. 2002) p1314]

This is a nice way of bringing out the idea that evolutionary processes occur with no eye to the future. Cancer cells proliferate with no long-term view to organismal well-being. Alberts et al. continue:

> In general, the rate of evolution in any population would be expected to depend on four main parameters: (1) the *mutation rate*, that is the probability per gene per unit time that any given member of the population will undergo genetic change; (2) the *number of individuals* in the population; (3) the *rate of reproduction*, that is, the average number of generations of progeny produced per unit time; and (4) the *selective advantage* enjoyed by successful mutant individuals, that is, the ratio of the number of fertile progeny they produce per unit time to the number of surviving fertile progeny produced by non-mutant individuals. These are the critical factors for the evolution of cancer cells in a multicellular organism, just as they are for the evolution of organisms on the surface of the Earth. [(Alberts et al. 2002) p1320-21]

Cancer is thus evolution writ small, occurring relatively quickly, within the milieu intérieur of a multicellular organism, in the course of its ontogeny.

For cancer to occur, it looks as though there must be several, independent, mutations occurring in the lineage of a given cell—perhaps as many as 10 mutations of the right kind. When these are in place, a cell can begin the process that leads to violation of the multicellular social contract. The requirement for multiple unlucky mutations in part explains why cancer rates rise precipitously with age. To use Greaves unhappy image, it takes time to get a "full house". Cancer incidence in humans is affected by environmental exposures (that reflect social, demographic and economic factors). The impact of these exposures reflects an individuals' immune, system, diet and inherited genetics. But mutations occur spontaneously—if they did not, there would be no genetic variation, and evolution would be impossible. We can make matters worse, but some background frequency of cancer incidence is probably inevitable in populations composed of multicellular organisms such as ourselves.

Then again, there are species-specific factors in our own evolutionary history. Thus, Greaves has recently noted:

> Many of the relevant exposures and negative modulators in cancer causation are the flip side of what were originally—for an emerging hominid species, subject over millennia to Darwinian fitness selection—advantageous attributes. These include selection for hormonally driven fertility, with persistently primed breasts, ovaries, uteruses, and prostates; some degree of promiscuity (in males at least); pale skins (for whatever reasons) in northerly folk; a propensity for periodic Bacchanalian bingeing and energy storage in fat; and longevity. And last but not least, an insatiably curious, risk-taking, tinkering and entrepreneurial personality. [(Greaves 2000) p216]

Moreover, it is worth pointing out that cancer is mainly (though not exclusively) a disease that afflicts the aged. Evolutionary mechanisms will only modulate features of individuals that bear on reproductive success. Most cancers occur after individuals have ceased reproducing. It is possible that to some extent the phenomenon of cancer is a result of the long-term effects of selection for traits contributing to youthful fecundity—thus bringing it in line with pleiotropic theories concerning organismal senescence.

While one needs to be cautious in forming generalizations about cancer cells within an organism, let alone between organisms of the same, or indeed different, species, the following list of properties seems to be relevant [(Alberts et al. 2002) p1325]:

(1) Cancer cells acquire the ability to violate the multicellular social contract by acquiring the genetic capability to ignore internal and external restraints on cell proliferation.
(2) Cancer cells avoid suicide by apoptosis.
(3) Cancer cells are genetically unstable. Poor replicative fidelity helps to generate much variation in the population of cancer cells.
(4) Cancer cells can invade their home tissues and neighboring tissues and are said to be invasive.
(5) Cancer cells can survive and proliferate at sites foreign to the site of their origination (they have metastatic capabilities).

Points (3), (4) and (5) deserve scrutiny since they are related. Genetic instability results in their being much heritable variation in the population of cancer cells—selection acts on these variants. Some of these variants have the ability to emigrate and proliferate beyond the tissues in which they arose, and an even smaller number have a constitution enabling them to establish new colonies at distant sites from the site of their origination. This is the origin of metastatic cancer. All it takes is one founder cell to establish a colony on a distant organic "island" for cancer to spread. Genetic variation in the resulting population of colonists, coupled with selection, will adapt a growing population of cancer cells to life in the new environment (say the liver or the brain). The process is akin to the adaptive radiations discussed by ecologists, whereby populations split up and insinuate themselves into a multiplicity of ecological niches, and comparisons with the island-hopping finches of the Galapagos Islands are not altogether inappropriate.

The precise genetic mechanisms behind metastatic cancer are not currently well-understood, but it seems likely that metastatic cancer cells proliferating in the liver, for example, will differ markedly from such cells proliferating in, say, the brain. The cells may be related through descent from a common ancestor—the original mutant cell that initiated the disease—but there has been much subsequent evolutionary modification by the time metastases have formed. There is thus much room for intra-patient variability—treatments successful at one site may be ineffective at others. We can also expect much inter-patient variability. In fact, this is what we see:

> Many types of cancers that have been analyzed genetically show a large variety of genetic lesions and a great deal of variation from one case of the disease to another. . . Different combinations of mutations are encountered in different patients and correspond to cancers that react differently to treatment. [(Alberts et al. 2002) p1355]

But intra-patient variability plays a role too. How?

Standard therapeutic measures involve the use of radiation or cytotoxins. But because of the high degree of variation displayed by cancer cell populations, cancer cells vary widely in their responses to these treatments. Cells that show even modest resistance to therapeutic insults will reproduce at the expense of more sensitive cells. The result over time will be the emergence of a population of cancer cells that is resistant to the cytotoxic effects of a given drug or to the effects of ionizing radiations. But as Alberts et al. go on to point out:

> To make matters worse, cells that are exposed to one anticancer drug often develop resistance not only to that drug, but also to other drugs to which they have never been exposed. The phenomenon

of *multidrug resistance* is frequently correlated with amplification of a part of the genome that consists of a gene called *Mdr1*. This gene codes for a plasma-membrane-bound transport *ATPase* . . . The overproduction of this protein or some other members of the same family can prevent the intracellular accumulation of certain lipophilic drugs by pumping them out of the cell. [(Alberts et al. 2002) p1357]

It is clear that the drug resistance story concerning cancer cells is very similar to that encountered in the context of host-parasite co-evolution— for example, in the context of the evolution of antibiotic resistance by microorganisms such as *Staphylococcus aureus*—even to the point of the exploitation of molecular pumps. Greaves [(Greaves 2000) p242] sees similarities with respect to the rapid evolution of pesticide resistance in insects. It is, then, not surprising that different humans react differently to cancers and the treatments for those cancers.

Cancer is not a disease unique to humans, though it has features in humans that are different from those found in other species. Thus Greaves has observed:

Most animals, both vertebrate and invertebrate, have benign tumours and in some cases invasive cancers; even plants can have bacteria-induced tumours. These, we can safely assume, have existed long before *Homo sapiens* stood up and strode forth on the planet. Accurate estimates of incidence rates, especially in feral species, are not available although one might expect smaller, short-lived animals to have a relatively low risk. Cancer has been recorded in most species of captive mammals in zoos, but the incidence rates, even in ageing primates, appear low in comparison with similar cancer types in humans. [(Greaves 2000) p210]

This invites the following question: are these differences between humans and other mammals (typically used as models of human disease) relevant for human biomedicine?

It turns out that there are both interspecific differences and intraspecific differences worthy of consideration. For example, discussing the challenges of using animal models to study human cancer, Williams and Weisburger point to well-studied intraspecific differences between different strains of mouse and they note:

. . . in animal models utilizing random bred animals or highly inbred animals, variations in sensitivity or in target organs occur to a given carcinogenic challenge. Even a cryptogenic cancer seen in an aging, untreated population of animals occurs to different extents in different species and strains. Among inbred strains of mice there are wide differences in the occurrence of cryptogenic neoplasms. Some, such as the strain *A* mouse, develop a high incidence of pulmonary neoplasms, and they are also very sensitive to the chemical induction of these tumors. The *C3H* strain readily develops cryptogenic or induced neoplasms of the liver. In contrast the *C57* black strain mouse is resistant to induced or spontaneous liver neoplasms, but cutaneous or subcutaneous cancer can be induced fairly readily . . . [(Williams and J Weisburger 1993) p147]

With relevance to human medicine, interspecific differences matter, too. Thus Alberts et al. point out:

The mouse is the most widely used model organism for the study of cancer, yet the spectrum of cancers seen in mice differs dramatically from that seen in humans. The great majority of mouse cancers are sarcomas and leukemias, whereas more than 80% of human cancers are carcinomas—cancers of epithelia where rapid cell turnover occurs. Many therapies have been found to cure cancers in mice; but when the same treatments are tried in humans they usually fail. [(Alberts et al. 2002) p1347]

The problem of similarity with respect to the manifestation of cancer does not merely concern mice. Our close evolutionary relatives, the nonhuman primates have not been good models for human tumors either. Thus, Beniashvili has observed:

Attempts to obtain malignant tumors in monkeys failed, since primates turned out to be highly resistant to certain blastogenic agents, carcinogenic for other animals... Spontaneous tumors in monkeys are very rare...Many researchers believe that monkeys have an inherent specific resistance to malignant tumors. The low incidence of spontaneous tumors in monkeys has been associated with difficulties in experimental induction of tumors in these animals...Thus unlike humans...in monkeys lung tumors are extremely rare....spontaneous tumors of the respiratory organs and mediastinum in monkeys, unlike in man, are extremely rare....spontaneous tumors of skin and soft tissue in nonhuman primates are comparatively rare...For spontaneous skin tumors in monkeys, recurrences and metastases were not characteristic....The above findings show that at present there have been just a few successful cases of the induction of soft tissue tumors in monkeys. (Beniashvili 1994)

Moreover, in a discussion of the human relevance of cancer treatments in mice, Gura has observed of work at the National Cancer Institute (NCI) that:

> The institute started by pulling together mouse models of three tumors: a leukemia, which affects blood cells; a sarcoma, which arises in bone, muscle, or connective tissue; and a carcinoma, the most common type of cancer, which arises in epithelial cells and includes such major killers as breast, colon, and lung cancers. Initially, many of the agents tested in these models appeared to do well. However, most worked against blood cancers such as leukemia and lymphoma, as opposed to the more common solid tumors. And when tested in human cancer patients, most of these compounds failed to live up to their early promise.
>
> Researchers blamed the failures on the fact that the drugs were being tested against mouse, not human, tumors, and beginning in 1975, NCI researchers came up with the xenograft models, in which investigators implant human tumors underneath the skin of mice with faulty immune systems. Because the animals can't reject the foreign tissue, the tumors usually grow unchecked, unless stopped by an effective drug. But the results of xenograft screening turned out to be not much better than those obtained with the original models, mainly because the xenograft tumors don't behave like naturally occurring tumors in humans--they don't spread to other tissues, for example. Thus, drugs tested in the xenografts appeared effective but worked poorly in humans. (Gura 1997)

The phenomenon of cancer in humans is complex. It is complex in other animals as well. Greek and Greek [(Greek and Greek 2002) p76-96] have discussed some of the ways in which cancer in nonhuman animals—including experimentally induced cancers—differs from cancer in humans.

Cancer and the Evolution of Mice and Men

We have just seen that mice and men are different in medically significant ways. The line leading to modern mice diverged from that leading to modern humans over 70 million years ago, for at least 140 million years of independent evolution. This fact is mentioned here because the vast majority of biomedical research involving animals to predict human

outcomes involves rats and mice (usually in highly inbred lineages). Biological evolution is a process involving descent from common ancestors
(hence divergence of lineages) coupled with subsequent evolutionary
modification (involving adaptations all the way down to the details of
regulation at the level of basic biochemistry and cell biology). One of
the issues we confront in this book centers on the use of one kind of
complex system—the mouse—to model another type of complex system—the human. Humans and mice are similar in some respects and
different in others. Humans and mice are similar with respect to a gene-
count (though without an analysis of issues pertaining to genomic regulation, the significance of this observation is highly unclear). Humans
and mice certainly differ with respect to degree of delegated complexity
in their respective nervous systems. A hypothesis to be introduced here
is that the physiological *milieu intérieur* of the mouse is not simply a
smaller version of that found in the human. Instead, it is a differently
complex physiological environment reflecting the long reach of evolutionary causes that culminate in mice and men having taken very different evolutionary trajectories. Rangarajan and Weinberg hit the nail on
the head when they observe:

> About 30% of laboratory rodents have cancer by the end of their
> 2-3 year lifespan and about 30% of people have cancer by the end
> of their 70-80 year lifespan. Moreover, although the incidence of
> cancer increases with age in both species, 30% of humans do not
> have cancer by the age of 3 years. So, a marked decrease in age
> specific cancer rates has accompanied the substantial increase in
> lifespan that has occurred during the last 80 million years of the
> mammalian evolution that led, via the primate lineage, to humans.
> This decrease in cancer susceptibility has been accompanied
> through the development of several distinct antineoplastic me-
> chanisms, many of which are intrinsic to human cells. (Rangarajan
> and Weinberg 2003)

Among the factors relevant to species differences with respect to cancer
susceptibility cited by Rangarajan and Weinberg (Rangarajan and Weinberg 2003) are the following:

[1] Organismal differences in basal metabolic rate (associated with differences in oxidation damage).
[2] Differences with respect to the metabolic pathways by means of which
carcinogens are broken down.

[3] Mouse tumors tend to be lymphomas and sarcomas, human cancers tend to be carcinomas—these are differences that may reflect basic differences between human and rodent tissues.

[4] Cells in human tumors often involve changes in chromosome number and other genetic changes such non-reciprocal translocations. Such changes are rare in mouse tumors.

[5] Differences with respect to telomere biology. Telomeres are specialized DNA "caps" at the ends of eukaryote chromosomes that modulate the tendency of chromosomes to shorten with each successive round of replication. The telomeres of mice are longer than those of humans, and telomerase—the enzyme responsible for the maintenance of telomeres, is active in mouse cells but difficult to detect in human cells. It is possible that these differences, observed *in vitro*, are relevant to an understanding of why mouse cells readily become cancer cells, and why human cells tend toward a state of replicative senescence. For instance, as noted by Rangarajan and Weinberg, telomere erosion triggers cellular senescence and places limits on the capacity for replication in human cells.

There are also other important differences with respect to mechanisms and pathways regulating cellular senescence, especially tumor-suppressor pathways. The *p53* pathway, the *RB* (retinoblastoma) pathway and the *Raf-MAPK* (Raf-mitogen activated protein kinase pathway) are important here. Rangarajan and Weinberg comment:

> These various observations show the existence of key differences in the signaling requirements for the transformation of mouse and human cells *in vitro*. In mouse fibroblasts, perturbation of just two signaling pathways—those involving *p53* and *Raf-MAPK*—seems to be sufficient to mediate tumorigenic conversion. In human fibroblasts, perturbations of six or more pathways—those involving *p53*, *RB*, telomerase, *PP2A*, *RAL-GEF* and one or more additional *RAS*-effector pathways—seems to be essential for achieving the same outcome. (Rangarajan and Weinberg 2003)

Again, as we saw in the earlier discussion of developmental biology, it is not so much that species differences require for their appearance the acquisition of novel components at the level of basic "parts" (so mice and humans differ with respect to basic constitution), rather, the differences lie in the evolution of *novel uses* of a conserved set of parts in different lineages. What do we mean by this?

The Unity of Biochemistry

In order to begin to get a handle on the issue of the evolutionary conceptualization of the physiological similarities and differences between mice and men, it will be helpful to briefly discuss one of the central pieces of molecular evidence supporting the theory of evolution: *the unity of biochemistry*.

Extant life is divided into three domains: Archaea, Bacteria and Eukarya. In terms of the tree of life, these can be thought of as three giant limbs branching off the central trunk where the last common ancestor of all contemporary life on Earth must have existed. In these terms, mice and men are at the ends of different twigs on the branch of a branch . . . of a branch coming off the Eukarya limb. Organisms are historical entities that carry with them records of their evolutionary pasts. Embodied in the unity of biochemistry, however, is the idea that biochemical features common to organisms are reflections of their descent from common ancestors. Over thirty years ago Szent-Györgyi put it this way:

> Life has developed its processes gradually, never rejecting what it has built, but building over what has already taken place. As a result the cell resembles the site of an archaeological excavation with the successive strata on top of one another, the oldest one the deepest. The older a process, the more basic a role it plays and the stronger it will be anchored, the newest processes being dispensed with most easily. [(Lahav 1999) p100]

To be interested in the unity of biochemistry is to be interested in the conserved features of organisms. These are the features of organisms that can be found in many, perhaps all, contemporary lineages. These are also the features of organisms that serve as evidence of descent from common ancestors. The unity of biochemistry is embodied in the following four claims concerning the biochemical features of life on Earth:

[1] *Common molecules.* The macromolecules found in all life on Earth are derived from a common set of monomeric subunits (for example, amino acids). Some macromolecules, such as the 16S ribosomal RNA, are found in all extant organisms with function apparently conserved across all lineages [(Lahav 1999) p105].

[2] *Common systemic features.* Organisms are not simply bags of molecules. At the biochemical level of description, organisms are complex, dynamic, integrated biochemical systems. These systems have similar features such

as the ribosome method of protein synthesis, and the catalysis of reactions using enzymes. There are also similarities with respect to central metabolic pathways. For example, the Krebs cycle is common to all aerobic organisms. Genes involved in the core components of biochemical systems tend to be conserved across many lineages. Such genes will include (1) those involved in the manufacture of enzymes for DNA replication and repair; (2) genes undergirding central metabolic pathways; (3) genes for ion channels, ion transport and ion and metabolite pumps; and (4) genes for the synthesis and degradation of intracellular structures [(Hochachka and Somero 2002) p14].

[3] *Pathway and network complexity.* The reaction networks in all biochemical systems are highly complex. Chemical reactions at a given location in such reaction networks are catalyzed by enzymes that are themselves products of reactions occurring at other locations in the common, integrated, interactive chemical network. Reactions are organized into pathways or sequences of reactions. Pathways are themselves components of more complex parts of the network, such as cycles. The multiplicity of pathways and cycles are then integrated to generate the entire biochemical network itself. Reaction networks are open-dissipative chemical systems far from equilibrium with their surroundings. They are, in fact, systems whose self-regulatory dynamics reflects the combined effects of various positive and negative feedback loops. Though this systemic complexity reflects the long course of evolution, it is sustained in every cell of all organisms by self-organizing dynamical processes driven by the flow of energy through those cells (Kauffman 1993).

[4] *Common conventions.* These are features of all biochemical systems that could have been different—features that represent conventions (like driving on the right hand side of the road rather than the left in the US). Examples include the triplet code and the use of left-handed amino acids. The common biochemical conventions represent features of organisms that were most likely already fixed in the common ancestor from which all contemporary life descends.

Taken together, the different aspects of the unity of biochemistry are powerful evidence that all contemporary organisms are descended from a common ancestor in the distant past. The acknowledgment of the *unity of biochemistry* is not the same, however, as the very different assertion of the *identity of biochemistry*, for along with the common biochemical features discussed under the heading of the unity of biochemistry are many biochemical differences between extant organisms. These differences arise from bio-chemical evolution as different organisms, confronting different evolutionary problems, found ways to adapt to a vast multiplicity of different environ-ments. Some biochemical differences between organisms represent biochem-ical adaptations, other differences may arise as by-products of selection for specific biochemical adaptations, or as the result of genetic drift (in the case of neutral molecular evolution).

Hochachka and Somero estimate that some 6,000 human genes are involved in central biochemical processes of the kind that are conserved across eukaryotes [(Hochachka and Somero 2002) p15]. What then are the biochemical (molecular) origins of physiological diversity in cells observed in multicellular organisms? Hochachka and Somero have re-cently suggested that to address this question it is helpful to examine genes involved in the sensing of changes in the extracellular environ-ment and in the communication of that information to intracellular tar-gets. They observe:

> The differential expression and differential evolution of these kinds of genes . . . supply the raw material for evolutionary change and species specificity. That this is one key point of departure for physiological diversity in multicellular organisms is strongly sup-ported by studies showing that genes involved in such functions are the most differentiated of all, that is, they are expressed in most numerous isoforms of all. We shall see . . . that the complex-ity of physiological systems in multicellular organisms requires ev-er more complex signaling, signal transduction, and communica-tion, as body plans attain higher levels of complexity. The control networks that have evolved are hugely complex by comparison with single-celled eukaryotes such as yeasts. [(Hochachka and So-mero 2002) p15]

In the light of these observations, Hochachka and Somero make the following proposal:

> . . . the hypothesis is that genes whose products are involved in such processes as interorgan communication, in cell-cell communication, in development and differentiation, in general sensing and signal transduction, in immune system defenses, and in host defenses against pathogens and parasites, are fundamental to the evolution of physiological diversity. [(Hochachka and Somero 2002) p15]

In this way, physiological differences between organisms (especially those belonging to different species) are reflections not so much of differences with respect to basic components or building blocks, but of differences with respect to the ways in which common, highly conserved parts are employed and organized (recall the piano analogy in our discussion of organismal development). A failure to appreciate this last point will be seen as a source of controversy in the next chapter, where the focus of the discussion is on the metabolism of xenobiotics.

CHAPTER 10

EVOLUTION AND THE METABOLISM OF XENOBIOTICS

Facts don't cease to exist because they are ignored.
—Aldous Huxley

In the context of research into the metabolism of xenobiotics aimed at benefiting humans, an ideal animal model should be a *CAM* satisfying further pragmatic constraints. The relevant causal analogies will be found at the lower levels of the biological hierarchy, since it is here that we expect extensive similarities between organisms of different species—it is here that we expect them to be, " . . .the same animal dressed up differently [(Cairns-Smith 1986) p50]." The pragmatic constraints will reflect the properties of organisms that appear at the organismal level of the hierarchy, where, it would appear, evolution has inadvertently bequeathed to some species properties that make them particularly convenient for experimental studies. However, as we saw in the last chapter, an appeal to the unity of biochemistry does not give warrant for the conclusion that, because organisms belonging to different species are rooted in a common set of building blocks, they are fundamentally the same animals *at the biochemical level of description* (so that one may be used straightforwardly as a *CAM* for another).

Methodological Issues

In the context of teratological investigations into developmental toxins, Schardein has suggested that we need to seek nonhuman animal species that satisfy the following *desiderata*:

- Absorbs, metabolizes, and eliminates test substances like man.
- Transmits test substances and their metabolites across the placenta like man.

- Has embryos and fetuses with developmental and metabolic patterns similar to those of man.
- Breeds easily and has large litters and short gestation.
- Is inexpensively maintained under laboratory conditions.
- Does not bite, scratch, kick, howl or scream. [(Schardein 1985) p19]

In other words we need nonhuman animal species that are similar in some respects to humans, but different in other respects and not just in teratology. Bahls et al. have recently listed similar criteria for the use of animal models generally, adding:

> Each model organism is distinctly suited, in its guise as a simplified model, to the study of complex aspects of biology. Researchers are repeatedly surprised that discoveries in simple organisms are relevant to human biology, which encourages transposition of results from one model system to another, and highlighting the extent of conservation and commonality of life forms. The differences hold value as well, as they provide important insights to understanding cell physiology and pathology. (Bahls, Weitzman, and Gallagher 2003)

In the context of basic biomedical research, similarities and differences can indeed be of great importance. But when animal models are used to predict human responses (as opposed to yield basic biological insights), the differences can be of great importance. The problem here is that mice are not simplified versions of humans, they are differently complex systems. The idea that mice are simplified versions of humans seems at best a holdover from discredited pre-Darwinian Lamarckianism (and its attendant view of an organismal struggle to achieve ever greater degrees of complexity, humans being exemplars of very complex creatures). So what are the scientific principles that enable us to cross the logical and biological chasms that seemingly separates mouse from man?

Both toxicology and pharmacology are branches of science that concern the effects of xenobiotics. Toxicology is concerned with adverse effects, whereas pharmacology is concerned with the use of xenobiotics to treat illnesses. Both branches of science make extensive use of nonhuman animals as CAMs of human toxicological and pharmacological phenomena.

According to Klaassen and Eaton, all animal toxicity testing rests on two basic principles, which they explain as follows:

The first is that the effects produced by the compound in laboratory animals, when properly qualified, are applicable to humans. This premise applies to all of experimental biology and medicine. On the basis of dose per unit of body surface, toxic effects in humans are usually in the same range as those in experimental animals. On a body weight basis, humans are generally more vulnerable than experimental animals, probably by a factor of about 10. With an awareness of these quantitative differences, appropriate safety factors can be applied to calculate relatively safe dosages for humans. (Klaassen and Eaton 1993)

This first principle is a clear manifestation of ideas about the relationship between humans and nonhuman animals that come straight from Claude Bernard. In this context, species differences are primarily differences with respect to quantitative factors, and not qualitative factors reflecting evolved organizational differences.

The standard way in which species differences are dealt with in toxicology is through consideration of quantitative scaling factors:

Scaling factors are a means for correcting species differences in cross-species comparisons. From a mechanistic perspective, scaling factors implicitly consider two independent physiological processes: (1) differences in pharmacokinetics, which determine the actual dose delivered to target tissues, and (2) differences in tissue sensitivity between species to an identical delivered dose. In practice scaling factors seldom incorporate such considerations explicitly but rather use dose adjustments across species based on some normalizing factor such as body weight or surface area. The most common form of scaling factor is body weight. This is largely empirical. The most appropriate scaling factor to use in interspecies extrapolations for carcinogenicity has been extensively debated. Surface area which is roughly equivalent to $(\text{body weight})^{2/3}$, may be a more accurate scaling factor between species than body weight directly. [(Klaassen and Eaton 1993) p44-45]

Having decided how to accommodate purely quantitative species differences, the problem is then to find a way to generate results about human populations on the basis of the study of nonhuman species. It is at this juncture that the second principle of toxicology plays a crucial role. Thus, Klaassen and Eaton continue:

> The second main principle is that exposure of experimental animals to toxic agents in high doses is a necessary and valid method of discovering possible hazards in humans. This principle is based on the quantal dose-response concept that the incidence of an effect in a population is greater as the dose or exposure increases. [(Klaassen and Eaton 1993) p32]

In other words, high doses are sometimes required to generate effects in small experimental populations that may be of concern at lower doses in large human populations.

In the famous Canadian saccharin-bladder cancer rodent study, the rats received saccharin equivalent to 800 sodas per day [(Giere 1991) p232]. The study was a 2-generation study in which, in the second generation, 14 of 94 rats so exposed developed bladder cancer. Of this result Giere offers the following commentary:

> . . . taking account of differences in dose and body weight, those fourteen cancers in ninety four rats translate into about 1200 cases of bladder cancer in a population of 200 million people drinking less than one can of soda a day. You can easily calculate that 1200 cases of bladder cancer in a population of 200 million means that any individual is facing a 6/1,000,000 chance of getting bladder cancer. So it may be argued that the risk is small and people ought to be allowed to decide for themselves whether to take this risk. That, however, is an entirely different question than the scientific question of whether there is any risk at all, and if so, how much. [(Giere 1991) p233]

The scientific question is by no means as simple as this case suggests.

Low dose risk assessment is a complex matter. Formaldehyde at 10 parts per million (ppm) in air is carcinogenic in rats, but has not been found to be so at 2 ppm. Extrapolation from high dose to low dose in the test species thus poses problems. There seems to be a threshold. Moreover, as Williams and Weisburger have pointed out, tests in mice (as opposed to rats) have been negative and, "Even though formaldehyde has been used extensively for decades as a tissue fixative in pathology and in embalming practices, so far there appears to be no documented record of a higher incidence of respiratory tract neoplasia in the many humans exposed to it occupationally or generally [(Williams and J Weisburger 1993) p181]."

To use an amusing example, consider caffeic acid (3, 4-dihydroxycinnamic acid). In rats, 50% develop cancer when exposed to 300 mg/kg of body weight. This is known as the TD_{50}. (For mice the TD_{50} value is 4900 mg/kg of body weight (Gold et al. 1992). Because both rats and mice develop carcinomas of the forestomach, the IARC considers caffeic acid to be a possible human carcinogen. In terms of the Human Exposure Rodent Potency (HERP) index, the rat, being the more sensitive species, is used in the calculation of possible cancer hazard. The method is as follows:

$$\text{HERP } (\%) = 100(\text{human dose in mg/kg}) \, / \, (\text{rodent } TD_{50} \text{ in mg/kg}).$$

A HERP value of 1% is a dose where humans are exposed to 1% of the TD_{50} for the sensitive rodent. Typical human exposure to caffeic acid in coffee alone has a HERP value of 0.1% (Gold et al. 1992).

A typical human dose of caffeic acid from all sources is about 70 mg per day (it is in the coffee, fruits and vegetables we enjoy). This is approximately 1 mg/kg of body weight. Using rat data employed in the HERP calculation, the risk of cancer at 300 mg/kg is 1 in 2. Making the assumption of linearity (direct proportionality of dose to response), we get an extrapolation of human cancer risk as (0.5 x 1/300), i.e. 1 in 600. Harris has wryly observed:

> If caffeic acid were to come under regulatory scrutiny what probable regulatory limit would be recommended? According to the linear model it was calculated that for our average daily dose of about 1mg/kg, the estimate of cancer risk would be 1 in 600. For a regulatory one-in-a-million risk target the safe level for humans would be a dose at least 1,000 to 10,000 times lower than the actual doses we receive ... Regulators are not mandating dose limits. The lack of attention is appropriate but for reasons that are not consistent with widely used but unrealistic linear dose to risk conversions. (Harris 1997)

In the IARC evaluation of caffeic acid (IARC 1993), it is noted that there is no relevant human data, the threat of carcinogenicity is derived from rodent studies and the additional claim that humans and rodents metabolize caffeic acid to the same metabolite. The IARC's line of reasoning is precisely what one would expect if rodents were being used as *CAMs* for the human condition. (Web searches actually reveal numerous sources, including the USDA, that list many healthful benefits forthcoming from caffeic acid in the diet. The USDA's agricultural research service actually points out that caffeic acid has useful antioxidant and antiviral properties.)

Never mind rodents and humans (whose lineages diverged over 70 million years ago), consider rats and mice. Tests for chemically-induced cancers in rats and mice yield the same results (non-site specific concordance) for about 70% of the substances tested. The figure drops to 51% for site-specific cancers (Gold et al. 1991). The divergence between rat and mouse lineages may be as recent as 15 million years ago. The divergence between humans and old world monkeys is estimated, by contrast, at approximately 25 millions years ago [(Li 1997) p224]. (For reference, the human and chimpanzee lineages are estimated to have diverged approximately 7 million years ago.)

The scientific question is very much at issue in a study such as this. More cautiously Gallo and Doull comment on the extrapolation problem as follows:

> Toxicology, like medicine, is both a science and an art. *The science of toxicology is defined as the observational and data-gathering phase, whereas the art of toxicology is the predictive phase of the discipline.* In most cases, these phases are linked since the "facts" generated by the science of toxicology are used to develop the extrapolations and hypotheses for the adverse effects of chemical agents in situations where there is little or no information. For example, the observation that the administration of 2, 3, 7, 8-tetrachlordibenzo-p-dioxin to female (Sprague Dawley) rats induces hepatocellular carcinoma is a fact. However, the conclusion that it will do so in man is a prediction or hypothesis. It is important to distinguish the facts from the predictions. *When we fail to distinguish the science from the art, we confuse facts with predictions and argue that they have equal validity, which they clearly do not.* (Gallo and Doull 1993) (Emphasis added.)

This is a particularly elegant way of making the basic claim about *CAMs* that is our concern in this book. Conclusions about humans rooted in nonhuman animal *CAMs* are simply predictions or hypotheses; they are not established facts. Moreover, in many cases, for legal reasons, such unconfirmed predictions or hypotheses will remain just that since it is unlikely that large-scale randomized, double-blind human studies are going to be performed with suspect carcinogens or developmental toxins. Here is the rub, for as Hau has argued:

Most of the regulating authorities require two species in toxicology screening, one of which has to be non-rodent. This does not imply that excessive numbers of animals will be used, because an uncritical use of one-species models may mean that experimental data retrospectively turn out to be invalid for extrapolation, representing a waste of animals. The appropriateness of any laboratory animal model will eventually be judged by its capacity to explain and predict the observed effects in the target species. [(Hau 2003) p8]

In predictive toxicology, two unconfirmed hypotheses rooted in tests in two species (possibly differing from each other) are no better than one such hypothesis. What matters is confirmation, and that requires data from the target species. Hau points to a general problem: "It is not possible to give reliable general rules for the validity of extrapolation from one species to another. This has to be assessed individually for each experiment and can often only be verified after first trials in the target species [(Hau 2003) p6]." Where the target species is human, such data will often *not* be forthcoming.

This issue also comes to the fore in the context of the issue of animal drug residues and their metabolites finding their way into human food. Here there are two problems: first, exposure to low doses of the animal drug; and second, exposure to the products of nonhuman animal metabolism which will typically alter the biological properties of the drug administered. However, if you believe that humans and nonhuman animals are the same animal dressed up differently, neither of these issues should be important, and the effects of human exposure to animal drugs and their metabolites ought to be forthcoming from what is known as a *relay toxicity study*.

Here, a drug is fed to one set of animals and then these animals are slaughtered and their products and tissues are fed to another set of animals (to model human consumption of contaminated products and tissues). But as Miller has observed:

Relay studies present problems associated with feeding animals diets that may be different from their natural diets, thus resulting in associated problems of unpalatability, altered digestion, and inadequate nutrition. These studies also have the inherent problems associated with all toxicologic evaluations such as extrapolation of animal data to humans. (Miller 1993)

The evidence forthcoming from a relay study tells us how one species responds to a diet consisting of contaminated tissues from another species. We are left, as before, with the untested hypothesis that humans will respond similarly.

Two issues emerge from the present discussion. First, claims about the direct relevance and applicability of results in mice, say, to the biology of humans are not typically rooted in experimental evidence (for this would require extensive human testing). Instead, they are claims rooted in bogus inferences from the unity of biochemistry to the effect that at the lowest (biochemical) levels of the biological hierarchy or organization, mice are basically the same as humans, differing only with respect to such quantitative factors such as mass or surface area—in essence, we are the same animals dressed up differently, and, since mice are relatively close to humans from a phylogenetic standpoint, this degree of phylogenetic closeness further bolsters the applicability or results in mice for the biology of humans. Second, there are issues about methods used in extrapolation from animal data to (often untested) hypotheses and predictions about the human condition. The facts of toxicological inquiry lie in the results of animal experiments. Unless the resulting hypotheses about humans that are rooted in these facts are actually tested rigorously, there are no further facts in the nature of the case, and all we have are theoretical inferences based on doubtful interpretations of the unity of biochemistry that turn out to be highly contentious. We now turn to examine this matter.

In the end, human data is forthcoming—as a consequence of studies of individuals exposed to drugs when they are consumers of pharmaceutical products, and as a consequence of studies into the effects of environmental exposures. *But these studies seem to take second place to studies involving nonhuman animals.*

The Place of Evolution

It is instructive to examine the scientific rationale for using nonhuman animals, typically rats and mice, as experimental models of humans. Most researchers will probably acknowledge that there are indeed biological differences between humans and the animals used to model them. However, as Sir William Patton has pointed out:

> But one should not over-estimate these differences. We have only to look at an account of evolution or at textbooks in comparative anatomy to see how much we have in common: hearts, lungs, kid-

neys, brains, endocrine glands, nerves, muscles, digestive systems, all built on the same plan. This homology goes back still further as one moves down to the biological elements like the nucleus, the mitochondria, or the cell membrane, out of which the higher organisms are built. The only differences [in drug actions] that appear are in dose required, duration of action, sometimes in the way action manifests itself, and sometimes in side-effects. [(Patton 1993) p166]

The similarities between organisms that Patton refers to are explained as the result of biological evolution, and the fact that humans and the animals used to model them in experiments share common evolutionary ancestry. Hau puts it this way:

The rationale behind extrapolating results to other species is based on the extensive homology and evolutionary similarity between morphological structures and physiological processes among different animal species and between animals and human . . . . [(Hau 2003) p6]

But how relevant is the mere fact of this common evolutionary ancestry?

The modern mechanistic physiologist does not deny the facts of evolution, but sees in its implications two methodological rules. Geneticist Richard Lewontin characterizes these as follows:

First, it is assumed that the detailed similarity between organisms increases as one goes to more and more basic cellular processes within organisms. So, a rat and a human may not look alike, but they are assumed to have similar general structures for their nervous systems, extremely similar chemistry for the actual firing of individual nerve cells, and identical chemistry for the copying of genes that code for the production of these chemicals. So the model maker feels comfortable carrying over the results from rats to people if it is a question of basic cellular processes. That is why rats are used for drug trials, at least to screen for harmful effects of new drugs. *Second, it is assumed that whatever differences exist at any level, these get smaller and smaller as one compares organisms that are more closely related in evolution.* Thus the model maker feels confident about carrying over nearly anything seen in chimpanzees to people. (Emphasis added.) [(Lewontin 1995) p9]

It is worthwhile to pause and examine these assumptions, for they do, in fact, play a role in the way in which researchers think of the similarities and differences between humans and nonhuman species.

To borrow words from biochemist A.G. Cairns-Smith [(Cairns-Smith 1986) p50], the first assumption implies that at the biochemical level humans and their nonhuman animal models are essentially "the same animal dressed up differently." Never mind humans and mice, for Alberts et al. observe:

> In molecular terms we have a more thorough knowledge of the working of *E. coli* than of any other living organism. Most of our understanding of the fundamental mechanisms of life—for example, how cells replicate their DNA to pass on their genetic instructions to their progeny, or how they decode the instructions represented in the DNA to direct the synthesis of specific proteins—has come from studies of *E. coli*. The basic genetic mechanisms have turned out to be highly conserved throughout evolution: these mechanisms are essentially the same in our own cells as in *E. coli*. [(Alberts et al. 2002) p27-28]

An important part of the rationale for using animals as models of human biomedical phenomena derived from the claim that there are levels in the hierarchy of organization where, despite the obvious differences at the organismal level of description, there are fundamental similarities. We agree that there are fundamental similarities of the kind Alberts et al. refer to. It does not follow from this, however, that there are relevant similarities in the ways in which humans and mice, for example, metabolize xenobiotics.

The second assumption referred to by Lewontin brings out traditional physiology's linear, quantitative estimate of the significance of species differences to a discussion of the significance of evolved differences. What do we mean by this? The basic idea is that while there are similarities between all organisms at the basal levels of the hierarchy of organization, differences between species with respect to other levels in the hierarchy diminish as one considers closer and closer phylogenetic relatives. Again, as Alberts et al. point out:

> In terms of genome size and function, cell biology and molecular mechanisms, mammals are nevertheless a highly uniform group of organisms. Even anatomically, the differences among mammals are chiefly a matter of size and proportions; it is hard to think of a human body part that does not have a counterpart in elephants and mice, and vice versa Evolution plays freely with quantitative fea-

tures, but it does not readily change the logic of structure. [(Alberts et al. 2002) p42]

Small wonder Alberts et al. continue:

> The mouse, being small, hardy, and a rapid breeder has become the foremost model organism for experimental studies of vertebrate molecular genetics. Many naturally occurring mutations are known, often mimicking the effects of corresponding mutations in humans. [(Alberts et al. 2002) p42]

If these two assumptions are an accurate reflection of the implications of evolutionary processes for differences between the species, *then* it is hardly surprising that researchers are inclined to think of mice as men writ small. We think that the study of the details of the metabolism of xenobiotics will reveal much about the validity of both of the assumptions that undergird the use of rodents as predictive models of human biomedical phenomena.

For example, in this book we are concerned (among other things) with species differences in the metabolism of xenobiotics. How might a deeper understanding of evolution affect our appreciation of this interesting subject matter? It is clear that animal studies employed in the screening of xenobiotics for human safety display little or no concern for evolutionary biology. All too often, ubiquitous species differences are seen as "noise" spoiling a clear "signal" about the human significance of the results of animal studies, rather than as being a "signal" that something is deeply flawed from a methodological standpoint, in the conception of mouse as man writ small.

Interestingly there is in fact a scientific tradition extending back at least forty years which has actively attempted to incorporate metabolic issues into sound microevolutionary science. We refer to *pharmacogenetics* and its recent conceptual descendant, *pharmacogenomics* (which uses genome-wide approaches to the study of inherited differences in drug metabolizing activity) (Evans and McLeod 2003). Vogel coined the term pharmacogenetics in 1959. As Evans et al. have recently observed:

> Although many nongenetic factors influence the effects of medications, including age, organ function, concomitant therapy, drug interactions, and the nature of the disease, there are now numerous examples of cases in which interindividual differences in drug response are due to sequence variants in genes encoding drug-metabolizing enzymes, drug transporters, or drug targets. Unlike

other factors influencing drug response, inherited determinants generally remain stable throughout a person's lifetime. (Evans and McLeod 2003)

One of the first drugs noticed to have significantly different effects among individuals was isoniazid. Isoniazid can be rapidly or slowly acetylated and this difference is under genetic control (Evans, Manley, and McKusick 1960). The fact that succinylcholine, a drug used to induce very rapid muscle relaxation could have a prolonged half life due to an enzyme abnormality (noticed shortly after the isoniazid difference), also reinforced the notion that drugs could have very different effects on different individuals.

Studies in this tradition of research focus on of the role of inheritance in shaping individual variation to drug response. These approaches represent major steps forward precisely because they focus on what is important in evolution: individual variation. Unlike predictive animal modeling, which typically involves the study of populations of genetically inbred strains of rodent, with subsequent extrapolation to the typical human (as though such a being really existed), pharmacogenetics and pharmacogenomics are disciplines which have sprouted from the human clinical setting, and observations of significant variability in response to exposure to xenobiotics. Weinshilboum has recently observed:

> The concept of pharmacogenetics originated from the clinical observation that there were patients with very high or very low plasma or urinary drug concentrations, followed by the realization that the biochemical traits leading to this variation were inherited. Only later were the drugs metabolizing enzymes identified, and this discovery was followed by the identification of the genes that encoded the proteins and the DNA-sequence variation within the genes that was associated with the inherited trait. Most of the pharmacogenetic traits that were first identified were monogenic—that is, they involved only a single gene—and most were due to genetic polymorphisms; in other words the allele or alleles responsible for the variation were relatively common. . . Today there is a systematic search to identify functionally significant variations in DNA sequences in genes that influence the effects of various drugs. (Weinshilboum 2003)

Weinshilboum (Weinshilboum 2003) identifies several important themes that occur in the context of pharmacogenetic research: (1) human populations display medically significant variation in alleles whose products influence responses to drugs; (2) striking differences between ethnic groups with respect to the frequencies with which these alleles are found; (3) a growing realization that variation affects not just the structure and functions of enzymes (e.g., the cytochrome P450 (CYP) superfamily which will be discussed in some detail later in this chapter) involved directly in metabolism, but also the structure of transporter molecules and target receptors; (4) a need to understand polygenic (many-gene) effects; (5) a growing interest in pharmacogenomics, a new discipline resulting from the fusion of traditional concerns in pharmacogenetics with the fruits of genomic science devoted to the study of whole organismal genomes.

Abnormalities of genes that lead to differences in drug action may be explained by single nucleotide polymorphisms (SNPs), gene regulation, or by differences in genetic networks. For example:

- Complex haplotypes of SNPs in the coding region and promoter of the B_2-adenoreceptor gene can influence a patient's response to bronchodilator therapy (Drysdale et al. 2000);
- Breast cancer susceptibility is influenced by the interactions of several different genes (Ritchie et al. 2001);
- A serine to glycine polymorphism in the dopamine D_3 receptor gene has been associated with a susceptibility to tardive dyskinesia when given antipsychotic drugs;
- A functional polymorphism in the cholesterol ester transfer protein has been associated with a greater probability of pravastatin to lower plasma cholesterol levels (Kuivenhoven et al. 1998).
- Asthma can be treated with B_2-adenorecptor (B2AR) agonists or ALOX5 (5-lipoxygenase) inhibitors but many patients treated thus fail to respond to therapy or suffer adverse drug reactions that limit their ability to take the medication. The gene for the B_2-adenoreceptor is *B2AR*. The polymorphism Arg16Gly of *B2AR* was studied and it was discovered that when a patient was homozygous for Gly16, he had significantly lower airway responsiveness for the B_2 agonist salbutamol (Kotani et al. 1999). Similarly, when the antiasthma drug ABT-761 was studied, it was discovered that when patients had a polymorphism in the promoter region of the gene *ALOX5*, they failed to respond to ABT-761 (Drazen et al. 1999).
- Polymorphisms in *CYP2C9* account for differences in warfarin metabolism (Aithal et al. 1999).

- The risk of thiopurine toxicity is determined by the presence of specific S-methyltransferase alleles. Genotyping is usually performed prior to beginning treatment (Relling et al. 1999; McLeod et al. 1999).
- The drug trastuzumab will be beneficial to only 25-30% of patients with late stage breast cancer (those who over-express the protein encoded by the *HER2* gene) (Menard et al. 2000).

Equally important is the fact that drug-metabolizing enzymes (DMEs) can vary intraindividually (within the individual). This fact is of greater importance when administering drugs for a prolonged period of time or when the therapeutic window is narrow. The drugs themselves may induce changes in DMEs or the changes may be due to other drugs, disease, aging, diet, or may be environmentally induced. In healthy volunteers, intraindividual variability in CYP1A2 ranged from 5 to 50%, CYP2D6 ranged from 12 to 140%, CYP 3A4 ranged from 5 to 21%, and N-acetyltransferase 2 from 2 to 27% [(Lerer 2002) p166].

Evolutionary medicine begins with the study of variation in relevant populations. This variation may be heritable variation, or it may arise from phenotypic plasticity due to different individuals encountering different environments in the course of developmental time. Evolutionary medicine sees adaptations in the context of their evolution (be they biochemical or morphological). The focus on this approach to biomedicine is inevitably species-specific. It rejects, because of irrelevancy to the human condition, the use of animals *as predictive models* of human biomedical phenomena, especially where such animals are genetically homogeneous inbred strains that manifest little or no variation. Transgenic animals or animals with xenografts do little to meet these objections, since tissues growing in an alien context, in an immunologically crippled organism, do not behave like tissues growing in the organism evolved to support them, and the expression of genes (hence their phenotypic effects) reflects the local environment (the organismal, regulatory context) in which the gene finds itself.

In the end, progress in biomedical research will involve a recognition that if the physicist Newton and his physiological descendent, Bernard, were right—that where there are qualitatively identical systems, and where these are similarly stimulated, same cause is followed by same effect—this is of little help to the predictive animal modeler of human biomedical phenomena. Because of the differential effects of evolution in divergent lineages, humans and rodents are not qualitatively identical, causally speaking, even after differences in mass and surface area have been allowed for. Being not merely different in size, but differently organized with respect to the *milieu intérieur*, there is no rational expectation that when the latter is used as a *CAM* for the former, same cause will be followed by same effect (and if it does, it does so for the same reason, following the same pathway). It is a consequence of evolution that the focus of biomedical research must move away from the study of organismal "sameness" to approaches that explicitly involve population studies of variability.

Ernst Mayr, one of the architects of the New Synthesis of evolutionary biology has emphasized variability and uniqueness as follows:

> No two individuals in a sexually reproducing species are identical. Among the millions of cells of an organism no two are probably exactly identical, owing to the diverse activity (suppression and activation) of regulatory genes. Even greater, of course, is uniqueness among species, higher taxa, and ecosystems. . . Uniqueness results in variability, and variability is characteristic of living systems, from the cells of the body through individuals to species and to still higher aggregations. (Mayr 1989)

Mayr continues:

> The consequences arising from uniqueness are many. It explains the almost incomprehensible diversity of the living world. It explains why in the course of evolution so often different organisms adopt different pathways to achieve the same adaptation (the principle of multiple pathways). It explains why the response to selection pressure is only probabilistic. Indeed it is one of the reasons why predictions in biology are so often impossible. (Mayr 1989)

The Metabolism of Xenobiotics

We have seen that according to predictive modelers who attempt to incorporate evolutionary considerations into their work, evolution has played freely with quantitative features of organisms, but has done little to alter the underlying biological systems, especially at the lower levels of the biological hierarchy. We would thus expect there to be extensive similarities between mammals not only with respect to the structures, but also the functions and substrate specificities of the enzymatic machinery involved in xenobiotic metabolism. We would also expect there to be—as Bernard had hypothesized—good ways to compensate for the ways in which evolution has played with the quantitative features of humans and their nonhuman animal models (for example, compensation for differences in body weight).

We have seen that the use of nonhuman animals as *CAMs* for human biomedical phenomena rests on two principles: first, that for a given nonhuman species, the differences between humans and members of that species became fewer and less significant as one considered the respective organisms at lower and lower level in the hierarchy of organization. At the biochemical level, they ought to be "the same animal dressed up differently;" second, that as nonhuman animals that were closer and closer to humans from a phylogenetic standpoint were considered, differences at any level in the hierarchy became fewer and fewer. Perhaps chimpanzees really are humans in ape suits.

In the remainder of this chapter we will examine the first assumption in some detail, and a discussion of the similarities between humans and nonhuman primates will be deferred to a later chapter. The groundwork for that discussion will be laid down here.

Briefly put, the metabolism of xenobiotics (a chemical foreign to the biological system) can occur in one or both of two *phases*. In phase I metabolism, a substance (the substrate) is metabolized through chemical reaction involving reduction, oxidation or hydrolysis. The resulting substance (product) is a phase I *metabolite*. In phase II metabolism, the original substance, or a phase I metabolite, is joined (conjugated) with an endogenous molecule (one produced by the organism), perhaps to enhance water solubility and hence ease of excretion. A substance may be excreted in unchanged form, or as a phase I or a phase II metabolite; or as a metabolite arising from phase I and phase II metabolism. A *metabolic pathway* is the chemical route—a series of reactions—by means of which a substance is metabolized.

The enzyme system playing an important role in phase I xenobiotic metabolism is the cytochrome P_{450} system. Some 500 P_{450} enzymes have been characterized in terms of DNA sequences; members of a given species may carry 40-50 (Guengerich 1997). Cytochromes P_{450} play a variety of roles in humans and other mammals:

> P_{450}s function in critical roles in the synthesis of steroids, and loss of function can be debilitating or lethal. P_{450}s are also involved in the oxidation of other compounds normally found in vertebrates or "endobiotics", e.g., fat-soluble vitamins, fatty acids, eicosanoids, and even alkaloids. However, most of the liver microsomal P_{450}s oxidize a wide variety of substrates and have the function of removing natural products (ingested in foodstuffs) from the body. In support of

this view, the levels of many of these P_{450}s vary dramatically among humans, as opposed to the levels of steroidogenic P_{450}s, which are usually tightly regulated. (Guengerich 1997)

Enzymes participating in phase II reactions include glutathione S-transferases (GSTs), N-acetyltransferases (NATs), sulfotransferases (STs), UDP-glucuronosyltransferases (UGTs), and NAD(P)H-quinone oxidoreductase (NQO1s). Also important are microsomal and soluble epoxide hydrolases that inactivate epoxides to dihydrodiols (mEHs and sEHs) (Gonzalez and Kimura 2001). We will return later to a fuller discussion of these enzyme systems.

Drug manufacturers and scientists are interested in the construction of better ADMET tests (Johnson and Wolfgang 2000; Monro and Mac-Donald 1998). (ADMET stands for absorption, distribution, metabolism, elimination and toxicity.) According to Bains, deficiencies in ADMET tests are the cause of approximately half of the failures of drugs in development to reach the market, and that half of the drugs that do make it to market still exhibit ADMET problems (Hodgson 2001). Further, Selick estimates that for every drug that is pulled from the market as a result of ADMET difficulties not revealed in clinical trials, there are ten more that remain on the market with labeled restrictions because of the potential for drug-drug interactions (Hodgson 2001). In this light, McLean has recently observed:

> Yes, I think it is very clear to all of us who are engaged in the business of assessing toxicity data that, when volumes of data are proudly presented to us after a carcinogenicity study [on animals] showing that there was a tumour in this organ or that, we look at it and we scratch our heads, and we wonder what on earth we can make of it. This is especially true when huge doses are given, with nothing to suggest what would be expected at low doses. I think very often the carcinogenicity studies are a waste of everybody's time and a fearful waste of animals. They are conducted partly because we are not sure what to do instead, and partly because they are a political gesture and a very miserable one at that. (McLean 1991)

There have been many problems associated with the practice of using animals in ADMET studies. Humans and animals may metabolize a substrate to different metabolites, or to the same metabolite but by a different pathway. Animals and humans may also differ in the way their organs and genes respond to the chemical. Different humans may even respond differently because of genetic differences. Because ADMET studies in animals are so predictably unreliable, when a drug enters clinical trials, the company has very little idea if it will damage humans.

One of the first attempts to evaluate the correlation between human and animal data was in 1962 by Litchfield (Litchfield 1962). He reported that the toxicities that occurred in rats rarely occurred in humans while toxicities reported in dogs also were rare in humans. But when the same toxicity occurred in both species it also occurred in humans about 70% of the time. He measured only true positives. Igarashi similarly showed that animals *failed* to predict 43% of toxicities that occurred in humans. The Igarashi study evaluated 139 drugs released in Japan from 1987 to 1991 (Igarashi 1994; Igarashi, Nakane, and Kitagawa 1995; Igarashi). Another study revealed that in only 4 of 24 cases did animals predict human toxicity, and in yet another, 6 out of 114 times (Spriet-Pourra. and Auriche. 1994; Heywood 1990). Many studies have been published outlining the many differences between species that impact on predicting toxicity (Oser 1981; Calabrese 1984; Zbinden 1993; Garattini 1985; Calabrese 1991).

As we know, a single family of isoenzymes known as CYP, are responsible for metabolizing, or breaking down in the body, about 95 percent all current drugs. Using this information, researchers can use cloned CYP enzymes in cell-based assays to test lead compounds for metabolic vulnerability. Liver cells from dogs or rats, or immortalized human liver cell lines, such as HepG2 are currently being used. However, Coleman has recently observed that data from the primary human cells are more informative than that provided by dog or rat livers, explaining that: "There are significant differences in the way that dog or rat livers metabolize compounds and differences in hepatotoxicity [liver toxicity] too (Hodgson 2001)."

In a study designed to compare enzyme activity across species, P450 isozymes were used as probes to study *in vitro* metabolism in horse, dog, cat, and human liver microsomes. The researchers found that there were "large interspecies differences in the way the selective P450 inhibitors affect the *in vitro* metabolism of the various substrates in horse, dog, and cat liver microsomes.... Overall, no one species behaved exactly like humans regarding the efficiency of the various inhibitors (Chauret et al. 1997)." In a similar vein, Weatherall has observed:

> Every species has its own metabolic pattern, and no two species are
> likely to metabolize a drug identically. Small differences in the rate

of conversion of drug to inactive, or toxic, metabolite can have large effects on the concentration of active substances at the point of action. Most experiments to seek toxic effects in whole animals involve oral administration; differences in diet, gut physiology, rate of passage and liver enzymes raise serious questions about the relevance of findings in rats or mice to man. Compounds which are not absorbed in laboratory animals are not, with minor exceptions, ever tested in man. Nobody knows how many drugs, which would be useful in man, may have been lost in this way. Similarly compounds toxic in laboratory animals at doses near the predicted therapeutic level do not receive trial in man, so it is never revealed whether they would actually have been harmful in man. Thus we lack the *evidence* of the false positive element in animal toxicology studies, so it is easy to give more weight to such studies than is justifiable. (Emphasis added.) (Weatherall 1982)

We have already cautioned in the context of our analysis of evolutionary biology, that similarities are not identities. The relevance of this observation for drug discovery has been discussed by Palfreyman et al. as follows:

Mice and humans have more than 95% of their genes in common, yet mice are not men, or women ... Although cell-based and animal models of disease have been the cornerstone of drug discovery it is increasingly apparent that they are of limited predictive value for complex disorders ... One of the major challenges facing the drug discovery community is the limitation and poor predictability of animal-based strategies. Over the last decade, drug discovery has largely been based on finding targets in animal models and then identifying the human homologue ... many drugs have failed in later stages of development because the animal data were poor predictors of efficacy in the human subject ... One of the overriding interests of the pharmaceutical and biotechnologies industry is to create alternative development strategies that are less reliant on poor animal predictor models of human disease ... Although the species [chimpanzees] share more than 98.9% gene identity [with humans], the expression of genes in the brain was more than five-fold greater in humans than in the chimpanzees ... Differences from mice were even greater. These differences reinforce the importance of using human disease models in drug discovery as a real predictor of human efficacy ... Discovery of drugs that act on the human central nervous system, are best studied in human-cell based systems. (Palfreyman, Charles, and Blander 2002)

In the context of the metabolism of xenobiotics, it would appear that the usual animal models are not the same animal dressed up differently at the basal levels of the biological hierarchy, nor does phylogenetic closeness (which seems to mean little more than "being a mammal") have the significance that predictive animal modelers accord to it, and researchers, as opposed to their public policy advocates, seem to be well aware of this. So we must now examine the causes and effects of metabolic variation.

The Causes and Effects of Metabolic Variation

In this section, we will take a closer look at similarities and differences between humans and their rodent *CAM*s with respect to the metabolism of xenobiotics. We will be concerned mainly, but not exclusively, with the cytochrome *P450* enzyme system that plays a central role in phase I metabolism. Cytochromes P450 will be abbreviated to CYPs for ease of reference.

First, some terminology. Alleles belonging to a group of repeated sequences in a genome are said to belong to a gene family. Members of the family, having arisen by gene duplication, are typically found in close proximity to each other on a chromosome. Such alleles exhibit similarities with respect to sequence and, possibly, function. (Note, however, that the evolutionary process of exaptation, whereby gene duplicates can diverge in function due to the accumulation of differences between a gene and its duplicate, along with the subsequent effects of selection acting on the resulting variation, means that it cannot be assumed that there is conservation of function.) When duplicates become sufficiently different in sequence structure or function, they are said to belong to a superfamily. For example the α-globins and the β-globins belong to separate families. Together with the myoglobins, they form a superfamily. Other superfamilies include the collagens, actins, immunoglobulins and serine proteases [(Li 1997) p280].

The CYPs form a superfamily of genes. The following example will help with the nature of CYP nomenclature. Consider CYP1A2. The first number designates the family the gene belongs to, and this is determined on the basis of at least 40% sequence similarity. The letter designates subfamily, determined on the basis of at least 59% sequence similarity. The last number identifies the gene (or protein). CYP 1A2 and CYP 3A4, for example, belong to different families constitutive of the CYP superfamily. By contrast, CYP2C9 and CYP2D6 belong to different subfamilies of the same family. Specific alleles may be denoted by an additional number and an asterisk. CYP2D6*10 refers to a specific allelic variant of CYP2D6, and so on.

For the animal modeler engaged in prediction of metabolic effects, the upside is that there are extensive similarities between mammalian species with respect to the CYP enzyme system. Consider caffeine. CYP1A2 carries out the first step in the caffeine metabolism pathway: 3-demethylation of 1, 3, 7-trimethylxanthine (caffeine) to yield 1, 7-dimethylxanthine. CYP1A2 null mice had a markedly longer plasma half-life than wild type mice thereby indicating the role played by CYP1A2 in caffeine metabolism (Gonzalez and Kimura 2001). Caffeine can thus be used as a metabolic probe to examine levels of CYP1A2 expression in humans, where the gene plays a similar role.

But similarities are not identities and there are indeed differences. But here is the rub, *for in complex, interactive systems such as organisms, small differences can have significant nonlinear effects so that two organisms exhibiting a high degree of quantitative similarity can nevertheless show very different effects when identically stimulated.* As always in discussions of evolution, the devil is in the details of the differences.

Organisms are very similar with respect to basic metabolic pathways, and this has led to the hope that they could be used to predict the threat posed to humans of chemical substances we are exposed to. But things are not quite so simple:

> It is a matter of common experience that the actions of the major classes of drugs, which in the main work by interfering with the normal function of physiological systems, are the same throughout mammals and most other living organisms. . . Despite this commonality of fundamental mechanisms of drug action, it is now appreciated that there are numerous situations where the effects of a drug or a chemical on the body depend on the animal species in question. (Caldwell 1992)

Species differences are ubiquitous and biomedically important. This is the down side to prediction of metabolic outcomes. How are we to think of such species differences?

The problem is generated in part by polymorphisms in human populations, giving rise to variation in chemical metabolizing activity, and also by polymorphisms in nonhuman animal populations (and the fact that the polymorphisms in the human and nonhuman animal populations may not be the same). These polymorphisms can affect catalytic activity and substrate specificities of enzymes. The polymorphisms are thus the root cause of intraspecific variation in human and nonhuman animal populations, respectively. These enzymatic polymorphisms give rise to a number of effects.

When considering species differences in enzymatic metabolism, toxicologists differentiate between *quantitative* differences and *qualitative* differences. Qualitative differences are generated by metabolic reactions that are unique to a given species. Perhaps members of one species are capable of achieving metabolic reactions not found in members of other species, or perhaps members of one species cannot achieve a particular reaction common to members of other species. For example, rats have very active N-acetyltransferases, whereas dogs completely lack this enzyme family (Collins 2001). For another example, consider a model substrate such as phenol (carbolic acid) and the transferases involved in phase II metabolism. Phenol can be excreted either through conjugation with glucuronic acid or conjugationwith sulfate. Cats are incapable of the former reaction, and pigs of the latter. Notice here that members of both species achieve the same metabolic function, but they do so using different causal pathways. Functional similarities do not imply underlying causal similarities.

Quantitative differences occur when members of different species use the same reactions to metabolize a given product, but differ with respect to the relative extents of these reactions. If we look at humans and rats, both species can metabolize phenol by conjugation with both glucuronic acid and sulfate. But if we look at the ratio of conjugation with sulfate to conjugation with glucuronic acid in terms of percent excreted in 24 hours, we see that it is 80:12 percent in humans whereas it is 45:40 percent in rats (Caldwell 1980). These sorts of differences are quite common in mammals. In a given organism, this sort of overlap of substrate specificity by different enzymes shows how the redundancy seen in some important biochemical systems (whereby similar functions—phenol metabolism in the present case—are achieved by distinct metabolic strategies within a given organism) can be important. Thus, blocking conjugation with sulfate will not prevent a rat from metabolizing phenol, though it may affect the rate at which this function is achieved.

A nice example of qualitative and quantitative species differences is afforded by a consideration of aflatoxin B_1. This is a mycotoxin produced by *Aspergillus flavus*. Common commodities that may be contaminated include corn, peanuts, pistachios, almonds, and walnuts. Human data suggests that aflatoxin is a potent hepatocarcinogen in humans [(Miller 1993) p846-847]. In humans, aflatoxin B_1 is transformed as a result of phase I metabolism employing *P450* enzymes, into an electrophilic metabolite, aflatoxin B$_1$-epoxide. This is a frameshift mutagen (one whose effects alter the meaning of the genetic code) that is associated in humans with development of liver cancer.

Mice, however, are resistant to aflatoxin-induced liver cancer, because even though they are effective producers of the reactive epoxide as a phase I metabolite, they express in phase II metabolism an α-class GST that has high activity towards this reactive metabolic intermediate. Rats, by contrast are sensitive to aflatoxin-induced liver cancer because they only express small amounts of an α-class GST with high activity toward the reactive intermediate. Human α-class GSTs, by contrast, have little or no activity with respect to the reactive intermediate. Unlike mice, we are indeed susceptible to aflatoxin-induced liver cancer. Interestingly, the nonhuman primate *Macaca fascicularis* expresses α-class GSTs that are up to 100-fold higher in activity than that observed in the human liver, but about 100-fold lower than that found in the mouse liver (Wang, Bammler, and Eaton 2002).

This example not only shows how species differences can be important at the biochemical level of the biological hierarchy, it also shows how differences here can have importance for differences elsewhere in the biological hierarchy, for cancer is a disease that affects systems at the organic and organismal levels of the biological hierarchy. Moreover, given the differences in metabolic activity between rats and mice and between macaques and humans, respectively, one might wonder whether relative phylogenetic closeness can really be considered to be a guide to closeness of metabolic activity, at least with respect to the activities of enzymes involved in detoxification.

It is possible to make some basic points of principle by considering some examples concerning differences with respect to the metabolism of xenobiotics.

(i) Intraspecific variation

We all too readily speak of mice and humans as if all mice and all humans were the same. This is an error from an evolutionary perspective, for there are metabolic differences between individuals constitutive of both human and animal populations that reflect heritable variation within those populations. These differences are often sufficiently well-marked to be significant from the standpoint of population genetics. There are both individual differences and strain differences. These will not be explicable, in general, by the usual *quantitative* suspects invoked in the context of scaling—differences in mass, etc. As Hau has observed:

> The validity of extrapolation may be further complicated by the question, "To which humans?" As desirable as it often is to obtain results from a genetically defined and uniform animal model, the humans to whom the results are extrapolated are genetically highly variable, with cultural, dietary, and environmental differences. This may be of minor importance for many disease models but can become significant for pharmacological and toxicological models. [(Hau 2003) p6]

For example, human sub-populations exhibit polymorphisms with respect to N-acetylating activity. Thus Sipes and Gandolfi point out:

> In humans, large individual differences exist in the acetylation of the antituberculosis drug isoniazid. The population is bimodally distributed into rapid and slow acetylators. . . The incidence of slow and rapid activators is not the same in all racial groups. Among Caucasians, a slightly higher percentage of slow acetylators predominate. . . . In contrast, in Orientals, rapid acetylators predominate. (Sipes and Gandolfi 1993)

There can also be metabolic variation reflecting gender differences. This is apparently not as marked in humans as it is in rodents and other mammals (though it is worth noting that of ten medications recalled between 1998 and 2001, eight were recalled because of side effects that occurred primarily in women (General Accounting Office 2001)). Gender differences may well be significant when mixed gender rodent populations are used to form hypotheses about human populations. For example, Manson and Wise give the following example:

> Chloroform is converted to a reactive intermediate (phosgene) ten times faster by microsomes obtained from the kidneys of male mice than those of female mice. Male mice are susceptible to chloroform-induced nephrotoxicity, whereas female mice are resistant. (Manson and Wise 1993)

For another example, in the IARC evaluation of caffeic acid discussed earlier, renal cell adenomas were observed in female, but not male mice, whereas the reverse was true in rats. In male mice, by contrast, there was an increase in the combined incidence of squamous cell papillomas and carcinoma of the forestomach—an effect that crossed gender lines in rats.

There can thus be serious gender differences with respect to the downstream effects of exposure to model substrates.

Recently, issues concerning race and ethnicity have entered studies of the metabolism of xenobiotics through the issue of statistical genetic differences between populations. As Holden has noted:

> No one disputes that some diseases strike disproportionately in some racial or ethnic groups—thalassemia in people whose ancestors came from the Mediterranean area, sickle cell anemia in people of African origins. . . Less clear-cut than these single gene disorders—but the subject of increasing research is the medical significance of a host of more subtle gene variants that appear in differing frequencies in various populations and that seem to influence a multitude of conditions. (Holden 2003)

Holden continues:

> The chief argument against the notion that biological race can be medically meaningful is that there are far more genetic differences among individuals that there are between different ancestral groups . . . however. . . if 30% of one population can't metabolize a certain drug, compared with 10% of another population, the between group variability is low because most people in both groups lack the metabolism polymorphism. Nonetheless, this variation is important when it comes to estimating the probability of response to treatment. . . (Holden 2003)

These effects have tended to be masked in clinical trials because over 80% of the participants in these trials are white. For example, nitric oxide is a chemical important for keeping blood vessels fit and toned. It is also important in the mechanisms of action of ACE inhibitors used to control hypertension. Studies have revealed that the variant of the gene making nitric oxide synthase (the enzyme important for vascular nitric oxide production) that ACE inhibitors work best with is found in 60% of whites but only in 30% of blacks (Holden 2003).

In humans, CYP polymorphisms can manifest themselves in the form of intraspecific differences in drug metabolism. Two genes, CYP2D6 and CYP2C19, are particularly important since they affect how people metabolize approximately 25% of the drugs on the market (Marshall 2003). Sipes and Gandolfi observe that with respect to the antihypertensive agent debrisoquine, some 3 to 10 percent of Caucasians are poor metabolizers because they are homozygous for 2 nonfunctional alleles for CYP2D6 (the source of debrisoquine 4-hydroxylase). Moreover, there appear to be more than 75 allelic variants of CYP2D6 circulating in human populations (Weinshilboum 2003). There are ethnic differences in the distributions of these alleles: individuals homozygous for the *10 allele, for example, have low CYP2D6 activity and make up nearly 20% of the Japanese population—a figure that differs from both Caucasian and Chinese populations (Tateishi et al. 1999). Studies in molecular genetics indicate that causality is complex among those exhibiting little or no CYP2D6 activity. Causal factors range from single nucleotide polymorphisms in the protein coding sequences, to deletions of the gene itself, to polymorphisms affecting the splicing of CYP2D6 (Weinshilboum 2003). On the other side of the coin, there are rapid metabolizers who possess duplicates of the CYP2D6 gene, and who require more than the standard doses of drugs to achieve therapeutic responses. Weinshilboum comments:

> Although the occurrence of multiple copies of the CYP2D6 gene is relatively infrequent among northern Europeans, in East African populations, the allele frequency can be as high as 29%. The effect of the number of copies of the CYP2D6 gene—ranging from 0 to 13—on the pharmacokinetics of the antidepressant drug nortriptyline. . . could hardly be a more striking illustration of how genetics influences the metabolism of a drug. (Weinshilboum 2003)

It hardly needs to be pointed out that these important differences could never have been revealed by nonhuman animal studies.

In the case of the antiepileptic drug mephenytoin, more than 20% of the Japanese population are poor metabolizers (compared to about 3% of the Caucasian population) [(Sipes and Gandolfi 1993)p95-6]. Enzymes in the CYP2C subfamily have been shown to be responsible for mephenytoin metabolism, with CYP2C19 being the main (S)-mephenytoin 4'-hydroxlase (Guengerich 1997). The poor metabolizers appear to make a stable, but defective protein [(Sipes and Gandolfi 1993) p96]. The presence of CYPC19*2 and *3 alleles account for 99% of oriental poor metabolizers and 87% of Caucasian poor metabolizers. These examples represent only a small sample of what is known about polymorphisms with respect to the specific enzymes and substrates mentioned. But they highlight the importance of paying attention to intraspecific variation when considering metabolic activity.

Sometimes these polymorphisms are masked by the effects of metabolism in humans. CYP3A4 and CYP3A5 are involved in oxidative metabolism of more than 50% of drugs on the market, and these two enzymes have considerable overlap in substrate specificity (including the protease inhibitor indinavir, the immunosuppressant cyclosporin, and oral contraceptives) (Moore, Goodwin et al. 2000; Ernst et al. 1998; Ernst 1999; Fugh-Berman 2000; Piscitelli et al. 2000; Ruschitzka et al. 2000). A nice example is afforded by Evans et al. who observe:

> About three quarters of whites and half of blacks have a genetic inability to express functional CYP3A5. The lack of functional CYP3A5 may not be readily evident, because many medications metabolized by CYP3A5 are also metabolized by the universally expressed CYP3A4. For medications that are equally metabolized by both enzymes, the net rate of metabolism is the sum of that due to CYP3A4 and that due to CYP3A5; the existence of this dual pathway partially obscures the clinical effects of genetic polymorphism of CYP3A5 but contributes to the large range of total CYP3A activity in humans. The CYP3A pathway of drug elimination in humans is further confounded by the presence of single nucleotide polymorphisms in the CYP3A4 gene that alter the activity of this enzyme for some substrates but not for others. (Evans and McLeod 2003)

The polymorphisms mentioned so far concern proteins directly involved in metabolism. But organisms are integrated, interactive complex systems, and polymorphisms not directly connected to the mechanisms of metabolism can have significant biomedical effects. Thus Evans and McLeod observe:

> . . . inherited differences in coagulation factors can predispose women taking oral contraceptives to deep-vein or cerebral-vein thrombosis, whereas polymorphisms in the gene for the cholesterol ester transfer protein have been linked to the progression of atherosclerosis with pravastatin therapy. Genetic variation in cellular ion transporters can also have an indirect role in predisposing patients to toxic effects of drugs. For example, patients with variant alleles for sodium or potassium transporters may have substantial morbidity resulting from drug-induced long-QT syndrome. (Evans and McLeod 2003)

What are the practical implications of the discovery of all these different types of human polymorphism? Collins has recently pointed out:

> In the field of metabolism, as well as some segments of toxicity and efficacy, there has been a major shift from animal-derived data to

human-based data. Except for comparative studies to assess interspecies differences, animal studies have declined in importance. Part
of this shift is driven by an appreciation for the uncertainty in cross-
species metabolic pathways. From the practical side, the well-
organized, readily available supply of human tissue has fueled this
shift (Collins 2001).

Observations of intraspecific metabolic variation of the kind just discussed have prompted some investigators to develop the science of
pharmacogenetics. The history of this branch of science goes back to the
middle of the 20th century. Thus Willyard has observed:

> It was in the 1950s that scientists found the first evidence that
> people's genetic makeup can alter their response to drugs. Alf Alv-
> ing at the University of Chicago had observed during the Korean
> War that black soldiers were more likely than white soldiers to de-
> velop debilitating anemia when given antimalarials such as prima-
> quine. In 1956, he found the cause: an enzyme deficiency—the re-
> sult of a rare allele on the X chromosome—that leaves red blood
> cells vulnerable to oxidative damage and causes them to burst.
> Based in part on this work, German geneticist Friedrich Vogel
> coined the term 'pharmacogenetics' in 1959 to describe the role of
> genetics in drug response. (Willyard 2007)

Pharmacogenetic investigations aim to identify complex patterns of
gene variation in an attempt to correlate these patterns to different drug
response genotypes.

Differences between groups of people exist not merely with respect to
the metabolism of xenobiotics, but also with respect to patterns of disease. For example, black women with certain breast cancers have a higher
incidence of developing metastatic cancer than their white counterparts
even when the same treatments are given. Further, black women under 50
years account for 31% of breast cancer while white women under 50 account for only 21%. The incidence of breast cancer that is estrogen recep-
tor-negative also varies with black women having an incidence of around
40% and white women 23%. Triple-negative- tumors—negative for estrogen receptors, progesterone receptors, and human epidermal growth factor receptor-2 tend to spread quickly and are far more common in young
black women (Couzin 2007). Incidence of diseases vary between the sexes. Males experience more schizophrenia while women experience more
depression, anxiety and eating disorders (Holden 2005). Myocardial infarc-

tion (MI) differs between sexes with MI occurring much later in women than men. Women have greater risk from MI if they suffer from diabetes and high blood pressure, but heart failure and sudden death are less frequent women. Women experience less typical angina but MI are more likely to be lethal. Left ventricular hypertrophy occurs later in women but is associated with greater risk.

Ober et al. have examined the genetic underpinnings of epidemiological differences with respect to gender, reporting that:

> In humans, sexual dimorphism is also observed in the prevalence, course and severity of many common diseases, including cardiovascular diseases, autoimmune diseases and asthma. Although sex differences in the endocrine and immune systems probably contribute to these observations, recent studies suggest that sex-specific genetic architecture also influences human phenotypes, including reproductive, physiological and disease traits. It is likely that an underlying mechanism is differential gene regulation in males and females, particularly in sex steroid-responsive genes. Genetic studies that ignore sex-specific effects in their design and interpretation could fail to identify a significant proportion of the genes that contribute to risk for complex diseases. (Ober, Loisel, and Gilad 2008)

A tendency to downplay evolved intraspecific gender differences in favor of pre-evolutionary typological thinking ("Male as representative human type") has led Simon to caution that:

> Most biomedical and clinical research has been based on the assumption that the male can serve as representative of the species. This has been true in spite of increasing awareness of significant biological and physiological differences between the sexes, beyond the reproductive ones. Women and men differ in their susceptibility to and risk for many medical conditions, and they respond differently to drugs and other interventions. The close of the previous decade saw 8 out of 10 prescription drugs withdrawn from the U.S. market because they caused statistically greater health risks for women than men. Thus, what is true and good for the gander does not seem to be necessarily good for the goose . . . NIH initiated a study of aspirin in the 1990s involving 39,876 women. In March 2005, this $30 million study reported that women were different from men. Women, unlike men, did not have a significantly lower risk of heart attack when taking aspirin, but they did have a somewhat lower risk of stroke.

> Gender differences in CVDs [cardiovascular diseases] definitely exist and are very likely to have an increasing role in therapeutic decisions in the near future. This should be supported by gender-specific research on existing drugs and gender-specific strategies in the development of novel agents. Research on oestrogen and testosterone and their receptors has begun, but gender in all its complexity is only starting to be recognized as a scientific category in medicine. We need a comprehensive approach to understand the entity of gender-related differences, including their genetic basis and gene–environment interactions, as well as the pathophysiology of sex hormones through all developmental stages to optimize pharmacological therapy for women and men. (Simon 2005)

For the evolutionary biologist, the study of intraspecific variation—the way characteristics vary within populations of a given species—is of paramount importance. But as Burggren and Bemis point out, "While comparative physiologists have made an art of avoiding the study of variation, such heritable variation nonetheless is the source for evolutionary changes in physiology as well as for all other types of characters [(Burggren and Bemis 1990) p201]." We will shortly see just how important intraspecific variation is, and the reason is simple: the existence of such variation *undermines* the view that species differences (which, in evolutionary biology, are viewed as the result of amplification, in historical time, of variation initially found in populations of a given species) can be understood using quantitative, mechanistic scaling formulae.

This issue comes to the fore in dramatic form in connection with the Human Genome Project itself, a project that has deliberately ignored heritable variation and which, in its present incarnation, reflects non-evolutionary typological thinking -- thinking that has unfortunately led scientists and others to think in terms of a genetic *gold standard*. As Alberts et al. themselves observe, while the Genome Project only sampled the DNA from a very small number of humans:

> The human genome—the genome of the human species—is, properly speaking, a more complex thing, embracing the entire pool of variant genes that are found in the human population and continually rearranged and reassorted in the course of sexual reproduction. Ultimately we can hope to document this variation too. Knowledge of it will help us understand, for example, why some people are prone to one disease, others to another; *why some respond well to a drug, others badly*. (Emphasis added.) [(Alberts et al. 2002) p44]

Indeed, J. Claiborne Stephens has stated that, if the findings of the Human Genome Project are accurate then "the functional complement of the human genome is going to be a repertoire of something like 400,000 to 500,000 gene [allele] versions (Durham 2001)." This brings us back to the basic ideas of evolutionary biology, one of the most important of which concerns the biological significance of heritable variation. If intraspecific variation and individual uniqueness is this important from a biomedical standpoint, why isn't the more significant interspecific variation—a longer term product of the same evolutionary mechanisms that generated the intraspecific variation in the human population—of more relevance to biomedical researchers using non human animals as *CAMs* for human biomedical phenomena? This issue is all the more acute since drug metabolizing activity is primarily biochemical activity at the lowest level of the biological hierarchy where the similarities between species (let alone within them) are deemed to be the greatest. To explore this issue, it will be useful to discuss interspecific differences with respect to the metabolism of xenobiotics.

We have just seen that intra-species extrapolation is problematic among humans. The metabolism of many drugs varies greatly among humans so genetic testing is performed to allow personalized treatment. Such drugs include: 6-mercaptopurine (Yates et al. 1997; Schaeffeler et al. 2004; Relling et al. 1999), warfarin, codeine (Caraco, Sheller, and Wood 1996), tamoxifen, imatinib, trastuzumab, antiretrovirals, drugs used to treat breast cancer, and the colon cancer treatment irinotecan.

(ii) Interspecific variation

There is also variation across species lines. The systematic study of this kind of metabolic variation brings content to the claim that humans and their animal models are differently organized complex systems, and not merely the same animal dressed up differently. Some examples of investigations into the causes of interspecific metabolic differences are instructive.

KW-4490 is a selective phosphodiesterase-4 inhibitor. It was studied in rats and monkeys and found to be metabolized to acylglucuronides and excreted in the urine in monkeys while in rats it was excreted in the bile. Fujita et al. found that "renal tubular secretion was significant in monkeys, whereas reabsorption was significant in rats. These species differences in urinary excretion of KW-4490 and its acylglucuronide metabolites are most likely due to substrate specificity of active transporters in rat and monkey kidney (Fujita et al. 2008)." 3,4-methylenedi-oxymethamphetamine or MDMA, also known as Ecstasy, is metabolized to 4-hydroxy-3-methoxymethamphetamine sulfate (HMMA-Sul) and 4-hydroxy-3-methoxymethamphetamine glucuronide (HMMA-Glu). Shima et al. found that sulfation "is quantitatively more significant than the glucuronidation for humans. In rats, in contrast, almost all the conjugated HMMA (>99%) was excreted as the glucuronide . . . It is concluded that a considerable interspecies variation exists in the conjugation of HMMA between humans and rats (Shima et al. 2008)."

ZD6126, a vascular agent, has been studied in the rats and dogs. Partridge et al.:

> In rat, the major metabolites were ZD6126 phenol its glucuronide and other metabolites consistent with O-demethylation and conjugation. ZD6126 was more extensively metabolized by male than female rats, and also in young compared with mature rats. In dog, metabolism occurred primarily via direct glucuronidation of the active species, ZD6126 phenol. (Partridge et al. 2008).

Examples such as these can be multiplied almost endlessly, the limiting factor being the willingness of the reader to search through the literature!

Cheung and Gonzalez summarize much of what is known about the various causes of species differences in the metabolism of xenobiotics as follows:

> The underlying reasons for interspecies differences in drug metabolism were originally proposed to be a species' inability to carry out a metabolic reaction (such as N-hydroxylation of aliphatic amines in rats), restriction of the occurrence of a reaction to a particular species, or competing reactions by which a compound may be metabolized. A classic example of species-specific difference in metabolism by competing reactions is the biotransformation of 2-acetylamino-fluorene, a potent carcinogen in a number of species. The carcinogenicity of 2-acetyl-amino-fluorene depends on the ratio of N-hydroxylation (bioactivation pathway) and aromatic hydroxylation (detoxification pathway). The degree of N-hydroxylation in rat, rabbit, and dog is considerable; therefore, this chemical is considered a carcinogen in those species. In guinea pigs, no N-hydroxylated metabolites are formed; thus, 2-acetyl-amino-fluorene is not carcinogenic in this species.
>
> It is now recognized that rodents metabolize xenobiotics differently from humans, due in part to species differences in the expression and catalytic activities of P450s. Functional orthologs of almost all human genes exist within the mouse genome . . . It can be

assumed that differences in P450 isoforms [variant forms of a given protein] between species are a major cause of species differences in drug metabolism. (Cheung and Gonzalez 2008)

The role played by small differences between molecules—a matter expected to be of importance from our earlier discussion of dynamical systems theory—is also illustrated here through consideration of the CYP2A subfamily. Despite considerable sequence similarities (>90%), these proteins show marked differences in catalytic specificity. According to Guengerich (Guengerich 1997), it was in this group of enzymes that substitution at a single amino acid residue was shown to change (not merely modulate) the specificity of a CYP enzyme.

The results of comparative studies of CYP enzymes suggest that for some subfamilies, cross-species extrapolation seems to be justified (2E1), but for other families there are serious problems (2D, 3A, 2A, 2B, 2C). Commenting on these observations Guengerich asks:

> Practically speaking, how can one deal with the possible differences in P450 activities across species when addressing questions about metabolism of specific chemicals? To an extent, this must still be done empirically. . . Nevertheless, with new developments in the availability of human tissues, recombinant P450 products, and new approaches to analysis, it has been possible to move from experimental animals to man in a relatively rapid manner. Even the best of in vitro knowledge about P450s, however, must be interpreted with reservations until tested in vivo. (Guengerich 1997)

Conducting metabolic studies "empirically"—i.e., employing "trial-and-error" animal studies of dubious relevance to the human condition—can indeed be usefully replaced by in vitro methods using human products. However, unless subsequent testing in vivo involves testing in a wide variety of human subjects (reflecting intraspecific variation), it is not clear how the problems with animal studies, antecedently known to be unreliable, can be evaded. This is especially true since the recent history of this branch of science suggests that the deeper the exploration of the causes of metabolic differences between species is conducted, the greater are the numbers of metabolic causal factors that must be taken into account. As relevant causal factors are multiplied, so too is the space for species differences with respect to those same factors.

In view of this latter observation, the CYP 2B and 3A families deserve closer scrutiny, since some of the differences between species that have been observed here reflect not just differences with respect to CYP enzymes, but also with respect to regulatory signaling systems. In this regard we will focus on nuclear receptors, which are ligand-activated proteins regulating gene expression.

Among the molecules involved in cellular signaling are steroid hormones, thyroid hormones, retinoids and vitamin D. Though widely different in chemical structure, these molecules all work by binding to nuclear receptor proteins at ligand binding domains (LBDs). The receptors then bind to DNA to regulate the transcription of specific genes. These receptors all belong to a group of related proteins called the nuclear receptor superfamily [(Alberts et al. 2002) p839].

For example, we know that toxicities mediated by TCDD (2,3,7,8-tetrachlorodibenzeno-*p*-dioxin) differ between the *h* (human) AhR (aryl hydrocarbon receptor) and the *m* (mouse) AhR; probably secondary to differences in the regulation of gene expression (Moriguchi et al. 2003). Species differences with respect to AhR, bringing out the role played by differential patterns of gene activation, are discussed by Flaveny et al. as follows:

> The aryl hydrocarbon receptor (AhR) is a ligand inducible transcription factor that exhibits interspecies differences, with the human and mouse AhR C-terminal transactivation domain sharing only 58% amino acid sequence identity. The AhR has a transactivation domain comprised of proline/serine/threonine-rich, glutamine-rich, and acidic amino acid subdomains. A truncated *m*AhR and *h*AhR containing only the acidic subdomain displayed widely differing transactivation potentials. Whether the glutamine-rich subdomain of the mouse AhR and the human AhR differentially recruit LXXLL-motif coactivators was investigated. Transiently expressed GAL4 DNA binding domain (GAL4DBD)-LXXLL-motif fusion proteins were used to map the critical LXXLL binding sequence of the *h*AhR to amino acid residues 663-688. Several LXXLL-motif GAL4DBD fusion proteins dramatically differed in their ability to influence the transactivation potential of the *m*AhR and *h*AhR. These findings suggest that the human and mouse AhR may display differential recruitment of coactivators and hence may exhibit divergent regulation of target genes. (Flaveny et al. 2008)

It has been shown that xenobiotics induce transcription of certain families of CYPs by activating nuclear receptors. CYP 3As are regulated by the pregnane X receptor (PXR) and CYP 2Bs are regulated by the constitutive adrostane receptor (CAR). PXR and CAR belong to the same nuclear receptor subfamily (NR1), sharing some 40% amino acid identity with respect to ligand binding domains (Moore, Parks et al. 2000). Studies have been performed on human (*h*) and mouse (*m*) orthologs of PXR and CAR with respect to model substrates such as phenobarbital (PB), 1,4,-bis[2-(3,5-dichloropyridyl-oxy)]benzene (TCPOBOP), and clotrimazole. Moore et al. commented upon the results of these studies as follows:

> Nuclear receptor orthologs generally share >90% amino acid identity in their LBDs. However, comparison of PXR from four different species shows that this receptor has diverged considerably in the course of evolution. The human, rabbit and rodent PXR are all roughly equally divergent and share only ~70% amino acid identity. This divergence in PXR is an important component of cross-species differences in the regulation of CYP3A expression by xenobiotics. Similarly, the human and rodent CAR share only ~70% amino acid identity in their LBDs, and demonstrate marked differences in their responses to xenobiotics. For example, clotrimazole is an efficacious deactivator of *h*CAR but has little or no activity on *m*CAR. Conversely, TCPOBOP is a potent activator of *m*CAR but lacks any activity on *h*CAR. The divergence in amino acid sequences across CAR orthologs undoubtedly contributes to cross-species differences in the physiologic effects of xenobiotics. (Moore, Parks et al. 2000)

Species difference are not just associated with protein evolution of the structures of CYP enzymes, they are associated with evolution in the molecules that regulate the expressions of the genes coding for those enzymes as well. This brings out the importance of the observation made earlier in the context of dynamical systems theory, that small differences can be very significant in the context of complex, interactive dynamical systems. It is suspected, in fact, that PXR and CAR are components of a regulatory network governing endocrine hormone homeostasis (Moore, Parks et al. 2000). Whether or not this is true, it has recently been shown that the regulatory role of PXR is indeed medically significant.

The CYP 3A family is particularly important in the context of xenobiotic metabolism because, as Jones et al. have noted:

> The CYP 3A gene products are among the most abundant of the monooxygenases in mammalian liver and intestine. In humans, CYP 3A4 is involved in the metabolism of more than 50% of all drugs as well as a variety of other xenobiotics and endogenous substances, including steroids. (Jones et al. 2000)

One drug that is of interest in this regard is troglitazone, marketed as Rezulin, and used in the treatment of type-II diabetes. Troglitazone was removed from the market in the US in March 2000. Despite the fact that it had been shown to be safe and effective in rodent studies (Jones et al. 2000), more than 65 people died (two thirds were women), and many others required liver transplants as a result of Rezulin toxicity. In clinical trials involving a total of 2500 human subjects, about 2% showed alanine aminotransferase (ALT) levels more than 3 times the upper limit of normal. ALT levels this high are an indicator of active liver disease (see (Greek and Greek 2002) for more details of how Rezulin came to market on the FDA "fast track)." Lauer et al. recently examined gene expression in rat and human hepatocytes after exposure to troglitazone and diclofenac:

> Furthermore, we were able to demonstrate marked species differences in gene expression patterns of troglitazone treated rat and human hepatocytes. In contrast to rat hepatocytes, human cells showed distinct upregulation of various CYPs, regulators of xenobiotic metabolism and marker genes for oxidative stress. In contrast, gene expression alterations in rat and human hepatocytes treated with Diclofenac were rather similar. (Lauer et al. 2008)

The class of drugs to which troglitazone belongs was developed using rodent models of insulin resistance, but without prior knowledge of the cellular target (Jones et al. 2000). It is now known that troglitazone achieves its therapeutic effects by binding to the PPARγ (peroxisome proliferator-activated) nuclear receptor. But at the concentrations required to activate PPARγ, it also activates PXR in humans—something it did not do in rats and mice (Jones et al. 2000). One immediate consequence of this species difference is that in humans, patients taking troglitazone experience increased CYP3A4 activity. Jones et al. comment:

> Our data showing that troglitazone activates human PXR at concentrations similar to those required to activate PPARγ provide an explanation for its interactions with other drugs, including oral contraceptives. Interestingly, the relative lack of activity of troglitazone on the mouse or rat PXR may explain why these effects were not reported in animal toxicology studies. Additional studies will be required to determine whether PXR also plays a role in the hepatotoxicity observed with troglitazone. In this regard it is interesting that that the PXR ligand rifampicin has also been associated with hepatotoxicity in humans. (Jones et al. 2000)

(Recently, it has been argued that the increased CYP3A4 activity associated with troglitazone activation of human PXR results in the metabolism of troglitazone to a reactive quinone which has been hypothesized as the cause of hepatotoxicity (Willson and Kliewer 2002).)

PPARs exhibit other differences between species. Rigamonti et al. 2008:

> In the last few years, PPARs and LXRs [liver X receptors] have emerged as key regulators of macrophage biology. Clear evidence has been provided that these nuclear receptors control transcriptional programs involved in macrophage lipid homeostasis. In addition, PPARs and LXRs negatively regulate macrophage-mediated inflammation. However, several important issues need further exploration. For example, important differences exist between human and mouse in term of gene regulation by PPARs and LXRs, especially in the field of macrophage lipid homeostasis. These observations, which call to caution in the extrapolation of the results obtained in a given species to the other, can be attributable to different reasons.
>
> First, the expression levels of PPARs and LXRs can be different between human and mouse macrophages. As an example, PPAR-α is abundant in human macrophages and is only barely detectable in mouse macrophages. Second, the expression of nuclear receptor cofactors (coactivators or corepressors) which participate in the regulation of target gene transcription, can be also different between human and mouse macrophages. Third, because the promoter regions of genes are not entirely conserved across species, the transcription factors that control gene expressionin one species might not be crucial regulators in another. Finally, synthetic ligands used in the pharmacological studies can display different affinity and selectivity for the human and mouse PPARs or LXRs proteins. (Rigamonti, Chinetti-Gbaguidi, and Staels 2008)

While examples like this could be multiplied, the point is made if it is understood that at the molecular level of the biological hierarchy there are significant differences between species. Differences that arise from evolved differences in catalytic activity of enzymes, or from evolved differences in the regulation of gene expression.

Observations such as those just discussed above do not necessarily faze the intrepid animal modeler. Thus as Collins has suggested:

> Generally, we want to have animal systems that mimic human experience. For some purposes, it might be useful to do the reverse: make humans look like animal systems. While rats have highly active arylamine N-acetyltransferases, dogs totally lack this enzyme family, and humans have intermediate amounts. Because the acetylated metabolites are sometimes more toxic than the parent molecule, it could be desirable to inhibit NAT activity in humans. Unlike the many enzymes in the CYP family, there are only two arylamine N-acetyltransferases, NAT1 and NAT2. Using cytosolic subfractions and cryopreserved human hepatocytes, we have demonstrated that either of these NAT enzymes can be inhibited by drugs at concentrations that are achievable in patients. Clinical studies to confirm these findings in vitro are underway. Thus, in a reversal of the usual paradigm for drug development, humans can be modified to appear more like dogs. (Collins 2001)

While it might be even more convenient to modify humans to resemble inbred mouse strains of the kind typically employed in drug development, we think Collins is barking up the wrong tree. A better strategy—discussed later in the conclusion to this volume—is that employed in pharmacogenetics and pharmacogenomics, and which takes intraspecific variation in the form of evolved metabolic polymorphisms at face value, and tries instead to tailor drugs to suit the genetic individuality of the patient.

(iii) In vitro studies

Another way to see the relevance of evolved metabolic differences between species is to consider the fate of xenobiotics *in vivo*, in an intact animal, and *in vitro*, in a test tube containing ground up animal cells. In the situation *in vitro*, the grinding up of the cells destroys some of the evolved, organizational differences between species. As Caldwell has pointed out:

> . . . in order to be metabolized *in vivo*, compounds must pass several membranes to reach the metabolizing enzymes, and in only a few cases are only one set of enzymes involved. Although the oxidation of foreign compounds occurs principally (but not exclusively) in the microsomes, as does the hydration of epoxides and glucuronic acid conjugationmany reductases are present in the cytosol, as are the sulfate conjugating enzymes. . . Additionally in the whole animal many organs other than the liver can contribute to metabolism, sometimes catalyzing reactions that cannot occur in the liver, e.g., dog kidney can conjugate benzoic acid with glycine, while the liver

cannot and gut flora can perform several reactions not carried out by the tissues. [(Caldwell 1980) p109]

The context in which enzymatic metabolism takes place is an evolved, complex system: the organism.

In the situation *in vitro*, where reactions might take place in a test tube containing an aqueous solution of ground-up liver cells, or indeed, in the context of purified enzymes acting on model substrates, the organismal, and indeed cellular, context is destroyed, and though such studies are important for biochemists:

> . . . it must always be remembered that the inside of a cell is quite different from the inside of a test tube. The "interfering" components eliminated by purification may be critical to the biological function or regulation of the molecule purified. *In vitro* studies of pure enzymes are commonly done at low concentrations in thoroughly stirred aqueous solutions. In the cell, an enzyme is dissolved or suspended in a gel-like cytosol with thousands of other proteins, some of which bind to the enzyme and influence its activity. Within cells some enzymes are parts of multienzyme complexes in which the reactants are channeled from one enzyme to another without ever entering the bulk solvent. . . In short, a given molecule may function somewhat differently within the cell than it does *in vitro*. [(Lehninger, Nelson, and Cox 1993) p48]

These differences blur, in the context of *in vitro* studies, evolved organizational differences between species. For example, in the context of *in vitro* studies of microsomal metabolism in rhesus monkeys, squirrel monkeys, tree shrews, pigs and rats, only minor differences were found when model substrates such as phenol or amphetamine were examined. However, in *in vivo* studies of the same species with respect to the same substrates, "Since it is clear from data quoted previously that enormous differences exist in the fate of foreign compounds in these species *in vivo*, it is important to consider the difficulties in extrapolation from *in vitro* to *in vivo* studies [(Caldwell 1980) p109]."

But it is not just an extrapolation problem from *in vitro* to *in vivo*. For what this highlights is the importance of the evolved organismal context in which metabolism takes place. Organisms are not passive vessels in which reactions occur with differences between species to be explained away by scaling for differences in mass, surface area, etc. In a test animal, the environmental circumstances in which it finds itself, along with its long and complex evolutionary history—implying broad similarities with members of other species, but also the cumulative effects of differences between this animal and members of other species—constrain the metabolism of xenobiotics in ways that are biomedically significant.

(iv) Symbiosis and its metabolic consequences

In addition to the observation that species may differ with respect to biochemical adaptations, there is a second point to be made. Organisms have not merely evolved in the context of a physical environment, they have evolved also in a biotic environment and have thus evolved complex relationships with other species. Some of these relationships are metabolically important. Many organisms are colonized by other organisms. Sometimes these relationships are parasitic, and associated with disease states, sometimes they are benign—and relations of co-dependence can evolve.

Especially important for our purposes are relationships with intestinal flora, for these reflect the effects of organismal co-evolution, and will not be mere quantitative differences to be scaled away. As Sipes and Gandolfi note, these evolutionary phenomena are metabolically important:

> An aspect of *in vivo* extrahepatic biotransformation if xenobiotics frequently overlooked is modification by intestinal microbes. It has been estimated that the gut microbes have the potential for biotransformation of xenobiotics equivalent to or greater than the liver. With over 400 bacterial species known to exist in the intestinal tract, differences in gut flora content as a result of species variation, age, diet, and disease states would be expected to influence xenobiotic modification. [(Sipes and Gandolfi 1993) p109]

Metabolic differences between humans and nonhuman animal *CAMs* that arise as consequences of co-evolutionary relationships with microorganisms will not be effects that will reflect differences in body weight or surface area, and will thus be resistant to the usual quantitative compensations for species differences. Metabolic activity is not simply a reflection of the organism and its genes, it also reflects the presence and activities of other organisms that are present and in an intimate co-evolutionary association with the given organism. This can be a source not only of interspecific differences, but also of intraspecific metabolic differences.

So far we have examined species differences and some of the ways in which a sharper understanding of the causes and effects of evolutionary processes contribute to the existence of both intraspecific variation and interspecific variation with respect to the metabolism of xenobiotics. The differences we have documented here are differences with respect to metabolic events at the lowest levels of the biological hierarchy, where the similarities between species ought to be most evident. It is apparent from the above discussion that humans and rodents are not "the same animal dressed up differently" with respect to metabolic activities. But what of the claim that differences between species become less significant as one examines species that are close from a phylogenetic standpoint. This issue requires special treatment and is deferred to a later chapter. Before turning to examine this issue, it is important to examine some issues about the meaning of *prediction* in the context of animal modeling. What are the implications of the existence of intraspecific variation and interspecific differences for the use of members of one species to predict outcomes for members of another species when both are similarly stimulated?

PART IV
◆
THE PREDICTION ISSUE

CHAPTER 11

♦

PREDICTIVE MODELS IN TOXICOLOGY AND ONCOLOGY

"When I use a word," Humpty Dumpty said in rather a scornful tone, "it means just what I choose it to mean—neither more nor less." "The question is," said Alice, "whether you can make words mean so many different things."
—Lewis Carroll in *Through the Looking Glass*, 1871

Attention to the meaning of words is very important in all areas of study but especially science. Take for example the word *prediction*. A biomedical research method need not be predictive of human outcomes, but if one claims predictive ability for the test or project, then one ought to mean something very specific. After all, if the word *prediction* can be used any old how, then the word loses its bite and its meaning. To say that a scientific research method is predictive is to say something quite specific about the method (Shanks, Greek, and Greek 2009).

This chapter addresses the use of the word *predict* as applied to animal models. It is our position that there has been insufficient attention to the meaning of prediction in the context of biomedical inquiry. Predictions are statements about expected future observations, but in science they are not, as they are in ordinary usage, merely lucky guesses. Scientific predictions are derived from hypotheses. Crudely speaking, a hypothesis is something that explains past and present observations and enables the investigator to form (under suitable conditions) expectations—predictions—about the course of future events. These predictions about the course of future events must be testable—they must be sensitive to the fruits of evidential inquiry (which may, in the biological sciences, involve carefully-controlled experiments, field observations, or some combination of both).

In the case of predictive animal modeling, what we are typically interested in is prediction of human outcomes. The animal model systems are stimulated (perhaps by some toxicological insult of interest) and animal data is gathered. This data derived from the animal model, in and of itself, settles nothing about the actual course of human phenomena. The animal data enables the investigator to form hypotheses—expectations—about what he or she thinks is likely to happen when humans are similarly stimulated (with all the due allowances for differences in dose and so on). At this point all the investigator has is a hypothesis about human responses. The business of science is the very business of the testing of hypotheses. In the present case this requires careful studies of humans so that the human data can be compared with the expectations rooted in animal model data, thereby confirming or falsifying the animal-based hypotheses (it is also possible that the evidence gathered does not settle the issue one way or another, and hence that there is a need for more detailed studies). We again point out—to forestall a species of fatuous criticism—*that not all tests and studies involving animals are done with prediction in mind. Nevertheless, those tests promoted as being predictive must be judged by how well they actually predict human response.* It also makes sense to ask whether a particular method has a track record of predictive success, and if so, how this was determined.

There may indeed be extra-scientific moral, political, and social objections to testing hypotheses rooted in animal models in humans. Fair enough, so long as investigators realize that in this case, a failure to experimentally validate or invalidate animal-derived hypotheses simply leaves those claims about human biomedicine in the realm of untested hypotheses. Investigators sometimes respond to this observation by noting that they are not just concerned with good science (involving the rigorous testing of hypotheses), but also with human health and well-being. Yet it is hard to see how the latter can be advanced without careful attention to the former. Perhaps this would not be so crucial if we had reason to believe that humans and their nonhuman models were indeed the same animal dressed up differently—perhaps this might alleviate some of the evidential burden involved in the gathering of human data. But as we have shown at various points in this book, and most notably in the last chapter, there is little reason to believe this claim about humans and their animal models. We leave for others the difficult questions of public policy raised by these observations. We rest content by observing that from the strict standpoint of good science, the realm of untested hypotheses has as its denizens not merely hypotheses that are testable (but whose testing is deemed morally, politically or socially unpalatable), but also hypotheses about the occult, the paranormal and the merely pseudoscientific! This is not a good neighborhood to be in.

The philosopher W.V.O. Quine has remarked:

> A prediction may turn out true or false, but either way it is diction: it has to be spoken or, to stretch a point, written. Etymology and the dictionary agree on this point. The nearest synonyms "foresight," "foreknowledge," and "precognition" are free of that limitation, but subject to others. Foreknowledge has to be true, indeed infallible. Foresight is limited to the visual when taken etymologically, and is vague otherwise. "Precognition" connotes clairvoyance. . . . *Prediction is rooted in a general tendency among higher vertebrates to expect that similar experiences will have sequels similar to each other* . . . [(Quine 2005)159] (Emphasis added.)

Predictions, generated from hypotheses, are not always correct. But if a modality or test or method is said to be predictively successful then *at the very least* it should get the right answer more often than not. Quine continues his essay on prediction by observing:

> Science, for all its refinement, does not lose the common touch. The observation categorical is still the touchstone. It says that if the experimental condition is set up, observable by the scientists concerned, then the predicted observation will ensue. *If the prediction fails, then the theory, which implied the observation categorical, is refuted.* (Quine 2005) (Emphasis added.)

If a modality consistently fails to make accurate predictions then the modality cannot be said to be predictive simply because it occasionally forecasts a correct answer. We will expand on this shortly. For the present Quine continues:

> The empirical meaning or content of the theory, we might say, is the set of all implied observation categoricals . . . *But prediction is always the bottom line. It is what gives science its empirical content, its link with nature. It is what makes the difference between science, however high-flown and imaginative, on the one hand, and sheer fancy on the other.*
>
> This is not to say that prediction is the *purpose* of science. It was once, we might say, when science was young and little; for success in prediction was, we saw, the survival value of our innate standards of subjective similarity. But prediction is only one purpose among others now. A more conspicuous purpose is technology, and an overwhelming one is satisfaction of pure intellectual curiosity—

which may once have had its survival value too. [(Quine 2005) 161-2] (Emphasis added.)

Quine may be right in saying that prediction is no longer the purpose of science, and we certainly do not challenge the importance of basic biomedical research aimed at satisfying pure intellectual curiosity, or, indeed, the business of developing new biomedical technologies.

However, in biomedicine, there is a pressing interest in the prediction issue: human health and well-being. If animal models are predictive of human outcomes, there ought to be solid evidence to support this claim. If they are not, then the public and their policy makers need to rethink the ways in which increasingly scarce biomedical research funds are directed since many grant applications and Food and Drug Administration (FDA) and Environmental Protection Agency (EPA) regulations imply or explicitly state that animal modeling is predictive of human outcomes. Society is entering a period in which severe cutbacks in science funding appear imminent therefore society has a very real—in fact a life-preserving—interest in assuring that only viable scientific research and practices are funded and/or mandated from a regulatory standpoint.

In the physical sciences prediction is easy to illustrate. According to Newton's second law, force equals mass multiplied by acceleration. The law tells us, in effect, that forces can be evaluated by observing the way that masses accelerate. Consider the case of two masses where the same force is identically applied. If the masses accelerate in the same way (all other things being equal), we may predict that the masses are identical. If, under these circumstances, the masses accelerate differently, we may predict that the masses are different. An implication of this latter observation is that having observed different accelerations, we may adjust the forces applied, so that the masses accelerate in an identical manner. In short we may scale for the difference to bring about similar effects in both masses. The differences are quantitative, not qualitative.

As we saw in chapter 2, based on the laws of physics we can also make predictions about things never seen before. Urbain Leverrier predicted another planet in the solar system based on deviations in the orbit of Uranus. In 1846, the planet Neptune was discovered. An observation (deviations in the orbit of Uranus) led to a hypothesis (another planet exists), which was then tested (by people with telescopes) and confirmed. The field of astronomy used the laws of nature (as they were then understood) to predict the location of a hitherto unobserved planet. The other side of this story is that scientists had to adjust their understanding of the laws of nature to accommodate observations of anomalous motions of the planet Mercury (the annual precession of its perihelion). The predicted existence of yet another planet—Vulcan—closer to the Sun than Mercury, was falsified. The result of the needed adjustment of the laws of nature was the general theory of relativity.

In biology and biomedicine, there are numerous examples of the scientific method functioning in this fashion. For example, in the mid 1800s Ignaz Semmelweis observed higher mortality rates in the maternity ward for patients being taken care of by medical students rather than midwives. He hypothesized that the medical student's had *dirty* hands from dissecting cadavers in the morgue and that this contamination was responsible for spreading disease in the maternity ward. He predicted the mortality rate from puerperal fever (what we would now recognize as bacterial infection) would decrease if physicians and medical students washed their hands before returning to the wards from the morgue. It did.

Whether or not an animal model of a given type can be used to predict human responses *can* be tested, and if the test results have a high enough sensitivity, specificity, and positive and negative predictive values, then the hypothesis that animals can predict human responses would be verified. If verified, then one could say that animal models (of the type tested) are predictive for humans and if refuted then one could say such models are not predictive for humans.

Now is an appropriate time to discuss where the burden of proof lies in science. The burden of proof—the evidential burden—lies with the person making a given claim (be it about human metabolism or the Loch Ness monster). Thus, those claiming animal models are predictive of human responses in the context of biomedical research must show that what they are claiming is true. The burden is not on us to prove that animal models of, say carcinogenesis or toxicity, are *not* predictive (though the previous chapters contain enough evidence and theoretical analysis to at least raise this very issue). After all, it is the modelers, not us, who are consumers of increasingly scarce research dollars. It is the job of those claiming that animal models are predictive of human responses to demonstrate that this is in fact the case. In the spirit of the motto of the Royal Society for the Advancement of Science—*Nullius in Verba*—we should not have to blindly take someone's word for it. Too much is at stake! This matter will require a consideration of what the evidence actually shows.

Investigators who make the untested claim that animal models predict human outcomes who assume (in the absence of human data) that their models are relevant to the human condition, often make the claim that drugs and chemicals that *would have harmed humans* have been kept off the market or out of the environment on the basis of results from animal tests. This is disingenuous unless we have *a priori* reason to assume animal models are predictive. The hypothesis was, in these cases, never tested. As we have already observed, it would in many cases be unethical to conduct on humans the sorts of carefully controlled laboratory studies that are regularly conducted on, say, rodents. However, there are other, ethical ways to gain human data in the context of epidemiology (for example retrospective and prospective epidemiological studies), *in vitro* research using human tissue, *in silico* research, and the recent technological breakthrough of microdosing (Xceleron). Further, it must never be forgotten that when industrial chemicals find their way into the environment, or drugs are marketed to the general population, large-scale human experiments are then underway on people unwittingly exposed—sometimes with truly awful consequences for exposed populations of humans (not to mention other organisms).

Moreover, as Altman (Altman 1998) has observed, there are many examples, both ancient and modern, where researchers, doubting the applicability or relevance of animal models to the human situation, have experimented on themselves—a practice that Altman points out continues to the present (recent Nobel laureate Barry Marshal being but one example). In any event, at the very least an evidence-based track record of success (vis-à-vis positive and negative predictive values) using specific animal models should be evident if society is to accept hypotheses from animal testing as predictive for humans.

Misunderstandings about who has the burden of proof are seen even among those claiming to be animal advocates. Kenneth Shapiro, currently Co-Executive Director Animals & Society Institute writing in *ATLA*:

> What are the odds that it will work—that the results also will obtain in the target? What do they need to be for this strategy of animal model to be effective? Opponents of animal research, arguing the science issue, often critically claim that the prediction rate is only 50%, a rate no better than tossing a coin. This is a misunderstanding of the situation. Let's say that the prediction rate is only 50%—this means that, half the time, the hypothesis developed in the model and tested in the target will not be confirmed in the target. The causal relation of treatment effect does not reach a statistical level of 0.05 even though it did in the model. This rate is not a strong critique of the animal model process unless there is a better method of generating hypotheses. The 50% success rate must be compared, not to a coin toss, but to other ways of generating hypotheses—against both non-animal models and

hypotheses from clinical observation. If these other ways only produce a 25% success rate, then clearly, animal models are a more efficient way of generating hypotheses. (Shapiro 2004)

The first observation here is that if Shapiro is correct about the 50% success rate, then regardless of the existence of even worse methodologies,
all other things being equal, we are literally faced with a choice as to
whether to do predictive animal-based research or toss a coin. All other
things are not equal. Coin tossing costs a tiny fraction of the cost of carefully controlled predictive animal studies. Coin tossing wins out here on
the basis of a cost-benefit analysis. But there is more.

Without assuming anything about human clinical trials (the figure
quoted by Shapiro is a fiction to make a point), we note here that the
poor predictive relevance of a given research method cannot be defended
by pointing to even worse methodologies. No doubt making guesses after
hitting oneself over the head with a hammer would yield an even worse
success rate than clinical observation. In the present context, this would
hardly justify, in and of itself, either clinical or animal-based research methods! The issue concerns one thing and one thing only: in this case, evidence for the validity of a widely employed research methodology. After
all, one can usually find support for the claim that fortune telling via ho-
roscopes is predictive. What destroys the evidential value of such claims
as evidence of astrological success are all the times astrology failed; the
occasional correct answer does not mean a modality is predictive or even
useful. Nor does the greater failure rate of, say, palmistry or the reading of
tea leaves advance the astrologer's case.

The comparison with coin tossing *is* appropriate in the sense that a
gambler with a taste for games of chance, might play a game with such odds
(50% chance of success), and may well prefer this to a game with only a 10%
chance of success (other things being equal). Gambling of this kind is not a
luxury for those whose business is human health, safety and well-being—
animal investigators themselves make this very point. Where the costs of
errors are high (in terms of death or serious injury, not to mention the
enormous social costs associated with the misclassification of substances as
harmful), a modality's success rate of 90% may fail to be adequate.

This leads us to a discussion of some common misuses of the term
prediction (see Figure 11.1). First, random guessing occasionally achieves
the right answer. Consider the following. A fortuneteller says you will get
married this year, have a baby next year, make $500,000 the next, discover
a cure for mesothelioma the next and retire the next year. Five years pass

	Example	Reason not predictive
Error #1	The New Zealand White rabbit responded to thalidomide as did humans therefore it predicted thalidomide's affects.	Retrospective simulation is not prediction.
Error #2	My horoscope predicted I would have a bad day and I did.	Prediction is more accurate than random chance or guesses. Past correct and incorrect answers must be taken into account before an entity can be assessed for predictive power.
Error #3	The school of biological sciences always picks the college football champion.	Betting on most or all contenders is not the same as predicting an outcome.
Error #4	If I predict the right outcome 51% of the time in Las Vegas, in the long run I would be rich.	The standards of gambling should not be confused with the predictive standards of medical practice.

Figure 11.1. Errors associated with *prediction*.

and in year three you did in fact make a $500,000. Did the fortuneteller predict this? In lay terms, one could certainly say so. But in science the concept of prediction has a very specific meaning and the above does not qualify. It is likely nothing more than a lucky guess grounded in information that the hapless client has inadvertently provided the teller. The lottery is another good example. Assume a lottery in Kansas is conducted in such as fashion that six numbers are picked at random every week. Mr Smith plays the lottery and every week picks six numbers based on his birth date. Suppose he wins on his twentieth trial. Did Mr. Smith predict what numbers would win? No, not in the scientific sense of the word. He simply got lucky.

Of more relevance to present concerns, prediction is sometimes confused with *retrospective simulation*. Here it is claimed that an animal model predicts human outcomes when the human data is antecedently known and a model is then found, after the fact of the human data coming to light, that simulates or mimics the human phenomenon. It is often possible to find some animal model that retrospectively simulates antecedently known human data.

While we will have much more to say about the thalidomide disaster later in this volume, an example derived from this tragedy is instructive. The teratogenic effects of thalidomide were first discovered in humans whose mothers took the drug while pregnant. The human teratogenic effects of thalidomide were antecedently known. *After the fact* of the human tragedy, animal studies were conducted to simulate these human results. Many species tested, even at very high doses, failed to mimic the observed human condition. In rodents (typical test species in teratogenic inquiries), moreover, this was true for many distinct strains of the same species. Finally, the New Zealand white rabbit was found to exhibit important features of the human condition. Is the existence of one successful species mimicking known human results, out of many distinct species tested, a vindication of predictive animal-based research? What we have here is a case of retrospective simulation. This is not the same as prediction.

There are also many contemporary examples of the confusion of prediction with retrospective simulation. Here is a representative case concerning the drug MPTP which is known to induce parkinsonism in humans. Gad, writing in *Animal Models in Toxicology* (Gad 2007), states that the neurological effects of MPTP were predicted by animals, specifically monkeys.

The first report of a chemical inducing parkinsonism appeared in 1979 from 4-propyloxy-4-phenyl-N-methylpiperidine, a meperidine analog (Davis et al. 1979). MPTP-induced parkinsonism was also first seen in humans:

> Four persons developed marked parkinsonism after using an illicit drug intravenously. Analysis of the substance injected by two of these patients revealed primarily 1-methyl-4-phenyl-1,2,5,6-tetrahydropyridine (MPTP) with trace amounts of 1-methyl-4-phenyl-4-propionoxy-piperidine (MPPP). On the basis of the striking parkinsonian features observed in our patients, and additional pathological data from one previously reported case, it is proposed

that this chemical selectively damages cells in the substantia nigra. (Langston et al. 1983)

This was then followed by the injection of MPTP into monkeys (Burns et al. 1983). The data in humans preceded the data from monkeys, thus reversing the usual account of events which are supposed to flow from monkey to man.

The reason retrospective simulation is not the same as prediction is straightforward. There are many possible animal models for a given phenomenon of interest (e.g., safety of some drug X in humans) and they typically make very different predictions concerning this matter (consider some of the examples of species differences in drug metabolism discussed in the last chapter). The question is which of these possible predictions are correct. The field of valid contenders is narrowed considerably once the human data is known, yet it is this very human data that we wanted the model to predict to avert a possible medical catastrophe in humans. It is cold comfort indeed to have the human catastrophe as a matter of fact, retrospectively simulate it in, say, a rodent species, and say, "There! I told you so."

The predictive power of an animal model must be assessed on the basis of hypotheses about the human biomedical condition of interest that are forthcoming from the model initially in the absence of relevant human data. Perhaps the data is known, but is withheld from the investigator for the sake of testing the model once the predictions are in, or perhaps the relevant human data has not yet been uncovered—either way, animal-based hypotheses about human biomedical phenomena cannot be tested until the human data is available. Recall that classical mechanics allowed astronomers to predict the existence and location of the planet we know as Neptune. They didn't know of its existence and location antecedently, and then reason backwards. The subsequent vindication of the predictions was one of the crowning glories of classical mechanics. Prediction involves risk of failure—a failure to get the phenomenon of interest right. Put another way, there is a world of difference between giving students the right answer in a test, and asking them to come up with a plausible question that matches the answer, and giving them a question to see if they can come up with the right answer when the test is subsequently graded!

The business of retrospective simulation of antecedently known results must not be confused with the scientifically respectable concept of *retrodiction*, whereby hypotheses are formed about the course of past events, and these are subsequently tested. In historical sciences such as evolutionary biology, comparative linguistics, and geology the testing of retrodictions is part and parcel of the sciences themselves. For example, if evolutionary biology is correct, simpler lifeforms ought to be found at lower levels of the geological column than are found at higher levels. And this is what we find. It would be devastating to evolutionary biology to find a "Johnny-come-lately" animal such as a fossilized Rottweiler, in pre-Cambrian deposits!

Prediction in science involves risk of failure. There are various ways of limiting risk of failure, but they come with a price. For example, a school of biological sciences at a large research university has 200 faculty and staff. This university also has a football team and hence the campus is very interested in football. At the beginning of the season, the 200 people in the school of biological sciences wager on who will be the ultimate #1 team at the end of the season. Considering the finite number of serious contenders for the prize and the number of people in the school of biological sciences, it is highly probable that someone in the department will pick the winner, provided each individual picks a different possible contender. There is a good chance that they will do better as a group than the Department of Philosophy with only four faculty members!

Can we then say that the school of biological sciences always (or often) predicts the winner? No, and for the following reason: if each member of the school picked the same team, it is much less likely that the school as a whole will predict the winner. They may even do worse that the Department of Philosophy. In a big school or department, if enough distinct bets are laid, someone will likely win. This does not make the school of biological sciences a "lean, mean prediction machine."

Consider a gambler at the horse track. His selections most often lose—he wins just often enough to maintain his interest in racing. It is a bad strategy to bet on the all the horses in the same race. Needless to say, this "get rich quick" scheme of assuring a winner by betting equal amounts on all the horses in the race, while it may likely result in the selection of a winner, comes at a terrible cost to the gambler! Selecting all the horses will result in the gambler picking a winner, but will also result in having bet on nine or so losers. Not only is this not an example of prediction but obviously (lacking "friends" in the criminal community) it is impossible to know in advance which one horse to wager on. The winner emerges when the race is over and bets can no longer be made.

Claims about the predictive nature of animal models

The issue of why prediction is important does not exist in a vaccuum. Lives are at risk. According to Salmon there are at least three reasons for making predictions:

> 1. because we want to know what will happen in the future;
> 2. to test a theory;
> 3. an action is required and the best way to choose which action is appropriate is to predict the future. (Salmon 1998)

In the case of carcinogenesis we want to know: (1) what will happen in the future (will the chemical cause cancer in humans?); and (3) an action is required (allow the chemical on the market or not?) and the best way to choose which action is appropriate is to predict the future. Neither (1) nor (3) is subtle. We want a correct answer to the question, "Is this chemical carcinogenic to humans?" or to similar questions such as, "What will this drug do to humans?" and "Is this drug a teratogen?" and "Is this the receptor used by HIV to enter the human cell?" But guessing correctly or finding correlations are not, as we have seen the same as predicting the answer. Neither is a high degree of sensitivity alone, as we shall see, the same as prediction.

We are about to argue in detail as to why the practice of animal modeling is not predictive of human biomedical outcomes. Both authors of this volume have encountered researchers who have tried to short-circuit our objections by claiming that animal models are either known not to be predictive models (so we are making much ado about nothing), or indeed were never intended to be—they are actually used in basic research whose aim is to increase the sum total of biological knowledge, not to predict human outcomes (so we have misunderstood the aim of animal-based research). These claims by researchers are not usually made in the public arena, for the public and their policy-makers have been repeatedly told that animal-based research does indeed predict human outcomes, and in this belief that such is the case lies the willingness of the public to support the allocation of research funds and tax dollars to the enterprise.

This is then a good opportunity to point out, once again, that the authors of this volume do not deny that there are many legitimate uses of animals in biomedical research, basic research being among the most important. It is false, however, to say that researchers do not think animal models are predictive of the human condition. We shall provide some examples to make our case in this regard. Moreover, not only are there investigators who believe animal models are predictive, they also believe that they are very good, too. This latter claim will be subject to critical scrutiny in what follows.

That animal models are predictive of human responses is an idea with a long history going back to the dawn of the science of toxicology and the emergence of scientific medicine in the 19th century. For example, as noted by Bernard in the 1860s:

> Experiments on animals, with deleterious substances or in harmful circumstances, are very useful and entirely conclusive for the toxicology and hygiene of man . . . for as I have shown, the effects of these substances are the same on man as on animals, save for differences in degree. (Bernard 1949)

To understand the prediction issue, we do not have to go back to ancient history (alas, all too easily ignored in contemporary discussions), for the intellectual legacy of Claude Bernard is evident in contemporary writings. Thus, Gad has recently written:

> Biomedical sciences' use of animals as models to help understand and *predict* responses in humans, in toxicology and pharmacology in particular, remains both the major tool for biomedical advances and a source of significant controversy . . .
>
> At the same time, although there are elements of poor practice that are real, by and large animals have worked exceptionally well as *predictive* models for humans-when properly used. . .
>
> Whether serving as a source of isolated organelles, cells or tissues, a disease model, or as a *prediction* for drug or other xenobiotic action or transformation in man, experiments in animals have provided the necessary building blocks that have permitted the explosive growth of medical and biological knowledge in the later half of the 20th century and into the 21st century. . . .
>
> Animals have been used as models for centuries to *predict* what chemicals and environmental factors would do to humans. . . . The use of animals as *predictors* of potential ill effects has grown since that time [the year 1792].
>
> Current testing procedures (or even those at the time in the United States, where the drug [thalidomide] was never approved for human use) *would have identified the hazard and prevented this tragedy.* (Gad 2007) (Emphasis added.)

In a similar vein Hau notes that the business of prediction, while not the only aim of animal-based research, is certainly one of them:

> A third important group of animal models is employed as *predictive* models. These models are used with the aim of discovering and quantifying the impact of a treatment, whether this is to cure a disease or to assess toxicity of a chemical compound. (Hau 2003)

While there are many motivations for the use of animals to predict human responses in toxicology, a very important motivation focuses on the issue of *intact organismal systems*. For all the emphasis we have seen investigators place on the importance of the lowest levels of the biological hierarchy of organization, where, it is hoped, humans and our animal models are the same animals dressed up differently, investigators also suspect that cell and intracellular studies may miss something important: that is, the organismal context in which the metabolism of xenobiotics takes place. Thus, Krewski et al. of the Committee on Toxicity Testing and Assessment of Environmental Agents observe:

> For the foreseeable future, some targeted testing in animals will need to continue, as it is not currently possible to sufficiently understand how chemicals are broken down in the body using tests in cells alone. These targeted tests will complement the new rapid assays and ensure the adequate evaluation of chemicals. (Committee on Toxicity Testing and Assessment of Environmental Agents 2007)

There are several observations worth making here (which we shall return to in various guises as our argument proceeds). First, we saw in the last chapter that there are reasons to doubt that, at *the basal biochemical and molecular levels of the biological hierarchy,* humans and mice (for example) are in fact the same animal dressed up differently. Second, standard justifications for predictive animal-based research see the effects of evolution primarily at the higher levels of the biological hierarchy (where mouse and man even look to the untutored eye to be different). The problem is that if humans and mice are *not* the same animal dressed up differently— if differences at the basal levels of the biological hierarchy of organization reverberate "upwards" to affect tissues, organs and ultimately the organism itself—it is simply not clear (no matter how much it may matter to mouse biology) that intact mouse studies have any more predictive relevance for human biology than did the cellular and intracellular studies in the first place. In point of fact, we know volumes about mouse biology and mouse medicine—we can even cure murine cancers! Such knowledge of mice can rightly be listed as belonging to the triumphs of basic biomedical research. What is much less clear—and is the concern of the present volume—is the predictive relevance of this work on mice for human medicine. It is here that opinions differ.

For example, in May 2005, one of us (RG) debated issues surrounding animal experimentation with Andrew Skolnick, at the University of Buffalo. Skolnick is a science and medical journalist who was then the Executive Director of the Commission for Scientific Medicine and Mental Health in Amherst, NY. A member of the audience asked Mr. Skolnick whether he had any statistical evidence relevant to the prediction question that had been debated. Skolnick replied:

> I don't have data to answer that question, but I do have this observation. . . there's inadequate predictability. When they do a clinical trial of 5,000 people here, there is inadequate predictability of how the drug is going to work in the marketplace . . . Animal studies are not perfectly predictable. It's not that accurate as prediction. But that's not what animal models are mostly used for. They're used to give insights and generate hypotheses for clinical testing. (Skolnick and Greek 2005)

This statement, emphasizing as it does the role of animals in basic research, is notable for its downplaying of the prediction question. Perhaps, then the prediction issue is, after all, a mere storm in a teacup, much ado about nothing.

Unfortunately, as we have seen in this section, there are prominent investigators who do take the prediction issue seriously, and as we shall see in the next section, there is relevant statistical data to help settle the matter. At some point, contrasting opinions must rub up against hard evidence.

Prediction in biological complex systems

So what *does* constitute prediction in biological systems? To forestall a kind of specious objection, let's be clear about what we are *not* saying. No reasonable person seriously doubts that experiments on animals can prompt new ideas and new approaches to issues in human medicine. Perhaps a genetic abnormality in a mouse with breast cancer might lead an investigator to look for the something analogous in humans. The very presence or absence of the abnormality in humans may then suggest ways forward with the human problem of interest. This is part and parcel of the business of basic research. This is not the same as prediction (for reasons to be clarified below), and the epistemological pathway from knowledge derived from basic research (of any kind) to practical human application is typically a long and tortuous one.

How then should we think of prediction in the context of toxicology, pathophysiology, and pharmacology? The 2x2 table for calculating sensitivity, specificity, positive predictive value and negative predictive value illustrates how predictability is assessed in these contexts (see Figure 11.2).

In biology, many concepts are best evaluated by using simple statistical methods involving probability. For example, in medicine, physicians can use a blood test to determine whether someone has liver disease. In order to ascertain how well this test actually determines the health of the liver we calculate the sensitivity and specificity of the test along with the positive predictive value (PPV) and negative predictive value (NPV). The sensitivity of a test is the probability (measured on a scale from 0.0 to 1.0) of a positive test among people whose test should be positive—those who do, in fact, suffer from liver disease. Specificity is the probability of a negative test among people whose test should be negative—those without liver disease. The positive predictive value of a test is the proportion of people with positive test results who are actually positive. The negative predictive value is the proportion of people with negative test results who are actually negative. This is all quite straightforward. Very few tests have a sensitivity or specificity of 1.0 or a PPV and NPV of 1.0, but in order for a test to be useful given the demanding standards of medical practice—in this case tell us if the patient actually has liver disease—it needs to have PPV and NPV in at least the .90 to 1.0 range.

The following illustrate how these statistics are used in medicine. Ravipati et al.:

> Sixty-four-multislice coronary computed tomographic angiography (CTA) and coronary angiography were performed in 145 patients (mean age 67 +/- 10 years), and stress testing was performed in 47 of these patients to determine the sensitivity, specificity, positive predictive value, and negative predictive value of coronary CTA and of stress testing in diagnosing obstructive coronary artery disease (CAD) in patients with suspected CAD. In 145 patients, coronary CTA had 98% sensitivity, 74% specificity, 90% positive predictive value, and 94% negative predictive value in diagnosing obstructive CAD. In 47 patients, stress testing had 69% sensitivity, 36% speci-

ficity, 78% positive predictive value, and 27% negative predictive value for diagnosing obstructive CAD, whereas coronary CTA had 100% sensitivity, 73% specificity, 92% positive predictive value, and 100% negative predictive value for diagnosing obstructive CAD. In conclusion, coronary CTA has better sensitivity, specificity, positive predictive value, and negative predictive value than stress testing in diagnosing obstructive CAD. (Ravipati et al. 2008)

We will shortly see how these statistical concepts can be used (and abused) to settle issues concerning prediction.

By definition, when we speak of animals predicting human response in drug testing and disease research we are addressing the risk of wrong predictions and how much risk society is willing to tolerate. Troglitazone (Rezulin) is a good example of the margin of error for medical practice tolerated in society today. Troglitazone was taken by well over 1 million people with less than 1% suffering liver failure, yet the drug was withdrawn because of this side effect (CNN 1999). (Interestingly, animal studies failed to reproduce liver failure from troglitazone (Masubuchi 2006).)

Rofecoxib (Vioxx) is another example of the small percentage of morbidity or mortality tolerated in the practice of medicine vis-à-vis introducing a new drug. Figures vary, and are controversial, but it now appears that apparently less than 1% of people who took rofecoxib experienced a heart attack or stroke as a result, yet it was also withdrawn (Topol 2004). This means that even if a test with a PPV of .99 had assured industry that rofecoxib and troglitazone were safe, the test would not have been accurate enough for society's standards.

This is an important point. Medical practice does not tolerate risks (probability of being wrong) acceptable in some experiments conducted in laboratories (where the stakes may be much lower)—and certainly not the sort involved in coin-tossing, for example. In basic research we might proceed with a study based on the outcome being more likely than not. For basic research this is acceptable. However, getting the answer wrong in medical practice has consequences; people die. Societal standards for medical practice today demand very high sensitivity, specificity, PPV and NPV from its tests. We will apply the above remarks to animal models shortly.

Cancer

So how well do animal models predict carcinogenesis? While it is true that that all known human carcinogens that have been adequately studied have

		Gold Standard	
		S+	S-
Test	T+	TP	FP
	T-	FN	TN

T+ = test positive
T- = test negative
T = True
F = False
P = Positive
N = Negative
S+ = standard positive
S- = standard negative

Sensitivity = TP/TP+FN

Specificity = TN/FP+TN

Positive Predictive Value = TP/TP+FP

Negative Predictive Value = TN/FN+TN

Figure 11.2. Statistics used in analysis of prediction.

been shown to be carcinogenic in at least one animal species (Wilbourn et al. 1986; Rall 2000; Tomatis and Wilbourn 2003), it is also true that an irreverent aphorism in biology known as Morton's Law states: "If rats are experimented upon, they will develop cancer." Morton's law is similar to Karnofsky's law in teratology, which states that any compound can be teratogenic if given to the right species at the right dosage at the right time in the pregnancy. The point here is that it is very easy to find positive results for carcinogenicity and teratogenicity; *indicating a high sensitivity.* But this is meaningless without also knowing *specificity, positive predictive value,* and *negative predictive value.*

The World Health Organization classifies chemicals according to carcinogenicity via the International Agency for Research on Cancer (IARC). Knight et al. discussed a study in 1993 by Tomatis and Wilbourn (Tomatis and Wilbourn 2003). Tomatis and Wilbourn surveyed the 780 chemical agents or exposure circumstances listed within Volumes 1-55 of the IARC

monograph series (IARC 1972-1992). They found that "502 (64.4%) were classified as having definite or limited evidence of animal carcinogenicity, and 104 (13.3%) as definite or probable human carcinogens . . . around 398 animal carcinogens were considered not to be definite or probable human carcinogens."
Knight et al. continue:

> . . . based on these IARC figures, the positive predictivity of the animal bioassay for definite probable human carcinogens was only around 20.7% (104/502), while the false positive rate was a disturbing 79.3% (398/502).
>
> More-recent IARC classifications indicate little movement in the positive predictivity of the animal bioassay for human carcinogens. By 1 January 2004, a decade later, only 105 additional agents had been added to the 1993 figure, yielding a total of 885 agents or exposure circumstances listed in the IARC Monographs (IARC). Not surprisingly the proportion of definite or probable human carcinogens resembled the 1993 figure of 13.3%. By 2004, only 9.9% of these 885 were classified as definite human carcinogens, and only 7.2% as probable human carcinogens, yielding total of 17.1%. (Knight, Bailey, and Balcombe 2006)

In animal tests, issues of gender turn out to be important (Haseman 2000). There are policy implications here. Giving consideration to such issues, Knight et al. conclude:

> If a risk-avoidance interpretation is used, in which any positive result in male or female mice or rats is considered positive, then nine of the 10 known human carcinogens among the hundreds of chemicals tested by the NTP [National Toxicology Program] are positive, but so are an implausible 22% of all chemicals tested. If a less risk-sensitive interpretation is used, whereby only chemicals positive in both mice and rats are considered positive, then only three of the six known human carcinogens tested in both species are positive. The former interpretation could result in the needless denial of potentially useful chemicals to society, while the latter could result in widespread human exposure to undetected human carcinogens. (Knight, Bailey, and Balcombe 2006)

These findings are not new. Discrepancies between animal-human studies and even animal-animal studies date back many decades. Percival Pott showed coal tar was carcinogenic to humans in 1776. Yamagiwa and Ichikawa showed it was carcinogenic in some animals in 1915. But even then, rabbits did not respond as mice (Shubick 1980).

In 1980 there were roughly sixteen hundred known chemicals that caused cancer in mice and rodents, but only approximately fifteen were known to cause cancer in humans (Coulston 1980). The Council on Scientific Affairs reported in 1981 that:

> *The Council's consultants agree that to identify carcinogenicity in animal tests does not per se predict either risk or outcome in human experience* . . . the Council is concerned about the hundreds of millions of dollars that are spent each year (both in the public and private sectors) for the carcinogenicity testing of chemical substances. The concern is particularly grave in view of the questionable scientific value of the tests when used to predict human experience. (Council on Scientific Affairs 1981) (Emphasis added.)

Again, in the early 1980s, David Salsburg of Pfizer, referring to a report by the National Cancer Institute that examined 170 chemicals, concluded that lifetime feeding studies using rodents lacked sensitivity and specificity. He stated:

> If we restrict attention to long term feeding studies with mice or rats, only seven of the 19 human non-inhalation carcinogens (36.8%) have been shown to cause cancer. If we consider long term feeding or inhalation studies and examine all 26, only 12 (46.2%) have been shown to cause cancer in rats or mice after chronic exposure by feeding or inhalation. Thus the lifetime feeding study in mice and rats appears to have less than a 50% probability of finding known human carcinogens. *On the basis of probability theory, we would have been better off to toss a coin.* (Salsburg 1983) (Emphasis added.)

Studies conducted prior to 1984 on mice and rats found that 46% of chemicals found to be carcinogenic in rats were not in mice (Di Carlo 1984) and that mice do produce tumors in response to chemicals that do not cause cancer in rats (Hoffman 1993). A 1987 study by Ennever (Ennever, Noonan, and Rosenkranz 1987) discovered that of twenty probable *non*carcinogens in humans, nineteen were known animal carcinogens. Conversely, of nineteen compounds known to cause oral cancer in humans, only seven caused cancer in mice and rats using the standard NCI protocol (Salsburg 1983). The findings of these studies have been supported by others (Meijers, Swaen, and Bloemen 1997; Monro 1996; Gold, Slone, and Ames 1998). Regardless of the era and the specific animal tests employed, the results have been the same: animals are not predictive for humans.

What should we do? Should we discard every drug that causes cancer in animals? Acetaminophen, chloramphenicol, and metronidazole are known carcinogens in some animal species (IARC 1972-2001; Sloan et al. 1983). Phenobarbital and isoniazid are carcinogens in rodents (Clemmensen and Hjalgrim-Jensen 1980; Shubick 1980; Clayson 1980). Does this mean they never should have been released to the market? Diphenylhydantoin (phenytoin) is carcinogenic to humans but not rats and mice (Anisimov 1987, 2003; Dilman and Anisimov 1980). Occupational exposure to 2-naphthylamine appears to cause bladder cancer in humans. Dogs and monkeys also suffer bladder cancer if exposed to 2-naphthylamine orally and mice suffer from hepatomas. It does not appear to have carcinogenic properties in rats and rabbits. These are qualitative differences due to differences in metabolism of aromatic amines (IARC 1974). It also appears that fewer genetic, epigenetic, or gene expression events are needed to induce cancer in rodents than are needed to induce cancer in humans (Anisimov, Ukraintseva, and Yashin 2005; Hahn and Weinberg 2002; Rangarajan and Weinberg 2003). (A good review of species differences in relation to carcinogenesis and why they exist is Anisimov et al. (Anisimov, Ukraintseva, and Yashin 2005).)

At least some of the differences we see with respect to carcinogenesis are apparently reflective of differences between species at the lower, cellular levels of the biological hierarchy of organization where standard accounts of predictive animal modeling suggest the similarities should be greatest. The differences in basic biology between human and nonhuman species just alluded to can be further illustrated by the case of tamoxifen. Arguably, the biggest breakthrough in the treatment of human breast cancer has been the addition of Tamoxifen to the medical arsenal. An editorial in *Nature Reviews Drug Discovery*, however, notes that:

> In Tamoxifen's case, a drug first developed as a potential contraceptive languished for many years before its present application was found. Furthermore, its propensity to cause liver tumours in rats, a toxicity problem that thankfully does not carry over into humans, was not detected until after the drug had been on the market for many years. If it had been found in preclinical testing, the drug would almost certainly have been withdrawn from the pipeline. (Follow the yellow brick road 2003)

Tamoxifen demonstrates diverse effects among different species. It is a full estrogen in mice, a partial estrogen/antiestrogen in humans and rats, and an antiestrogen in chicks (Jordan and Robinson 1987). It induces hepatocellular carcinoma in rats after being given orally and acts as a promoting agent in some models of rat carcinogenesis. It did not induce cancers in mice when using the same protocol used in the rat studies (Wogan 1997; Tannenbaum 1997). In hamsters, it protected against estrogen-induced hepatocarcinogenesis (Coe et al. 1992). Myriad other differences also manifest (Hengstler et al. 1999).

Smoking and Cancer

Finally we come to smoking. One of the great failures of the animal model approach (as opposed to clinical and epidemiological studies of humans) in cancer research was its prediction that smoking was safe. Smoking usually does not cause cancer in lab animals.

> For decades the clinical observation of an association between cigarette smoking and bronchial carcinoma was subject to unfound doubt, suspicion, and outright opposition, largely because the disease had no counterpart in mice. There seemed no end of statisticians craving for more documentation, all resulting in the fateful delay of needed legislative initiatives. (Clemmensen and Hjalgrim-Jensen 1980)

Given the havoc wrought by smoking for human health and well-being, this case deserves further examination. Predictive animal modelers, when they admit the deficiencies in their models, often claim that nevertheless they are doing the best they can to promote human safety. The case of smoking in relation to cancer, then, should serve as a warning of the ways in which deficient modeling practices may be exploited for commercial gain (especially in the present case by predictive modelers serving the interests of "Big Tobacco").

For a variety or reasons, it is hard to simulate the long-term effects of exposure to tobacco in nonhuman animals. There are basic biological differences, including, but not limited to the fact that typical test species have much shorter lifespans than human smokers. Dr. Sigismund Peller, an internist and medical statistician, was, in 1937, among the first researchers to link cigarette smoking to lung cancer. Peller stated: "The overestimation of animal experiments is so rampant that the issue is of general interest (Peller 1952)." Concerning this unhappy state, Utidjian observed in 1988:

Surely, not even the most zealous toxicologist would deny that epidemiology, and epidemiology alone, has indicted and incriminated the cigarette as a potent carcinogenic agent, or would claim that experimental animal toxicology could ever have done the job with the same definition. (Utidjian 1988)

Readers inclined to think that these issues are of mere minor academic interest may wish to reflect on the following: William Campbell, president and CEO of Phillip Morris testified under oath in 1993:

Q. Does cigarette smoking cause cancer?
A. To my knowledge, it's not been proven that cigarette smoking causes cancer.
Q. What do you base that on?
A. I base that on the fact traditionally, there is, you know, in scientific terms, there are hurdles related to causation, and at this time there is no evidence that they have been able to reproduce cancer in animals from cigarette smoking. (Janofsky 1993)

If we were to do weighted averages for harm done, the fact that animals did not suffer from cancer secondary to smoking—and consequently policy makers ignored human data consistent with carcinogenesis—would surely outweigh all other considerations in a historical analysis of using animals as predictive models. The fact that the tobacco industry continues to quote animal data when justifying their product is evidence that society has somehow been led to believe that animal data is not only predictive for humans, but that it trumps human epidemiological studies (which would ultimately settle the matter in favor of human beings, not rats and mice).

The point to be made here is that there are serious scientific professionals (rather more than we have space to mention) who question the predictive and/or clinical value of animal-based research, and *history* is on their side. Of course, the opinions of scientists prove nothing in and of themselves. But this sword cuts two ways, for it also includes the opinions of scientific apologists for the practice of predictive animal-based research.

Further, all we have presented could be dismissed as anecdotes but this would be a mistake. First, the studies we have referenced are just that, scientific studies not anecdotes. They constitute publicly checkable sources of information. Our examples are not undocumented *tittle-tattle*! We find ourselves asking once again, where lies the burden of proof? If the animal model community claims the practice of animal modeling is generally predictive (we readily concede that there are uses of animal models for purposes other than prediction), then they must explain the examples and studies that reveal it was not. These cases of predictive failure are too numerous to be dismissed as anomalies, outliers or mere evidential "noise" muddying an otherwise clear picture of predictive success.

Drug Development

Drug testing with animals has traditionally focused on ADMET studies. How the drug is *A*bsorbed into the body. How it is *D*istributed to the tissues. How the body *M*etabolizes it. How the body *E*liminates it and the *T*oxic effects. Toxicity is determined mostly by how the chemical is metabolized by the body. Many different genes influence how the drug is metabolized, and as Kehrer has stated, "Small differences in gene structure can make large differences in function (Kehrer 2001)." This is not unsurprising, given what is known about mathematical models of complex biological systems.

Drugs are theoretically supposed to hone in on a specific single target/molecule and leave all other targets alone. The problem is that this is often more true in theory than practice. 6.7% of hospitalized patients suffer severe adverse drug reactions (Lazarou, Pomeranz, and Corey 1998). *Severe* being defined as an event that prolongs hospitalization or results in death or disability. William Bains, chief scientific officer of Amedis Pharmaceuticals in the UK estimates that 50% of all drugs in development fail to progress to the market because of problems associated with ADMET and 50% of all drugs that do make it to market have problems associated with ADMET. Barry Selick, formerly with Glaxo Wellcome and now with Camitro, a company specializing in ADMET, believes that for every drug that is withdrawn from the market, 10 remain available to people even though they have ADMET-associated problems. He also says that for every drug that is on the market with ADMET-associated problems, there is an additional 10-50 that will fail before they reach the general public (Hodgson 2001).

When a drug enters human clinical trials, the drug company typically has very little idea if it will damage humans. Because ADMET studies in animals are so unreliable, Tom Patterson, chief scientific officer at Entelos, likens the current practice of drug testing in humans during clinical trials to making airplanes, trying to fly them, and marketing the one that does not crash (Hodgson 2001).

Some history is relevant. Back in 1966, Theodore Koppanyi and Margaret A. Avery wrote an article about species differences, observing:

Many useful drugs have been introduced into therapeutics without previous animal screen; digitalis, ipecac, cinchona bark, and the earliest inhalation anesthetics may suffice as examples. It could be argued, however, that the discovery of these drugs was by serendipity rather than rational planning. Had these drugs first been tested in animal experiments for their safety, some of them might never have reached clinical trial. The animal screen might have revealed that the margin of safety between the toxic and truly effective doses of digitalis, ether, and chloroform is narrow; ipecac would have been shown to cause myocardial damage; and cinchona bark would have been found to produce blindness if certain species had been chosen as the test animal. Deringer and other authors discovered genetic as well as sex differences in mice in the response to chloroform. Strains C_3H, C_3H_1, A, and HR were highly susceptible to chloroform, while strains C_57BL, C_57BR/cd, C_57L, and ST were resistant to the accidental exposure. If it had been screened in guinea pigs (Harare et al. 1943) or Syrian hamsters (Schneierson and Perlman 1956), penicillin might have been discarded. (Koppanyi and Avery 1966)

These remarks were but a prescient foretaste of what would be revealed in subsequent decades of predictive animal-based research devoted to drug discovery and development. Given the enormous role that "Big Pharma" has come to play, not just in shaping health policies in the developed world, but also in influencing the contours of the global economy, it behooves us to pay some attention to this matter.

Perhaps in no other area of animal modeling is prediction a more important issue than in drug development. Young writing in *Drug Discovery World* 2008:

> The success rate of this heuristic approach [to drug development] is very low. For example, the average probability that a candidate emerging from lead optimisation will not make it to be a drug is above 99.8%. Building suspension bridges, cars or TVs with this sort of failure rate would not be thought reasonable by investors in those businesses, and searching questions are now being asked of the processes that are intended to build value in pharmaceuticals . . . It is evident from current attrition rates that the processes implemented in the recent past, and in the present if they differ little from what has been traditionally employed, are not sufficiently accurately predictive to yield the required productivity across the industry. This can be a little dissonant for those used to drawing together extensive preclinical packages for IND [Investigational New Drug] submissions, but, in the average, the PK, tox, *in vitro* and *in vivo* studies required for an

IND plainly do not predict efficacy, safety and deliverability in human
patients even nearly well enough. (Young 2008)

In a similar vein, Mankoff et al. also draw our attention to deficiencies in
animal-based drug discovery processes by pointing out:

> The overwhelming majority of Phase I-III human clinical trials fail
> for one of two main reasons: either the medication doesn't do what
> it is promised to do, or else, the drug is so toxic to humans that its
> other effects cannot be determined. In all cases, the experimental
> drug has proven safe and effective in preclinical studies using ani-
> mals of at least two species. (Mankoff et al. 2004)

The list of drugs used safely in humans but dangerous to animals is almost endless. Furosemide, commonly called Lasix, is a diuretic used to treat
high blood pressure and heart disease. Mice, rats and hamsters suffer liver
damage from furosemide, but humans do not. The drug is metabolized differently in each species (Walker and McElligott 1981). Corticosteroids,
drugs like prednisone and cortisone, were found to cause cancer in some
rodents. Other rats and monkeys were studied and found not to react to the
corticosteroids in the same way. Corticosteroids can cause birth defects in
some animals but not humans (Ward and Green 1988; Sidhu 1983).

We will now present specific toxicity and bioavailability studies, where
the results from animal testing were compared with human data, so we can
be better placed to judge whether animals are generally predictive of human
outcomes. Figures 11.3 and 11.4 illustrate graphically important data relevant to the animal prediction issue. Both figures chart bioavailability data
from three species of animals and compare it to data from humans. (Bioavailability is usually defined as the fraction of a drug that reaches the systemic circulation and reflects a number of different variables. Regardless of
the correlation, or lack thereof, of the variables, the bioavailability of the
drug is the final determinant of how much drug presents to the receptor or
active site.) Some of the drugs that showed high levels of bioavailability in
dogs had very low levels in humans and vice-versa. This was true regardless
of drug or species. Some levels did correlate between species but as a whole
there was no correlation between what a drug did in humans and what it
did in any given animal species or any combination thereof.

As can be seen, there is little correlation between animal and human
data. In some cases human bioavailability is high when bioavailability in
dogs is high but in other cases dogs and humans vary considerably. The
patterns exhibited by both are what are frequently referred to as a shot-

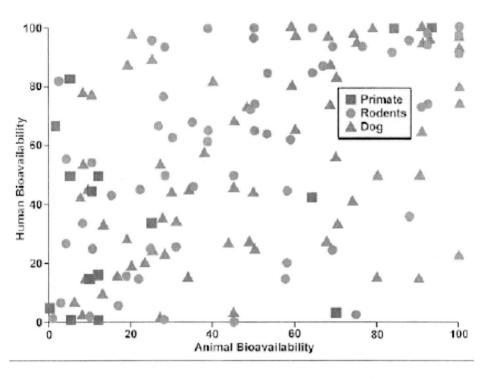

Figure 11.3. Bioavailability 1.

Figure was generously provided by James Harris PhD, who presented it at the Center for Business Intelligence conference titled *6th Forum on Predictive ADME/Tox* held in Washington, DC September 27-29, 2006 and is adapted from data that appeared in Grass GM, Sinko PJ. Physiologically-based pharma-cokinetic simulation modelling. *Adv Drug Deliv Rev.* 2002 Mar 31;54(3):433–5.

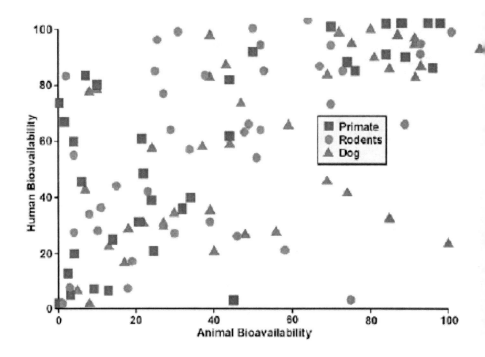

Figure 11.4. Bioavailability 11.

Figure was generously provided by James Harris PhD, who presented it at the Center for Business Intelligence conference titled *6th Forum on Predictive ADME/Tox* held in Washington, DC September 27-29, 2006 and is adapted from data that appeared in Arun K Mandagere and Barry Jones. Prediction of Bioavailability. In (Eds) Han van de Waterbeemd, Hans Lennernäs, Per Artursson, and Raimund Mannhold. *Drug Bioavailability: Estimation of Solubility, Permeability, Absorption and Bioavailability (Methods and Principles in Medicinal Chemistry)* Wiley-VCS 2003. P444-60.

gun pattern; meaning that if one fired a shotgun full of bird shot at a target one would see the same pattern. No precision and no accuracy. The pattern is also referred to as a scattergram, in this case meaning that the pattern is what one would expect from random associations.

The data on multispecies bioavailability illustrates some of the basic biological problems confronting those involved in the drug discovery process. Sankar offers the following gloomy assessment of where things stand with respect to predictive uses of animals in the drug discovery process:

> The typical compound entering a Phase I clinical trial has been through roughly a decade of rigorous pre-clinical testing, but still only has an 8% chance of reaching the market. Some of this high attrition rate is due to toxicity that shows up only in late-stage clinical trials, or worse, after a drug is approved. Part of the problem is that the toxicity is assessed in the later stages of drug development, after large numbers of compounds have been screened for activity and solubility, and the best produced in sufficient quantities for animal studies.
>
> Howard Jacob notes that rats and humans are 90% identical at the genetic level. However, the majority of the drugs shown to be safe in animals end up failing in clinical trials. "There is only 10% predictive power, since 90% of drugs fail in the human trials" in the traditional toxicology tests involving rats. *Conversely, some lead compounds may be eliminated due to their toxicity in rats or dogs, but might actually have an acceptable risk profile in humans.* (Sankar 2005) (Emphasis added.)

In the end, the profit motive may prompt drug companies to reassess the use of animals as predictive models. Such changes as might be forthcoming are likely to be slow and gradual. The public and more importantly jurists and policy-makers, have been sold on a methodology that has shown itself resilient in the face of the sorts of "incomplete validations" (we prefer "empirical refutations") illustrated above. Such is the power of institutional inertia!

CHAPTER 12

♦

PREDICTIVE MODELS IN TERATOLOGY

If you torture data sufficiently, it will confess to almost anything.

—Fred Menger

Developmental Toxins

We continue our discussion of the predictive nature of animal models with a discussion of developmental toxins, not least because here the standards of adequate prediction are particularly rigorous. The study of birth defects is known as teratology. Substances inducing birth defects are referred to as teratogens. Exactly how many drugs and other factors can result in birth defects is not known but the estimates are all very low. Only 40 or so drugs have been shown to be teratogens in humans. Early studies revealed over 800 to be so in animals (Shepard 1995); (Schardein 1976). The 40 known human teratogens include: infectious agents, physical factors (for example, ionizing radiation), maternal metabolic imbalances, drugs, and environmental chemicals. This is consistent with numbers cited in Schardein (Schardein 2000). Regardless of when the issue has been examined, the number of chemicals found to cause birth defects in humans was smaller by an order of magnitude than those causing birth defects in animals.

For a drug to cause birth defects, a number of criteria must be fulfilled. The drug exposure must take place at a critical stage of pregnancy and the dose must be high enough to cause a threshold of exposure for an appropriate duration of time. For most of the known human teratogens, more than 90% of pregnancies exposed during the first trimester result in normal offspring (Webster and Freeman 2003).

As you might expect, there are numerous examples of species differences with respect to teratogenicity. In this connection, Schardein observes:

Not all species are equally susceptible or sensitive to teratogenic influence by a given chemical…an agent that is teratogenic in some species may have little or no teratogenic effect in others…Because substances cross the placental membrane by a number of mechanisms, some differences in species reactivity to teratogens may be due to accessibility of the drug to the embryo…There are many experimental studies in which differences exist in teratogenic response when a chemical is given by different routes of administration…Chemicals that are teratogenic in animal tests may or may not indicate danger to humans, and negative results may incorrectly imply the absence of risk. As Fraser so aptly stated, '…final proof of whether a chemical is likely or not likely to be teratogenic in man must be sought in man'… A number of different animal species have been used in teratological research in an attempt to determine the most satisfactory model for predicting the hazard to humans…No single species thus far evaluated, however fulfills all criteria. [(Schardein 1992) p2-5, 12, 37]

What can we expect in terms of animal models predicting human teratogenicity?

Once again, the issue raised in the previous two chapters of the confusion of *prediction* with *retrospective simulation* raises its ugly head here with a vengeance. Thus, Newman et al. point out:

Some in the health community are concerned about the relevance of using animal data for predicting toxicologic scenarios that would interfere with normal in utero development of humans. This concern arises, at least in part, *because virtually all human teratogenic scenarios were not identified prospectively by tests in pregnant animals.* (Newman, Johnson, and Staples 1993) (Emphasis added.)

The fact is that virtually every medication used today has been shown to cause birth defects in some animal at some time. The cholesterol-lowering drug Lovastatin causes birth defects in rats, as did mevinolinic acid. Chondroitin sulfate caused birth defects in mice. Acetazolamide caused birth defects in rats, mice, rabbits, hamsters and furosemide caused birth defects in rats, mice, and rabbits. Spironolactone is also commonly used but was shown to cause feminization of male offspring in rats. Clonidine was shown to cause birth defects in mice; diazoxide in sheep and mice; hydralazine in mice and rabbits; reserpine in rats; and guanabenz in mice.

On the other hand, the ace inhibitors Captopril and Enalapril are strongly suspected of causing birth defects in humans but not mice or rats just as Minoxidil appears to cause human birth defects but not animal birth defects. Diltiazem is widely used and was found to cause birth defects in rats, mice and rabbits but not humans. Nifedipine and other vasodilators and calcium channel blockers have also been used safely in humans despite teratogenicity in animals. Most of the analgesics including codeine, hydrocodone, hydromorphone, meperidine, morphine, oxymorphone, phenazocine, propoxyphene, and others have been found to produce birth defects in animals but not humans. Acetaminophen is metabolized to a chemical that causes birth defects in hamsters. Cortisone produces birth defects in almost all animals normally tested, but not humans. Many antibiotics, antifungal medications and antiviral medications are teratogenic in animals, but not humans (Schardein 1992).

Rats have been shown to suffer birth defects from every chemical that causes birth defects in humans and therefore have been heralded as sentinels for chemicals that will cause birth defects in humans. But this is misleading since rat strains tested have high false positive rate so the results from rats are suspect. There is a growing awareness of the nature of the different types of mechanisms that account for species differences in this field of medicine. Again, Schardein has observed:

> The reaction to a specific teratogen may also be due to differences in a species' rate of metabolism, as well as to qualitative differences in metabolic pathways. In the human fetus, for example, drugs are metabolized both in the liver and at extrahepatic sites, such as the adrenal gland, whereas extrahepatic activity is negligible or absent in fetal rats, guinea pigs, rabbits, and swine. Chlorcyclizine, an antihistamine drug, is teratogenic in the rat, but is not so in the human. Although the drug is metabolized to norchlorcyclizine in both species, the steady-state level of the drug is three time higher in the rat than in the human, a factor perhaps accounting for the teratogenic difference between the species...Even when two species metabolize a drug at the same rate, differences in metabolic products may cause different teratogenic responses in the two species. For instance, the drug imipramine, which is teratogenic in some species of animals, is metabolized into different metabolites in various species, further emphasizing the problems inherent in extrapolating results from animal studies to humans. [(Schardein 1992) p3]

Lest anyone think the issues here are of minor academic interest, we draw the reader's attention to Hau, who in the *Handbook of Laboratory Animal Science,* summarized teratogen testing in animals nicely when he wrote:

> Uncritical reliance on the results of animal tests can be dangerously misleading and has cost the health and lives of tens of thousands of humans, as in Ciba Geigy's clioquinol scandal, the Opren disaster of Distra Products Ltd., or ICI's Eraldin calamity. Such counteraction in interspecies reactivity is bilateral: what is noxious or ineffective in nonhuman species can be innoxious or effective in humans. For example, penicillin is fatal for guinea pigs but generally well tolerated by human beings; aspirin is teratogenic in cats, dogs, guinea pigs, rats, mice, and monkeys but obviously not in pregnant women despite frequent consumption. [(Hau 2003) p4]

Furthermore, all of this should be considered in light of the fact that common table salt (Nishimura and Miyamoto 1969) and even water (Turbow, Clark, and Dipaolo 1971) have been shown to be teratogenic in the right circumstances. In the field of teratology, the failure of typical animal models to serve as predictive models with biomedical relevance to humans is widespread and acknowledged by key investigators. For further review of teratogens, animal testing and nonanimal-based testing methods see (Bailey, Knight, and Balcombe 2005). The discovery of the failure of these models to live up to the predictive promises made on their behalf by lobbyists is in fact a reflection of good, objective science by animal modelers themselves.

We will now examine some specific teratogens in more detail to bring out the nature of the problems encountered in the context of predictive modeling.

(i) retinol (vitamin A)

Retinoic acid is an important signaling molecule crucial to normal development. Excess or a deficiency can cause birth defects. Indeed, retinoids (derivatives of vitamin A) have been demonstrated to be potent human teratogens. Thus, Ross et al. have observed:

> Maternal ingestion of 13-cis-RA (isotretinoin or Accutane), prescribed for the treatment of severe cystic acne, has resulted in spontaneous abortions and the manifestation of severe malformations in the offspring. Abnormalities such as microtia/anotia, micrognathia, cleft palate. Conotruncal heart defects and aortic arch abnormalities, thymic defects, retinal or optic nerve abnormalities, and CNS malformations (e.g., hydrocephalus), were observed in the fetuses of women ingesting therapeutic doses of 13-cis-RA (0.5-1.5 mg/kg). (Ross et al. 2000)

This case is interesting since during embryogenesis there are causal connections between retinoic acid (RA), retinoic acid receptors (RAR) and the activity of specific homeobox genes via retinoic acid response elements (RARE). Thus, Ross et al. comment:

> The availability of retinoid response elements in *Hoxa1*, *Hoxb1*, and *Hoxd4* genes suggests that retinoids act as an early developmental signal, possibly conditioning the posterior-to-anterior gradient in the gastrula and providing positional specification of the A/P axis in the developing vertebrate embryo. . . The *Hoxa1* and *Hoxb1* genes appear to be modulated directly via the RARE, and the protein products of these genes may activate the proximal 5'-genes in these clusters, that is, *Hoxa2* and *Hoxb2* respectively. This scenario would allow for a sequential activation of *Hox* genes. (Ross et al. 2000)

There are, thus, good reasons to believe that the teratogenic effects of an excess of retinoic acid are mediated through disruption of normal *Hox* gene expression. When the dose of a retinoid is sufficient to act as a developmental toxin, the substance evidently works its mischief by disrupting identifiable genetic pathways that are themselves components of more complex circuits and networks.

Though it has proven possible to find animal models of the teratogenic effects of an excess of retinoic acid, significant species differences exist. Thus, as noted by Ross et al., not only are some retinoids more potent than others:

> . . . the potency of the retinoid can vary with different animal models. Humans, for example, are more sensitive to 13-*cis*-RA than monkeys and rabbits, whereas mice and rats are relatively insensitive to this retinoid. Species differences in response to the teratogenic potency of 13-*cis*-RA can be explained by variation in pharmacokinetics [i.e., ADME]. In many of the animal studies described above, the dosages tested were generally far in excess of likely human exposures. However, the spectrum of malformations observed in animal models are similar to those observed with human exposure to pharmacological doses of 13-*cis*-RA and, although seen less frequently, retinol. (Ross et al. 2000)

(ii) DES

Diethylstilbestrol (DES) is a synthetic estrogen that was given to women to prevent miscarriage starting in the 1940s. Ironically, the drug failed to do that which it was designed to do. This case is sometimes viewed as one of the triumphs of predictive animal modeling in the fields of teratology and oncology. For example, Vineis and Melnick have recently observed:

> We believe it to be irresponsible to ignore health effects data derived from animal studies. Public health protective strategies are needed to avoid repeating mistakes of the past similar to that of the DES tragedy *in which adverse health effects in animals were ignored and human use of this drug not banned until 1971,* after the discovery of high rates of rare, clear-cell adenocarcinomas of the vagina and cervix in DES-exposed daughters. (Emphasis added.) (Vineis and Melnick 2008)

Things are seldom as they appear, and this is true in the present case. Despite human clinical data available by 1953 that the drug was dangerous, it was indeed administered to humans until the early 1970s. The ultimate outcome was that DES increased the risk of vaginal and cervical cancer in the patients' daughters. Even the granddaughters of patients are affected (Langmuir 1971; Herbst, Ulfelder, and Poskanzer 1971; Blume 1992; Dieckmann et al. 1953).

So how valuable were the animal studies performed prior to the 1970s? Schardein and Macina have reported:

> In studies carried out in laboratory animals prior to the discovery in the early 1970s of the developmental problems in humans, researchers reported neither developmental toxicity nor teratogenicity in the three species tested [mice, rats, hamsters] . . . As stated above, developmental studies conducted in animals prior to the 1970s failed to demonstrate developmental toxicity, teratogenicity, or frank carcinogenicity.(Schardein and Macina 2007)

Could animal tests, then, have predicted the affects of DES?

Walker 1989 reports the mixed results forthcoming from animal testing and also draws out attention to the issue of retrospective simulation versus prediction proper:

> Animals of several species exposed perinatally to diethylstilboestrol (DES) have been evaluated for anomalies and tumours. In male offspring, anomalies of the testis and epididymis have been reported, but evidence for tumours has been very limited. *Many anomalies and tumours have been recorded in female offspring, and some of these duplicate the anomalies and tumours reported in DES-exposed women,* whereas others either have not yet been discovered or else do not occur in the human species. A variety of abnormal physiological responses have been identified in animals exposed perinatally to DES . . .

Most research on the effects of perinatal exposure to diethylstilboestrol (DES) in animals was performed in response to the report that prenatal exposure to DES was associated with adenocarcinoma of the vagina in young women. The experimental work has focused on female offspring; however, some investigators have reported abnormalities and tumours in male offspring. Testicular and epididymal abnormalities have been recorded in mice, rats, and hamsters. Adenocarcinoma of the rete testis in mice, squamous-cell carcinomas of the prostate region in neonatally castrated rats and granulomas of the epididymis and testis in hamsters have also been described . . .

The earliest papers on the effects of perinatal exposure to DES, by Greene et al. (1939) and by Dunn & Green (1963), were exploratory and reported morphological abnormalities. *The next series of papers focused on morphological duplication of the conditions that had been reported to occur in women exposed prenatally to DES* . . .

The effects of perinatal exposure to DES on hormone and receptor levels vary according to species, organ, stage of oestrus cycle, whether the animal was ovariectomized postnatally and whether exogenous hormones were administered. (Emphasis added.) (Walker 1989)

In fact, animal studies led to the notion that DES could be used to prevent miscarriage in the first place. Smith et al.:

There was some reason to suppose that diethylstilbestrol, as well as being inexpensive and effective by mouth, might be a more useful therapeutic agent for this particular purpose than the naturally occurring estrogens. Experimentally, it has been found to be 100 times as active as estrone in stimulating increased secretion and release of gonadotropic hormones from the pituitary of the intact rat. It has also been shown by Pencharz and others to have an augmentative effect upon the ovarian response of hypophysectomized rats to chorionic gonadotrophin, synergizing with this hormone to give pronounced enlargement of follicles and corpus luteum formation.

As a consequence of the above considerations and findings, we were led to investigate the therapeutic possibilities of diethylstilbestrol as a preventive measure against the premature decadence of placental secretion which appears to characterize late pregnancy toxemia and its associated accidents.(Smith, Smith, and Hurwitz 1946)

In this case, early animal studies suggested effects—maybe even beneficial effects—of DES exposure. But as so often happens in this field, the effects discovered were not the ones that would ultimately count from the standpoint of human epidemiological studies.

Karnofsky's Law states any substance can be teratogenic if given to the right species, at the right phase in development, in the right dose. That means all medications can cause birth defects in some species (Scialli 1991). An immense amount of experimentation has proven Karnofsky right. The unsolved problem here, as elsewhere, is that of which species and at what dose actually predicts human risk in advance of the emergence of human data. As we have seen here, and as we shall shortly see below, generating *some* developmental abnormality with a substance is fairly straightforward. Generating the one seen in humans is a different matter.

(iii) Thalidomide

Why are unborn babies so vulnerable to toxicity? Until the drug thalidomide manifested its horrific effects in humans, the fetus was considered by many to be immune to adverse effects of drugs taken by the expectant mother (Dally 1998). We now recognize that, by their very nature, drugs alter living systems. A good drug will treat the source of a disease without disturbing overall well-being. However, no living system is as precarious as that growing *in utero*. The cells are dividing rapidly, and the body is growing faster than it ever will again. With all this activity the cells are more prone to the adverse effects of toxins than they will be after the child is born. Drugs that are tolerable in adults—such as thalidomide—can result in a baby being born with malformations. The cells are at maximum vulnerability at the fetal stage.

Let us then take an in-depth look at one drug and the animal tests that *could* have been performed, and evaluate what we *would* have learned from them. There are many examples of animal models giving results at extreme variance from humans and even from each other—thalidomide being but one—but thalidomide occupies a special place in history so we will use it. Thalidomide was a sedative prescribed to pregnant women in the late 1950 and early 1960s. The children of some of these women were born without limbs, a condition known as phocomelia. There is much uncertainty concerning the mechanism of action of this teratogen, but species differences with respect to teratogenic effect are striking.

Notwithstanding this, Gad could recently write in *Animal Models in Toxicology*:

The use of thalidomide, a sedative-hypnotic agent, led to some 10,000 deformed children being born in Europe. This in turn led directly to the 1962 revision of the Food, Drug and Cosmetic Act, requiring more stringent testing. *Current testing procedures (or even those at the time in the United States, where the drug was never approved for human use) would have identified the hazard and prevented this tragedy.* (Emphasis added.) (Gad 2007)

Others in the scientific community also claim the thalidomide tragedy could have been prevented with appropriate animal testing (which was not, as a matter of routine, performed in advance of human epidemiological data suggesting a teratological tragedy had occurred in the exposed human population). The case of thalidomide assumes, therefore, the characteristics of a counterfactual inquiry (of the kind, what *would have* happened if, contrary to fact, *we had* done X, Y or Z?) In this connection, Walker et al. observe:

> *A failure to investigate the potential teratogenic effects of thalidomide in a primate model prior to exposing pregnant women resulted in tragic consequences in the 1950s and early 1960s.* Unfortunately, the teratogenicity of thalidomide could not be demonstrated in rodents prior to human exposure, but was subsequently shown to cause fetal abnormalities in primates . . . the humane use of nonhuman primates as a physiological, pharmacological, and toxicological research model is critical for safety assessment of new drugs and biotechnology products, increasing the need for prudent use of primate inventories. (Emphasis added.) (Walker, Nelson, and Bernard 2007)

Could the thalidomide tragedy have been prevented by more animal testing?

Consider the evidence. Schardein who has studied this tragedy has observed:

> In approximately 10 strains of rats, 15 strains of mice, 11 breeds of rabbits, 2 breeds of dogs, 3 strains of hamsters, 8 species of primates and in other such varied species as cats, armadillos, guinea pigs, swine and ferrets in which thalidomide has been tested, teratogenic effects have been induced only occasionally. (Schardein 1976)

We remind the reader that these results, and those below were from tests performed *after* thalidomide's affects had, as a matter of fact, been observed in humans. We have just seen Walker et al. sing the praises of primate models of thalidomide teratogenicity. But even here there is cause for caution. For as Schardein observes:

> It is the actual results of teratogenicity testing in primates which have been most disappointing in consideration of these animals' possible use as a predictive model. While some nine subhuman primates (all but the bushbaby) have demonstrated the characteristic limb defects observed in humans when administered thalidomide, the results with 83 other agents with which primates have been tested are less than perfect. Of the 15 listed putative human teratogens tested in nonhuman primates, only eight were also teratogenic in one or more of the various species. (Schardein 1985)

Manson and Wise summarized the thalidomide testing as follows:

> An unexpected finding was that the mouse and rat were resistant, the rabbit and hamster variably responsive, and certain strains of primates were sensitive to thalidomide developmental toxicity. Different strains of the same species of animals were also found to have highly variable sensitivity to thalidomide. Factors such as differences in absorption, distribution, biotransformation, and placental transfer have been ruled out as causes of the variability in species and strain sensitivity. (Manson and Wise 1993)

Could the use of animal models have predicted thalidomide's adverse affects? Even if all the animals mentioned above were studied the answer is no. Different species showed a wide variety of responses to thalidomide. Once again, if you bet on enough horses you will probably find a winner, or if you cherry pick the data after the fact you will find a winner. In the present case of thalidomide, human effects were already known, so cherry picking is easy. The animal models for thalidomide discussed above were aimed at retrospectively simulating *known* human effects. Even then, not many animal models succeeded. If the human effects were *unknown*, what would the case have looked like from the standpoint of prediction? In this case, to pursue the horse racing analogy, we would have numerous horses to bet on without any idea which one would win. Certainly one will win (which is not a given when testing on animals in hopes of reproducing or guessing human response), but which one? We cannot know that until after the fact so how do we judge *prospectively* which horse to wager on or which animal model to choose? Which model species were relevant to the human case in advance of the gathering of human data? This is by no means a trivial question, as evolutionary closeness does not increase the predictive value of the model.

Even if we retrospectively picked all the animals that reacted to thalidomide as humans did, we still could not say these animals predicted human response, as their history of agreeing with human response to other drugs varied considerably (for a given model that gets a human prediction correct, there is typically little or no track record of predictive success in *that* model with *other* substances to give even a *prima facie* reason to select the given model on the basis of antecedently known predictive success). Prediction vis-à-vis drug testing and disease research implies a track record. Single correct guesses are not predictions. Nonhuman primates are a good example of this. They more or less reacted to thalidomide as humans did (so we will give them the benefit of the doubt as corresponding to humans in this case). However, when tested with other drugs they *predicted* human response about as well as a coin toss. (We defer to a later chapter in this book a fuller discussion of the predictive powers of nonhuman primate models.)

Add to all this the fact that all the animals whose offspring exhibited phocomelia consequent to the administration of thalidomide did so only after being given doses 25-150 times the human dose (Runner 1967; Keller and Smith 1982; Staples and Holtkamp 1963) and it does not appear that any animal, group of animals, or the animal model *per se* could have been used to predict thalidomide's teratogenicity in humans. (Ironically, it was the thalidomide tragedy that ushered in many of the regulatory requirements for using animals, notwithstanding the facts of the case.)

Again, one must not confuse *sensitivity* with *prediction*. As of 1985, of 37/38 compounds known to be teratogenic in humans were also teratogenic in some animal species. However, of 168 compounds known not to be teratogenic in humans, 41% did cause birth defects in some animal species (Frankos 1985). Once again we see that animals (when taken as a whole) are very sensitive but not very specific. Add to this the fact that today some drugs that are known to be safe for most pregnant women have been associated with birth defects in others. Gene-based medicine is probably the only way we will ever be able to confidently prescribe drugs to pregnant women.

Karnofsky's law is relevant here. Any drug is teratogenic if given to the right animal in the right dose at the right time. Given thalidomide's profile today, physicians would advise pregnant women not to take the drug, which is what physicians advise every pregnant woman about almost every non-life-saving drug anyway, regardless of the results of animal tests.

As Teeling-Smith observed three decades ago:

There is at present no hard evidence to show the value of more extensive and more prolonged laboratory testing as a method of reducing eventual risk in human patients. In other words the predictive value of studies carried out in animals is uncertain. The
statutory bodies such as the Committee on Safety of Medicines that
require these tests do so largely as an act of faith rather than on
hard scientific grounds. With thalidomide, for example, it is only
possible to produce specific deformities in a very small number of
species of animal. In this particular case, therefore, it is unlikely that
specific tests in pregnant animals would have given the necessary
warning: the right species would probably never have been used.
(Teeling-Smith 1980)

There is a fallacy that humans seem to be particularly prone to. In the
present case it manifests itself in the claim that because animal testing was
mandated after the thalidomide disaster, and there has never again been
another incident comparable to thalidomide, it must be the case that animal testing practices instituted post-thalidomide have prevented further
tragedies. The fallacy is an example of the *post hoc ergo propter hoc* fallacy
(after this hence because of this). (A notorious example of this fallacy—
Cheney's fallacy—states that because the United States has not been attacked during the Bush Presidency after 9/11, it must be the direct result
of the security policies adopted by that administration. It may be that
such policies prevented attacks but simply a temporal relationship is not
proof and should not be heralded as such.)

It is true that society has not experienced a tragedy on par with the thalidomide tragedy. But this is not because of animal testing since, as we have
seen, animals testing vis-à-vis teratology is not predictive for humans. Occurring simultaneously with the post-thalidomide requirements for animal
testing was the education of society to the potential for all drugs to cause
harm to the fetus. Frankos gets close to the heart of the matter:

The public and the scientific community were generally unaware of the
potential of chemical agents to cause human malformations and thus
remained complacent. The lesson of thalidomide lies in how it shook
the public and the scientific community from its complacent sleep and
made it painfully aware of the "potential" harm chemicals can cause to
our unborn children. For toxicologists, the myth that teratology only
occurs after maternal toxicity was shattered. (Frankos 1985)

Drugs have not been prescribed to pregnant women in the same cavalier
way since the early 1960s.

CHAPTER 13

♦

EVIDENCE AND OPINION

*I cannot give any scientist of any age better advice than this:
the intensity of the conviction that a hypothesis is true has no bearing on
whether it is true or not. The importance of the strength of our conviction
is only to provide a proportionately strong incentive to find out if the
hypothesis will stand up to critical evaluation.*

—Peter Medawar

In biomedicine, we do not have the mathematician's luxury of modeling humans and rodents by beginning, "let humans and rodents be spheres." If only it were that simple. Instead, as we have seen in earlier chapters on evolution and genetics, what we do have are a lot of theoretical grounds for questioning the predictive utility of animal models. But of course, such theoretical reasoning may be dismissed as being just that. The real question is one of evidence. Our critics have drawn our attention to the famous Olson study to show that, appearances to the contrary, animal models (at least in toxicology) are genuinely predictive of human outcomes.

Predictive vs Useful

At this point in the debate, some will state that animal models can be useful in science and scientific research and then conflate the word *predict* with the word *useful*. This is disingenuous for many reasons. First, *useful* is too ambiguous to mean anything. Useful to whom? Useful how? Almost anything can be useful in some sense of the word. If someone gets paid to engage in fortune telling, then fortune telling may be very useful to that person. Whether it can be used to predict the future is an entirely different question. We do not deny animal models can be quite *useful* in certain circumstances, but this has nothing to do with whether they are predictive. Second, this is an example of bait and switch; sell animal models as predictive for humans then justify their use, since they are not predictive, because they are *useful*. Freeman and St Johnston illustrate the strategy as follows:

> Many scientists who work on model organisms, including both of us, have been known to contrive a connection to human disease to boost a grant or paper. It's fair: after all, the parallels are genuine, but the connection is often rather indirect. More examples will be discussed in later chapters. (Freeman and St Johnston 2008)

Finally, we again acknowledge that studying animals can lead to new knowledge, for example, in the context of basic research. This point is not in dispute. Accusing us of denying that particular, legitimate value of animal models merely because we criticize their use as predictive models is to erect a straw man. Once again we are drawn back to evidential matters.

The Famous Olson Study

The Olson study (Olson et al. 2000) (Figure 13.1) purports (and has certainly been cited in this regard) to provide evidence of the vast predictive utility of animal models in assessing human toxicity. In response to an article on the prediction problem confronting animal modelers by Shanks et al. (Shanks et al. 2007) Conn and Parker quoted the Olson study stating:

> The authors have simply overlooked the classic study (Olson, Harry, et al., 2000. "Concordance of the Toxicity of Pharmaceuticals in Humans and in Animals." that summarizes the results from 12 international pharmaceutical companies on the predictivity of animal tests in human toxicity. While the study is not perfect, the overall conclusion from 150 compounds and 221 human toxicity events was that animal testing *has* significant predictive power to detect most—but not all—areas of human toxicity. (Conn and Parker 2007) (Emphasis theirs.)

We encourage the reader to examine the Olson Study in its entirety (see Appendix 4). Here we include some important representative paragraphs from the Olson study and our commentary will follow.

Regulatory Toxicology and Pharmacology **32**, 56–67 (2000)
doi:10.1006/rtph.2000.1399, available online at http://www.idealibrary.com on **IDE𝐀L**®

Concordance of the Toxicity of Pharmaceuticals in Humans and in Animals

Harry Olson,[1] Graham Betton,[2] Denise Robinson,[3] Karluss Thomas,[3] Alastair Monro,[1] Gerald Kolaja,[4] Patrick Lilly,[5] James Sanders,[6] Glenn Sipes,[7] William Bracken,[8] Michael Dorato,[9] Koen Van Deun,[10] Peter Smith,[11] Bruce Berger,[12] and Allen Heller[13]

[1]*Pfizer Inc., Groton, Connecticut;* [2]*AstraZeneca Pharmaceuticals, Macclesfield, England;* [3]*ILSI-HESI, Washington, DC, 20036;* [4]*Pharmacia & UpJohn, Kalamazoo, Michigan;* [5]*Boehringer Ingelheim Pharmaceuticals, Ridgefield, Connecticut;* [6]*Rhone-Poulenc Rorer, Collegeville, Pennsylvania;* [7]*University of Arizona, Tucson, Arizona;* [8]*Abbott Laboratories, Abbott Park, Illinois;* [9]*Eli Lilly and Co., Greenfield, Indiana;* [10]*Janssen Research Foundation, Beerse, Belgium;* [11]*Monsanto-Searle Laboratories, Skokie, Illinois;* [12]*Sanofi-Synthelabo, Inc., Malvern, Pennsylvania;* and [13]*Bayer Corporation, West Haven, Connecticut*

Received January 22, 2000

This report summarizes the results of a multinational pharmaceutical company survey and the outcome of an International Life Sciences Institute (ILSI) Workshop (April 1999), which served to better understand concordance of the toxicity of pharmaceuticals observed in humans with that observed in experimental animals. The Workshop included representatives from academia, the multinational pharmaceutical industry, and international regulatory scientists. The main aim of this project was to examine the strengths and weaknesses of animal studies to predict human toxicity (HT). The database was developed from a survey which covered only those compounds where HTs were identified during clinical development of new pharmaceuticals, determining whether animal toxicity studies identified concordant target organ toxicities in humans. Data collected included codified compounds, therapeutic category, the HT organ system affected, and the species and duration of studies in which the corresponding HT was either first identified or not observed. This survey includes input from 12 pharmaceutical companies with data compiled from 150 compounds with 221 HT events reported. Multiple HTs were reported in 47 cases. The results showed the true positive HT concordance rate of 71% for rodent and nonrodent species, with nonrodents alone being predictive for 63% of HTs and rodents alone for 43%. The highest incidence of overall concordance was seen in hematological, gastrointestinal, and cardiovascular HTs, and the least was seen in cutaneous HT. Where animal models, in one or more species, identified concordant HT, 94% were first observed in studies of 1 month or less in duration. These survey results support the value of *in vivo* toxicology studies to predict for many significant HTs associated with pharmaceuticals and have helped to identify HT categories that may benefit from improved methods. © 2000 Academic Press

INTRODUCTION

A vitally important theme in toxicology is the search for and the assessment of *in vitro* and *in vivo* models that are predictive for adverse effects in humans exposed to chemicals. The conduct of toxicology studies in laboratory animals is driven by experience, historical precedence, and governmental requirements, and the results of these studies usually, and reasonably, lead to restrictions on the use, or method of use, of the chemicals concerned. Such a process must be based on the assumption that the current choice of animal models and the design of the studies are truly predictive of human hazard. The reliability of this assumption has far-reaching repercussions in terms of the potential for inappropriate use of animals and the unnecessary deprivation of, or restrictions in the use of, valuable chemicals including pharmaceuticals. Identification of any weaknesses in the assumption could lead to revisions of existing regulations and stimulate the search for better methods for the safety evaluation of chemicals in the future.

There have been relatively few attempts to methodically assess the correlation between the toxicity caused by chemicals in animals and in humans. This is not surprising, given that the toxicity of many chemicals observed in humans is after accidental exposure, the quantitative details of which in terms of duration and intensity are often not known. Chemicals, which are components of the diet, either macro- or micro-, are more susceptible to evaluation of their toxicity in animals and in humans, provided that the means to carry out epidemiological studies are available. However, a rich source of relevant information is pharmaceutical chemicals. For these, the human exposure is controlled and measured accurately. In addition, clinical studies of drugs employ systematic clinical examinations and

Figure 13.1. Olson study

We apologize for the length of the quoted material, but due to the importance many place on the Olson study, we believe a thorough examination is justified.

> This report summarizes the results of a multinational pharmaceutical company survey and the outcome of an International Life Sciences Institute (ILSI) Workshop (April 1999), which served to better understand *concordance of the toxicity* of pharmaceuticals observed in humans with that observed in experimental animals. The Workshop included representatives from academia, the multinational pharmaceutical industry, and international regulatory scientists. *The main aim of this project was to examine the strengths and weaknesses of animal studies to predict human toxicity (HT). The database was developed from a survey which covered only those compounds where HTs were identified during clinical development of new pharmaceuticals*, determining whether animal toxicity studies identified concordant target organ toxicities in humans. . . .
>
> The results showed the *true positive HT concordance rate of 71%* for rodent and nonrodent species, with nonrodents alone being predictive for 63% of HTs and rodents alone for 43%. The highest incidence of overall concordance was seen in hematological, gastrointestinal, and cardiovascular HTs, and the least was seen in cutaneous HT. *Where animal models, in one or more species, identified concordant HT, 94% were first observed in studies of 1 month or less in duration.* These survey results support the value of in vivo toxicology studies to predict for many significant HTs associated with pharmaceuticals and have helped to identify HT categories that may benefit from improved methods. . . .
>
> *The primary objective was to examine how well toxicities seen in preclinical animal studies would predict actual human toxicities for a number of specific target organs using a database of existing information.* . . .
>
> *Although a considerable effort was made to collect data that would enable a direct comparison of animal and human toxicity, it was recognized from the outset that the data could not answer completely the question of how well animal studies predict overall the responses of humans. To achieve this would require information on all four boxes in Fig. 1* [all boxes in our Figure 11.1], *and this was not practicable at this stage.* The magnitude of the data collection effort that this would require was considered impractical at this stage. *The present analysis is a first step, in which data have been collected pertaining only to the left column of Fig. 1: true positives and false negatives.* By definition, therefore the database only contains compounds studied in humans (and not on those that never reached humans because

they were considered too toxic in animals or were withdrawn for reasons unrelated to toxicity). Despite this limitation, it was deemed useful to proceed in the expectation that any conclusions that emerged would address some of the key questions and focus attention on some of the strengths and weaknesses of animal studies. . . .

A working party of clinicians from participating companies developed criteria for "significant" HTs to be included in the analysis. For inclusion a HT (a) had to be responsible for termination of development, (b) had to have resulted in a limitation of the dosage, (c) had to have required drug level monitoring and perhaps dose adjustment, or (d) had to have restricted the target patient population. The HT threshold of severity could be modulated by the compound's therapeutic class (e.g., anticancer vs anti-inflammatory drugs). In this way, *the myriad of lesser "side effects" that always accompany new drug development but are not sufficient to restrict development were excluded.* The judgments of the contributing industrial clinicians were final as to the validity of including a compound. The clinical trial phase when the HT was first detected and whether HT was considered to be pharmacology-related was recorded. HTs were categorized by organ system and detailed symptoms according to standard nomenclature (COSTART, National Technical Information Service, 1999) . . .

Concordance by one or more species: Overall and by HT. *Overall, the true positive concordance rate (sensitivity) was 70% for one or more preclinical animal model species (either in safety pharmacology or in safety toxicology) showing target organ toxicity in the same organ system as the HT* (Fig. 3) [see Figure 13.2 labelled Olson Figure 3]. . . .

This study did not attempt to assess the predictability of preclinical experimental data to humans. What it evaluated was the concordance between adverse findings in clinical data with data which had been generated in experimental animals (preclinical toxicology). (Emphasis added.) (Olson et al. 2000)

The Olson Study, as noted above, has been employed by researchers to justify claims about the predictive utility of animal models. However, we think there is much less here than meets the eye. Here's why:

1. The study was primarily conducted and published by the pharmaceutical industry. This does not, in and of itself, invalidate the study. However, one should never lose sight of the fact that the study was put together by parties with a vested interest in the outcome. If this was the only concern, perhaps it could be ignored, however, as we will now show, there are some rather more serious flaws.

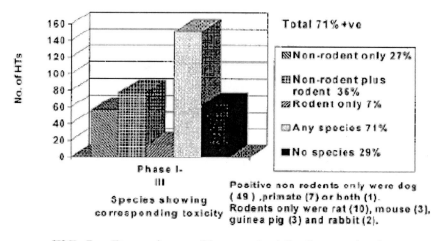

FIG. 3.   Concordance of human toxicity from animals.

Figure 13.2. Olson Figure 3.

2. The study says at the outset that it is aimed at measuring the predictive reliability of animal models. Later the authors concede that their methods are not, as a matter of fact, up to this task. This makes us wonder how many of those who cite the study have actually read it in its entirety.

3. The authors of the study invented new statistical terminology to describe the results. The crucial term here, unqualified at the beginning of the article, is "true positive concordance rate" which sounds similar to "true positive predictive value" (which is what should have been measured, but was not). A Google search on "true positive concordance rate" yielded twelve results (counting repeats), all of which referred to the Olson Study (see Figure 13.3). At least seven of the twelve Google hits qualified the term "true positive concordance rate" with the term "sensitivity"—a well-known statistical concept. In effect, these two terms are synonyms. Presumably, the authors of the study must have known that "sensitivity" does not measure "true positive predictive value," for later in the middle of the article they qualify the term "true positive concordance" with the term "sensitivity." In addition to "sensitivity" you would need

information on "specificity" and so on, to nail down the crucial concept of "true positive predictive value," and *this* the authors did not do! If all the Olson Study measured was sensitivity, its conclusions are largely irrelevant to the great prediction debate. Given the weight placed on the Olson study by friends of predictive modeling, we are left wondering how many of those citing the study got beyond the first page, and of those who did, how many understood elementary statistics.

4. Any animal giving the same response as a human was counted as a positive result. So if six species were tested and one of the six mimicked humans that was counted as a positive. The Olson Study was concerned primarily not with prediction, but with *retrospective simulation of antecedently known human results.*

5. Only drugs in clinical trials were studied. Many drugs tested do not actually get that far because they fail in animal studies.

6. ". . . The myriad of lesser 'side effects' that always accompany new drug development but are not sufficient to restrict development were excluded." A *lesser* side effect, so physicians have been known to facetiously say, is one that affects someone else. While hepatotoxicity is a major side effect, lesser side effects (which actually matter to patients) concern profound nausea, tinnitus, pleuritis, headaches and so forth. We are also left wondering whether there was any independent scientific validity for the criteria used to divide side effects into major side effects and lesser side effects.

7. Even if all the data is good—and it may well be—sensitivity (i.e. true positive concordance rate) of 70% does not settle the prediction question. Sensitivity is not synonymous with prediction and even if a 70% true positive prediction value is assumed, when predicting human response 70% is inadequate. In carcinogenicity studies, the sensitivity using rodents may well be 100%; the specificity, however, is another story. That is the reason rodents cannot be said to predict human outcomes in that particular biomedical context.

The Olson Study is certainly interesting, but even in its own terms it does not support the notion that animal models are predictive for humans. We think it should be cited with caution.

A citation search (also performed with Google on 7/23/08) led us to 114 citations for the Olson paper. There is a wide range of opinion about the actual significance of the Olson study by toxicologists themselves. The following are examples. Claude:

7/18/08 1:41 PM

Web   Images   Maps   News   Shopping   Gmail   more ▾   Sign in

| "true positive concordance rate" | Search | Advanced Search
Preferences |

Web   Books   Results 1 - 12 of 12 for "true positive concordance rate". (0.30 seconds)

Concordance of the Toxicity of Pharmaceuticals in Humans and in ...
Overall, the **true positive concordance rate** (sen-. sitivity) was 70% for one or more
preclinical animal. TABLE 2. Distribution of Compounds by Therapeutic ...
linkinghub.elsevier.com/retrieve/pii/S0273230000913990 - Similar pages

> **European Journal of Pharmaceutical Sciences : Early microdose drug ...**
> There was a **true positive concordance rate** of 71% for comparable target organs in
> rodent plus non-rodent studies and identified human toxicities for 150 ...
> linkinghub.elsevier.com/retrieve/pii/S092809870300040X - Similar pages

**Histopathology of Preclinical Toxicity Studies: Interpretation and ... - Google Books
Result**
by Peter Greaves - 2007 - Medical - 953 pages
Overall, the **true positive concordance rate** (sensitivity) is of the order of 70% with 30% of
human toxicities not predicted by safety pharmacology or ...
books.google.com/books?isbn=0444527710...

First dose of potential new medicines to humans: how animals help ...
Overall, the **true positive concordance rate** (sensitivity) of the data derived from
conventional studies is of the order of 70%, with 30% of human toxicities ...
www.nature.com/nrd/journal/v3/n3/full/nrd1329.html - Similar pages

> **First dose of potential new medicines to humans: how animals help ...**
> Overall, the **true positive concordance rate** (sensitivity) of the data derived from
> conventional studies is of the order of 70%, with 30% of human toxicities ...
> www.nature.com/nrd/journal/v3/n3/full/
> nrd1329.html;jsessionid=AC7F28279806834829F7F100EEEE4E6E - Similar pages

7100048a 440..447
showed the **true positive concordance rate** of 70%. for rodent and non-rodent species,
with non-rodents. alone being predictive for 63% of HTs and rodents ...
www.ingentaconnect.com/content/sage/ het/2000/00000019/00000008/7100048a?
crawler=true - Similar pages

> **7100048a 440..447**
> showed the **true positive concordance rate** of 70%. for rodent and non-rodent species,
> with non-rodents. alone being predictive for 63% of HTs and rodents ...

http://www.google.com/search?q=%22true+positive+concordance+rate%22&num=30&hl=en&safe=off&client=safari&rls=en&filter=0   Page 1 of 2

Figure 13.3. Results of Google search for "true positive
concordance rate."

www.ingentaconnect.com/content/ am/het/2000/00000019/00000008/7100048a -
Similar pages

First dose of potential new medicines to humans: how animals help ...
Overall, the **true positive concordance rate** (sensitivity) of the data derived from
conventional studies is of the order of 70%, with 30% of human toxicities ...
www.naturereprints.com/ nrd/journal/v3/n3/full/nrd1329.html - Similar pages

"Pre-clinical Safety Evaluation". In: Pharmacovigilance
Overall, the **true positive concordance rate** was. 70% for the pre-clinical animal species
to show. target organ toxicity in the same organ system as ...
doi.wiley.com/10.1002/0470853093.ch5 - Similar pages

First dose of potential new medicines to humans: how animals help ...
Overall, the **true positive concordance rate** (sensitivity) of the data derived from
conventional studies is of the order of 70%, with 30% of human toxicities ...
www.emboj.org/nrd/journal/v3/n3/full/nrd1329.html - Similar pages

First dose of potential new medicines to humans: how animals help ...
Overall, the **true positive concordance rate** (sensitivity) of the data derived from
conventional studies is of the order of 70%, with 30% of human toxicities ...
intl.emboj.org/nrd/journal/v3/n3/full/nrd1329.html - Similar pages

[PDF] FIRST DOSE OF POTENTIAL NEW MEDICINES TO HUMANS: HOW
ANIMALS HELP
File Format: PDF/Adobe Acrobat - View as HTML
true positive concordance rate (sensitivity) of the data. derived from conventional studies
is of the order of. 70%, with 30% of human toxicities not ...
npg.nature.com/nrd/journal/v3/n3/pdf/nrd1329.pdf - Similar pages

---

"true positive concordance rate"   (Search)

Search within results | Language Tools | Search Tips | Dissatisfied? Help us improve |
Try Google Experimental

Google Home - Advertising Programs - Business Solutions - Privacy - About Google

Figure 13.3. (cont.) Results of Google search for "true positive
concordance rate."

> Taken together, these elements [from the Olson study] support the value of toxicology studies to predict many human toxic events associated with pharmaceuticals. (Claude 2007)

Referring to the Olson study, Khor et al. note:

> The fact that only 71% of all human toxicities can be accurately predicted by using animal models indicates the existence of species-specific differences upon exposure to drugs. (Khor, Ibrahim, and Kong 2006)

In a similar vein Kaplowitz comments:

> This is not to say that animal toxicology studies are worthless— certainly, many chemicals that would have been dangerous to patients have been identified and development of the compound suspended (Kaplowitz 2005)

Again, with reference to the Olson study Foster observes:

> Evaluation of the relative predictivity of the rodent versus nonrodent studies for the prediction of human hepatotoxicity suggests that nonrodent species had a concordance of 63% between laboratory results and those obtained in the clinic, while rodent studies alone were only able to predict approximately 43% of subsequent hepatotoxic events in the clinic (Olsen [sic] et al., 2000). However, together studies conducted in both nonrodent and rodent species were able to predict approximately 71% of subsequent hepatotoxic events in man. (Foster 2005).

Referencing the Olson study, Sistare and DeGeorge state:

> A retrospective study of test animal experience, involving 150 compounds with toxicities documented in clinical trials, revealed that rodent toxicology studies alone "predicted" 43% of human toxicities, whereas non-rodents predicted 63%, and together both species captured only 71%. (Sistare and DeGeorge 2007)

Clearly the Olson study is cited as evidence that animal models are predictive (to varying degrees, it is true). The problem is that the study, in the nature of the case, does not even come close to dealing with the prediction issue, it merely conflates the issue of prediction with that of statistical sensitivity.

Other Voices

We are not the only ones concerned about the predictive power of animal models. The scientific community itself is not marching in lock-step when it comes to the predictive utility of animal models. We will take a moment to examine what some of these scientists actually say about the power of animal models to predict human responses. The following quotes from scientists, of course, prove nothing in the sense of mathematical proof, they nevertheless provide a window into the thinking of people well-versed in the field and as such a reasonable person, even one who disagrees, should give them consideration. They should give pause to those who think that the prediction issue is one where there is no reasonable controversy.

In the 2006 report *The use of nonhuman primates in research* chaired by Sir David Weatherall, the authors state: "It's undoubtedly the case that all animal models are limited in their predictability for humans (Weatherall 2006)." In what follows we provide some examples drawn from a wide variety of biomedical fields to reveal how investigators and commentators express concerns regarding the limited informational value of predictive animal models.

Immunology is one of the fields where mice have played a central role in research. Davis has recently observed that technology and materials available to conduct research sometimes shape the research actually done:

> How did we arrive at this state of affairs [where the clinical developments of immunology have lagged so far behind the basic science breakthroughs]? A good case can be made that the mouse has been so successful at uncovering basic immunologic mechanisms that now many immunologists rely on it to answer every question. Where it was once common to use a variety of species, there is now such an abundance of reagents available in mouse immunology that one has to have an overpowering reason to work in any other species, including humans. It also has raised the bar of evidence required for journals and grant reviews, as pointed out by Steinman and Mellman and by Hayday and Peakman. This has skewed the field so much that most clinically trained immunologists keep at least a few (and usually a lot more) mice in the "back room" so that they can have a steady flow of papers, grant funding, etc., and some have abandoned human work entirely as a lost cause. But this is just the price of progress, no? Well, except that mice are lousy models for clinical studies. This is readily apparent in autoimmunity and in cancer immunotherapy, where of dozens (if not hundreds) of pro-

tocols that work well in mice, very few have been successful in humans. Similarly, in neurological diseases, the mouse models have also been disappointing. (Davis 2008)

There are of course implications for differences between the immune systems of mice and humans. Brady (see Figure 13.4):

> The effort to develop drugs that interact with human immune system (whether by accident or design) has been dogged by mismatch between the data derived from animal models (mice in particular) and that found in man. Although the mouse provides the most common models for many aspects of the human immune system, the 65 million years of divergence has introduced significant differences between these species, which can and has impeded the reliable transition of pre-clinical mouse data to the clinic. The industry is littered with examples of delays, reiterations or even abandoned drug programmes arising from poor translation of animal responses to man. This article highlights some of the species differences and forwards the rationale to utilise high resolution human immune cell assays to improve the successful transition from pre-clinical project to proof-of-concept in clinical trial. (Brady 2008/9)

A similar theme can also be found in the study of neurodegenerative disorders. In this regard Schnabel comments:

> The results of drug tests in mice have never translated perfectly to tests in humans. But in recent years, and especially for neurodegenerative diseases, mouse model results have seemed nearly useless. In the past year, for example, three major Alzheimer's drug candidates, Alzhemed (3-amino-1-propanesulphonic acid), Flurizan (tarenflurbil) and bapineuzumab, all of which had seemed powerfully effective in mouse models, have performed weakly or not at all in clinical trials involving thousands of human Alzheimer's patients.
>
> In the case of ALS, close to a dozen different drugs have been reported to prolong lifespan in the SOD1 mouse, yet have subsequently failed to show benefit in ALS patients. In the most recent and spectacular of these failures, the antibiotic minocycline, which had seemed modestly effective in four separate ALS mouse studies since 2002, was found last year to have worsened symptoms in a clinical trial of more than 400 patients . . . (Schnabel 2008)

	MOUSE	HUMAN
Altered peripheral blood cell make up eg:		
Lymphocytes	~80%	~40%
Neutrophils	~20%	~60%
CD1 genes	One (CD1d)	Multiple (CD1a-e)
CD2-ligand interaction:		
T cell dependence	Low	High
Ligand	CD48	CD58 (LFA-3)
Affinity	Low	High
CD4 on macrophages	Absent	Present
EC present Ag to CD4⁺ T cells	No	Yes
CD5 and CD23 on B cells	Mutually exclusive expression	Co-expression
CD8 on DC	Present	Absent
CD28 expression on T cells	By 100% of CD4⁺ and CD8⁺ T cells	By 80% of CD4⁺ and 50% of CD8⁺ T cells
CD33 expression	Granulocytes	Monocytes
CD38 expression on B cells	Low on GC B cells, absent in plasma cells	High on GC B cells and plasma cells
CD40 on EC	Absent	Present
CD45 expressing cells	Purging extends graft survival	Purging does not extend graft survival
CD52 expression	Absent	Present
CD58 expression	Absent	Present
IL-10	Th2 cytokine	Th1 and Th2 cytokine
P-Selectin expression	Up-regulated by inflammatory mediators	Unresponsive to inflammatory mediators
TLR2 expression on PBL	Low (induced on many cells including T cells)	Constitutive (but not on T cells)
TLR3	Induced by LPS	Not induced by LPS
TLR10	Pseudogene	Highly expressed in lymphoid tissues
Hemotopoiesis in spleen	Continues into adulthood	Terminates prenatal
Hemotopoietic stem cells	c-kithigh	c-kitlow
Presence of Bronchus-associated Lymphoid Tissue (BALT)	Present	Absent in healthy tissue
Leukocyte defensins	Absent	Present on neutrophils
fMLP receptor affinity	Low	High
Fc RI	Absent	Present
Fc RIIA, C	Absent	Present
IL-13 effect on B cells	None	Induces switch to IgE
Thy1 expression	Thymocytes, peripheral T cells	Absent from all T cells, yet expressed by neurones
Caspase 10	Absent	Present
IFN-α promotes Th1 differentiation	No	Yes
Th expression of IL-10	Th2	Th1 and Th2
GlyCAM	Present	Absent
MHC II expression on T cells	Absent	Present
Kv1.3 K channel on T cells	Absent	Present
MUC1 on T cells	Absent	Present
Granulysin	Absent	Present
Chemokine receptor CXCR1	Absent	Present
Chemokines:		
CXCL7 CXCL8 CXCLII CCL13 CCL14 CCL15 CCL18 CCL23 CCL24/CCL26	All absent	All present
CCL6 CCL9 CCL12 CXCL15	All present	All absent
MRP-1/2, lungkine, MCP-5	Present	Absent
Passenger leukocytes	Account for graft immunogenicity	Do not account for graft immunogenicity

Ag = antigen, DC = dendritic cell, EC = endothelial cell, GC = germinal centre, LPS = lipopolysaccharide, N = neutrophil, PBL = peripheral blood leukocytes, Th = T helper cell, TLR = Toll-like receptor

Figure 13.4. Immune differences.

Reprinted with kind permission from *Drug Discovery World* Winter 2008/9 Volume 10 Issue 1

There is no doubt that some of the science involving the use of animals in biomedical research is truly spectacular. The point is made if it is granted that there are profound limitations to the use of animal models to predict human clinical outcomes—regardless of how useful animal modeling may be in non-predictive contexts!

Drug discovery and the study of disease are both time-consuming and expensive. The usual approaches to drug discovery involve extensive animal testing (if only to try and predict human safety). Do typical animal tests deliver the goods? In this book we have cited many sources that are frankly skeptical of what appear to be extravagant claims made on behalf of predictive modeling. (These matters are touched on in more detail in the appendices accompanying this volume. It is not our place to examine the motives of individual investigators when assessing why a practice continues in the face of evidence to the contrary. This would be a complex matter indeed. But some factors referred to in the literature are at least worthy of mention here, not least of which are the comments of Goldberg and Hartung in *Scientific American* January 2006:

> In reality, representatives of nine multinational companies revealed to Goldberg that all the firms use petri dish or nonmammalian tests, usually involving fish or worms, to decide if a chemical is safe enough to produce. Only then do they perform the life-span feeding studies [on animals]—to satisfy the company's lawyers and regulatory agencies. (Goldberg and Hartung 2006)

Not for the first time in this book, the issue of institutional inertia arising from extra-scientific legal and regulatory requirements rears its ugly head. Can anything be done to improve the predictive utility of animal models? This matter is explored in the next two chapters in connection with genetically modified animals and primate models.

CHAPTER 14

◆

IMPROVING PREDICTION: GENETICALLY MODIFIED ANIMALS

The first principle is that you must not fool yourself—
and you are the easiest person to fool.

—Richard Feynman

New technology often drives the process of scientific inquiry by opening up potential avenues for novel research into old problems of interest. In the field of biotechnology, this is especially true of the emergence in the last three decades of transgenic technologies in which it is hoped that animals can be genetically modified (*humanized*) to more closely resemble the vulnerable human target population. This turns out to be an area where promissory notes abound. Thus, commenting on breast cancer, Hutchinson and Muller note that:

> While no single genetically engineered mouse can offer a complete model of the wide assortment of human neoplasms found in human breast cancer, it is hoped that these multiple approaches will enable us to develop insights into the complex molecular events involved in tumorigenic progression of the breast . . . Fortunately, researchers now have many models available to them to study these steps in a controlled and rational manner. Furthermore these models provide the opportunity to study many various aspects of the pathogenesis of this disease, from hormonal effects to responses to chemotherapeutic drugs. It is hoped that through the combined use of these models, and the further development of more relevant models that a deeper understanding of this disease and the generation of new therapeutic agents will result. (Hutchinson and Muller 2000)

Garattini stated in 1990:

> Animal research should actually gain fresh impetus from the new technologies available in the area of genetic engineering.

> The development of transgenic animals is a powerful tool for re-
> producing human diseases with more accuracy, allowing the expres-
> sion or the suppression of enzymes, proteins and hormones charac-
> teristic of certain human diseases. [(Garattini 1990) p1]

Liggitt and Reddington stated in 1992:

> Twelve years ago the idea that exogenous genetic material could be
> injected into a pronucleus of an embryo, integrated, retained and
> expressed in a newborn animal was so novel that it lacked a name.
> Since then the proof of concept of what has become known as
> transgenic animals has been fulfilled and the technology has erupted
> into an extremely active area of investigation with significant and
> broad potential. As with any new technology, initial investigatory ef-
> forts have been predominantly directed toward exploration of the
> varied techniques, breadth and tolerances of the system rather than
> applications. Hence, currently there is more promise than proof
> available that transgenic mice and other species will prove to be use-
> ful particularly in the area of drug development/ evaluation. [(Liggitt
> and Reddington 1992) p1043]

But what exactly has our actual experience been with the fruits of transgenic technologies? And does what we actually know give reason to be
optimistic about the future prospects of these fruits of the biotechnological tree?

First, some background. Earlier in this book, when discussing the
theoretical motivations behind predictive animal modeling, we drew attention to two important claims made by the modelers in defense of the predictive prospects of their science. The first claim was that at the basal levels in the biological hierarchy of organization (e.g., the biochemical,
intracellular and cellular levels) there were profound similarities *relevant* to
the use of animals as predictive models (there are obviously similarities
between mice and men, say, that are not relevant to the prediction question—humans and mice are both composed of atoms consisting of protons, neutrons and electron; protons and neutrons are made of more basic
things such as quarks, and so on).

Perhaps when the focus is on the relevant basal similarities, from a
predictive point of view, mice and men are the same animal dressed up
differently. This claim, we have argued, both on theoretical grounds (evolutionary biology and developmental genetics) and empirical grounds, is
false. For instance, as we have seen in the geneticists' response to the an-

cient metaphysical puzzle of *unity in diversity*, there are enormous genetic similarities between mice and men, but the devil's differences lie in the details of the regulation of genes in the pathways, circuits and networks to which they belong. It is here that hopes for prediction at the basal levels of the biological hierarchy are confounded.

These observations invite the following question: could we not genetically modify typical animal test subjects to make them more like humans from a genetic point of view? We will devote this chapter to a discussion of some of what is known about genetically modified animals. We happily concede that such humanized animals are important tools in basic research, where they may be very suggestive about the process of pathogenesis in humans. But do they solve the prediction problem? Here you need more than mere suggestions.

(The next chapter is devoted to a discussion of what has emerged from predictive studies rooted in the biology of our close primate relatives. The reason for this is that the second claim that researchers have made to bolster the predictive prospects of their science involves an appeal to the principle of phylogenetic continuity of extant species—they are all descended with modification from common ancestors in the past. The possibility was floated that by using close phylogenetic relatives as models, predictive power of the resulting models would be enhanced because, with relatively short histories since divergence from common ancestors—perhaps as little as seven million years since the divergence of human and chimpanzee lineages—there has been less time to accumulate prediction-destroying species differences).

Prediction Through Genetic Modification

The explosive growth of genetic engineering technologies has given rise to *transgenic disease models*. Hau describes the status of the current situation as follows:

> The rapid developments in genetic engineering and embryo manipulation technology during the past decade have made transgenic disease models perhaps the most important category of animal disease models. A multitude of animal models for important diseases have been developed since this technology became available, and the number of models seems to be increasing quickly. Mice are by far the most important animals for transgenic research purposes, but farm animals and fish are also receiving considerable interest. [(Hau 2003) p4]

Gondo describes the aims of transgenic research this way, "The primary goal of mouse mutagenesis programmes is to develop a fundamental research infrastructure for mammalian functional genomics and to produce human disease models (Gondo 2008)."

First, a very brief reminder of some terminology we shall employ throughout this chapter. Much of our focus is on the use of animals as *causal analog models* (*CAMs*) of the human condition (see LaFollette and Shanks 1996). This use of animal models was introduced (and differentiated from other uses of animal models) in Chapter 6. For an animal model to be a *CAM*, several conditions need to be satisfied, and much of what follows hinges on these matters. Of particular relevance are prediction-destroying causal disanalogies between model and subject modeled.

In the context of monogenic diseases, transgenic models may be valuable heuristic devices capable of stimulating new insights about problems of interest—for example, in the context of basic research. But are they necessarily going to serve as *CAMs*? Many people uncritically assume that transgenic animals simply solve all the problems we have raised about prediction. All assumptions must be examined in this field of endeavor. The new biotechnological innovations that have allowed the construction of genetically modified animals are interesting and important in their own right.

We will argue, however, that biotechnological innovations of the kind discussed here do not solve the issues about prediction that we have raised in this book because, at the genetic level, the conceptual effort to make the technologies relevant relies on a fundamentally atomistic view of genetics according to which genes operate largely independently of each other and have consequences for their organismal bearers that are largely unconnected with the activities of other genes or biological subsystems present. Crudely put, the idea is that if a human gene is inserted into, say, a mouse genome, or a mouse ortholog of a human gene is deleted, the effects of these interventions are largely unmodulated by the other mouse genes present, and the effects of gene insertion or deletion reflect only (or mainly) the intrinsic properties of the inserted (human) gene (or the deleted mouse counterpart of a human gene) and nothing else in the rest of the organism. Consideration of the genetic background is often acknowledged by predictive modelers, but is also downplayed, either as being insignificant (a claim we can examine), or as being something to be solved in the future (lacking crystal balls, this is a claim neither we nor anyone else can confirm or refute).

This kind of *genetic atomism* has been abandoned by most geneticists who have, in turn, come to see genomic systems as consisting of interactive genetic networks where pathways and circuits (hence other genes) are crucial to a proper account of the integrated behavior of the organismal system as a whole, including its constituent genes. There is no reason to believe that the native genes present play second fiddle to the inserted human genes. For related reasons, functional deficits consequent upon knocking out mouse orthologs of human genes do not necessarily tell you what the human gene does *in humans*, and thus what its absence portends *for humans*. Once again, no gene is an island, and the evolved context in which it is found is often crucial to a discussion of functions and functional deficits.

It should also be recalled that the context of expression of a transgene is important. Simply inserting a human gene does not guarantee a relevant human result. The genetic background against which the transgene is regulated and expressed is important. The problems here are well-known and documented. As Houdebine has observed:

> However, despite the generation of several transgenic and knockout models, obtaining relevant models still faces several theoretical and technical challenges. Indeed, genes of interest are not always available and gene addition or inactivation sometimes does not allow clear conclusions because of the intrinsic complexity of living organisms or the redundancy of some metabolic pathways. In addition to homologous recombination, endogenous gene expression can be specifically inhibited using several mechanisms such as RNA interference . . .

> Models based on the use of gene knockout are sometimes disappointing. In up to 30% of the knockouts, no phenotypic effects are observed, perhaps because of insufficient observations of the animals or because redundant mechanisms mask the knockout effects. In other cases, the interference of the gene, which was knocked out with other genes, is too complex and cannot be analyzed easily, leading no clear conclusion.

> Gene knockdown by using shRNA also may face problems. . . Although mice and humans are mammals, some functions are too different in the two species, e.g., lipid metabolism and arteriosclerosis. Thus transgenic rabbits are extensively used to study human diseases resulting from disorders of lipid metabolism . . . *This species has been selected mainly for breeding.* The available lines of rabbits show relatively high heterogeneity in their genetic background. This characte-

ristic significantly reduces the relevance of the models, particularly when lipid metabolism is being studied. (Emphasis added.) (Houdebine 2007)

Transgenic animals have failed as models for human diseases for essentially 5 reasons (see (Lin 2008) for more detail on this):

1. The genes, proteins, protein-protein interactions, and the expression and regulation of genes varies with the species in question.
2. Transgenic animals still have organs and systems that influence the intact organism apart from the modified gene in question.
3. Animals are complex systems—so that slight modifications of the system, barring redundancy (which is known to be an important confounding factor in genetic knockout studies) can have unintended and unforeseen consequences for the system taken as a whole (pleiotropic effects are relevant here). In complex systems, effects of small manipulations are rarely simply proportional—in any straightforward linear fashion—to the magnitude of the causes bringing them about.
4. Polygenic effects are hard to study using current technologies. As observed by Hau:

    > Many physiological functions are polygenic and controlled by more than one gene, and it will require considerable research activities to identify the contribution of multiple genes to normal as well as abnormal biological mechanisms. (Hau 2003)

5. Intraspecific variation in the genetics of modified animals can be a confounding factor masking the effects of genetic manipulation:

    > Mice in any given inbred background are genetically identical; however, when mice are not on a pure background, phenotypic differences may be related to background strain differences rather than the genetic manipulation itself. Littermates in the mixed background help but, depending upon the phenotype, might not suffice. (Shapiro 2007)

Linn 2008 points out the difference between using genetically altered mice for basic research and predicting human response:

> Genetically modified mouse models in which a specific gene is removed or replaced have proven to be powerful tools for identifica-

tion/validation of target gene and scientific understanding of molecular mechanisms underlying drug-induced toxicity through mechanistic studies. In spite of the advantage, there are significant limitations of genetically modified mouse models. Modification of a given gene does not always result in the anticipated phenotype. In some instances, phenotypes of targeted mouse mutants were not those predicted from the presumed function of the given genes, while other null mutants revealed no apparent defects. Furthermore, the phenotypic outcome can be influenced by many environmental and genetic factors. Therefore, interpretation of the significance of the findings from studies using genetically modified mouse models is not always as straightforward as one would expect, especially when desire is to extrapolate the findings to humans. Interestingly, many humanized mouse models have been generated for evaluating the function and regulation of cytochrome P450 (CYP) enzymes . . . Although the creation of humanized animals that carry a particular human CYP gene provides useful tools for scientific understanding of the function and regulation of the CYP enzyme, these humanized mouse models are not so useful in prediction of human pharmacokinetics in a quantitative sense Accordingly, it is important to keep in mind that an animal engineered to express a human gene and its protein is still an animal. (Lin 2008)

Some of the mechanisms that complicate the predictive uses of genetic knockouts are well-known. Thus Regenmortel points out:

> The disappointing results of knockout experiments are partly caused by gene redundancy and pleiotropy, and the fact that gene products are components of pathways and networks in which genes acting in parallel systems can compensate for missing ones. As many factors simultaneously influence the behaviour of a system, one part might function only in the presence of other components. The essential contribution of other genes in achieving a particular function will therefore be missed, which will further encourage the reductionist view that a single gene has adequate explanatory power. (Van Regenmortel 2004)

While there is certainly a *prima facie* case to be made for transgenic models—if one gene influences a disease e.g., cystic fibrosis, then placing that gene in another species may shed light on the disease—the devil resides in the details. Notwithstanding anticipated similarities at the basal levels of the biological hierarchy, investigators frequently remind us of the importance of studying intact organismal systems. But this only serves to bring to the fore the relevance for predictive modeling, of confounding factors found at higher levels of the biological hierarchy. Shapiro makes the point this way:

> The greatest concern, of course, is the utility of mouse studies in predicting human pathophysiology and pathogenesis. The mouse lung has the same general structure and physiological mechanisms as the human lung; however, there are notable exceptions that make translation to humans difficult. For example, the mouse airway has few submucosal glands and only six to eight branches until the terminal bronchiole is reached, which goes directly to the alveolar duct. Humans have 20 branches before becoming the respiratory bronchiole, a structure not present in mice, which is the site of initial inflammation and genesis of centriacinar emphysema. In addition, mice do not always express proteins identical to those in humans, as seen in the previous example of MMP-1. (Shapiro 2007)

While public policy advocates supporting animal-based research (advocates who frequently fail to distinguish between questionable predictive modeling practices and legitimate uses of animals in science) sometimes claim that researchers know how to compensate for the differences we have just alluded to, they rarely if ever explain how, nor do they offer any explanation as to why researchers themselves are worried by these issues. It is not merely that transgenic models are *incompletely validated* (they work with only a few minor wrinkles to be ironed out), rather the so-called "wrinkles" seem to be marks of serious underlying conceptual problems for the predictive animal modeler.

Consider the example of cystic fibrosis. This case is important since mice modified with the human *CF* gene have become a paradigm case illustrative of the potential for the predictive modeling of monogenic diseases through genetic manipulation of model subjects.

The bronchial epithelium in mice that have had the human *CF* gene inserted is quite different from that in humans; in particular it does not contain serous glands, which express the majority of *CFTR* (cystic fibrosis transmembrane conductance regulator) in human beings. Further, the animals did not exhibit the same changes in the pancreas that humans do, neither did they face the lung infections that cystic fibrosis patients do. Considering the fact that the pancreas and lungs are the main human organs affected by the disease, one can say that the transgenic model of cystic fibrosis is a failure as a *CAM*, interesting as it may be in other ways. Moreover, the mice died of intestinal disorders before the lung abnormalities manifested. As Mepham et al. wrote in this regard, "None of the [transgenic] strains are ideal, with either the genotype and/or the phenotype of the mouse failing to accurately model the human condition (Mepham et al. 1998)."

Once again, however, it is important to examine the whole, intact animal, in this case the human subjects modeled—for the clinical characterization of cystic fibrosis is well-known in humans. In this regard Dorfman et al. point out:

> Although cystic fibrosis (CF) is a monogenic disease, its clinical manifestations are influenced in a complex manner. Severity of lung disease, the main cause of mortality among CF patients, is likely modulated by several genes. The mannose-binding lectin 2 (MBL2) gene encodes an innate immune response protein and has been implicated as a pulmonary modifier in CF. . . This MBL2 effect was amplified in patients with high-producing genotypes of transforming growth factor beta 1 (TGFB1). Similarly, MBL2 deficiency was associated with more rapid decline of pulmonary function, most significantly in those carrying the high-producing TGFB1 genotype. These findings provide evidence of gene-gene interaction in the pathogenesis of CF lung disease, whereby high TGF-B1 production enhances the modulatory effect of MBL2 on the age of first bacterial infection and the rate of decline of pulmonary function. (Dorfman et al. 2008)

Monogenic diseases may have modifier genes in the background that complicate the disease phenotypes the modified systems are aimed at modeling. Theorists now distinguish between monogenic disorders at one end of the spectrum of genetic causation, and polygenic (many-gene) disorders at the other. Somewhere in between these two extremes are oligogenic disorders. Thus, Agarwal and Moorchung note:

> It is now increasingly apparent that modifier genes have a considerable role to play in phenotypic variations of single-gene disorders. Intrafamilial variations, altered penetrance, and altered severity are now common features of single gene disorders because of the involvement of several genes in the expression of the disease phenotype. Oligogenic disorders occur because of a second gene modifying the action of a dominant gene.
>
> Oligogenic disorders remain primarily genetic in origin, in contrast to polygenic traits, which are believed to occur because of a complex interaction between the genes and the environment. These disorders require the synergistic action of a small number

of mutant alleles at a small number of loci. The position along this continuum depends on three main variables. It depends on whether a major locus is involved, the number of loci involved, and the extent of environmental participation. (Agarwal and Moorchung 2005)

Concerning the cystic fibrosis paradigm case, Agarwal and Moorchung draw out the relevance of modifier genes in the background of pathogenesis as follows:

> The pulmonary phenotype is altered by several modifier genes including the gene, associated with the migration of the *CFTR* protein to the cell surface. Low levels of mannose-binding protein also influence the severity of the disease. This occurs because mannose-binding protein is a lectin involved in the opsonization and phagocytosis of microorganisms. A decrease in the destruction of microorganisms would lead to an increase in infection and a consequent increase in the severity of symptoms. Polymorphisms in cytokine genes are also important in influencing the severity of the disease, probably by modulating the severity of the inflammatory response. Different alleles of the $\alpha_1$antitrypsin (*A₁AT*) gene are known to influence the severity of the disease. Pulmonary function was better in patients with *CF* bearing the *S* and *Z* alleles of the *A₁AT* gene. (Agarwal and Moorchung 2005)

Horrobin summarizes the current state of the art of enhanced prediction through genetic modification of organisms as follows:

> First, most human disease is highly unlikely to be due to a single abnormal gene. It may well be that the consequences of catastrophic failure of a single gene can be partly understood with the assistance of appropriate genetically modified mouse models. But such diseases are for the most part rare and tend, in any case, to be reasonably well understood from direct human studies.
>
> Second, consistent phenotypes are rarely obtained by modification of the same gene even in mice. The disruption of a gene in one strain of mice may be lethal, whereas disruption of exactly the same gene in another strain of mice may have no detectable phenotypic effect. If this is true of the impact on one gene of the rest of the mouse genome, how much more is it likely to be true of the impact of the rest of the genes in the human genome?

Third, the great majority of human diseases that affect large numbers of the population are likely to be the result of the interaction of several different genes. If one mouse gene is so difficult to understand in a mouse context, and if the genome of a different inbred strain of mouse has so much impact on the consequences of that single gene's expression, how unlikely is it that genetically modified mice are going to provide insights into complex gene interactions in the non-interbred human species? At the least one must conclude that most predictions of near term human benefit are not only overblown but are actually fraudulent. (Horrobin 2003) (See Appendix 5 for full Horrobin article.)

Genetic modification technologies are certainly interesting and important, but their promise to rescue predictive animal modeling by modifying basal features of the hierarchy of organization in nonhuman animals to more closely mimic the human condition is at best overly simplistic, given what is known about the complex nature of genetic causation, and at worst simply false. Our judgement here reflects the current state of the art of modification technologies. Perhaps future research will change this somewhat gloomy state of affairs. But then again, perhaps not: neither we, nor the advocates of the predictive value of genetically modified organisms possess crystal balls!

Two decades have passed since Garattini, Liggitt and Reddington, and Hutchinson and Muller (and others) promised society medical advances due to genetically altered mice. These promises have gone unfulfilled as far as predicting human response. Although we have not discussed it in this volume, there has been a trend in animal-based research for the proponents to excuse past failings by appealing to future successes of a new animal model—such as the genetically altered mouse. Robles and Varticovski echo this when they say:

Recent studies cast doubt on the value of traditionally used models as tools for testing therapies for human cancer. Although the standard practice of xenografting tumors into immunocompromised mice generates reproducible tumors, drug testing in these models has low predictive power when compared to the clinical responses in Phase II trials . . . In spite of their contributions to cancer biology, there are few examples of the use of GEM in preclinical testing because of significant obstacles which prevent widespread use . . . No single preclinical modality will provide adequate guidance for clinical development of new anticancer agents and existing and additional GEM will be increasingly used for preclinical testing. To bring GEM into the mainstream of preclinical testing, many obstacles need to be overcome. Tumors originating in GEM or after transplantation into naive recipients need to be validated for specific genes and pathways that mirror human disease, as well as for sensitivity and resistance to common chemotherapeutic agents. Although history often repeats itself, we hope that hard-learned lessons from xenografts will prove valuable in preventing inappropriate use of GEM in drug development. (Robles and Varticovski 2008)

One of us (NS) once saw a sign in a pub that read "cash today, checks tomorrow" you pay now but tomorrow never comes.

Because of the failure of animal models, historical and current, to predict human response to drugs and disease, there is now talk of making chimeras (mixed cell models): geeps—goat and sheep— (actual) and "himps"—humans and chimps—(hypothetical). The usual promissory notes have been proffered. Behringer:

Laboratory animals are routinely used to model human biology and disease but are not human and therefore cannot fully replicate human physiology. Thus, the primary goal of human-animal chimera research is to produce human cellular characters in animals. The animal carrying the human tissue can then be examined or treated to investigate human-specific biological processes and disease without experimentation on human individuals. . . . Model organisms offer in vivo systems to study fundamental biological processes, providing insights into human physiology. However, these animals are not human and have limitations for studying specific human cellular characters. Practical and ethical concerns preclude direct studies on humans. Thus, human-animal chimeras provide an in vivo system for studying human tissues without experimentation on human individuals . . . However, current discussions about the potential biological outcomes of hES cell-animal chimeras should consider the long heritage of human animal chimera research that has provided important insights into human physiology, disease, and drug discovery. (Behringer 2007)

Kneteman and Mercer 2005:

Mice with chimeric human livers should be useful in the evaluation of drug metabolism within human livers . . . Studies of potential toxicity of drugs to the human liver are also integral to drug development. Toxicity data from other species can be notoriously poor at predicting outcomes in subsequent human clinical trials. A mouse with a chimeric human liver should prove a helpful bridge to clinical development . . . The scid/Alb-uPA mouse carrying a chimeric human liver has been a significant step forward in studying the basic biology of human hepatitis viruses.[18] This work only begins to scratch the surface of what the ultimate utility of this model can be to investigators across a wide range of fields. *The detailed characterization of the uPA-SCID mouse chimera reported by Meuleman al. characterizes the type of study that will help propel the model to successful development in a broad range of applications in health and science that hold significant promise for improvement of the human condition.* (Kneteman and Mercer 2005) (Emphasis added.)

Notwithstanding the promises of new chimeric technologies, there are reasons for a healthy skepticism. Arguably, chimeras will not be predictively fruitful unless it turns out (a) that the human part of the chimera is organizationally and interactively separate from the nonhuman animal part, and (b) that the human part functions normally in the absence of human parts normally present in an intact human organism. Otherwise, all you have is a new messy system with novel synergistic effects found neither in humans nor the animals whose cells they have been combined with. After all, from the standpoint of basic physiology, responses to stimuli (be they toxic insults or other environmental perturbations) require complex interorgan and intercellular communication—communication of the sort that modulates the behavior of intracellular components, including genes and the products they make. There is every reason to believe that human cells in human-animal chimeras will be subject to novel signaling modulation of the kind not typically found in (variable) human populations. As always, the proof of the pudding is in the eating, and we could be wrong. But the poor track record of transgenic technologies in yielding predictive models should at the very least, be a cause for caution when it comes to the great expectations currently being uncritically touted for these new model systems.

Predictive Obligations?

Some have suggested we should not criticize animal models unless we have better suggestions for research and testing (Vineis and Melnick 2008). It is not incumbent upon us to postpone our criticisms of current (inadequate) predictive modeling practices until we ourselves have better predictive models (perhaps *in silico, in vitro* or *in vivo*). Our aim is to provoke critical thinking about the validity of extant scientific practices, and this aim can be pursued independently of the business of constructing new predictive methodologies. Put another way, astrology does not predict the future but can usefully be criticized even though no one knows (certainly not us) how to go about inventing a reliable fortune-telling device!

The matters discussed in this volume, however, are not merely of academic interest (unlike the astrology case). Lives are indeed at stake. And it is a sad fact that, viewed historically, medical research has made very little real progress in curing heart disease, cancer, AIDS, Alzheimer's, Parkinson's, diabetes and many other diseases. Given the coming scarcity of research funds, the question of whether animal models are predictive is a vital issue. We simply cannot afford to fund everything. Research proposals that are funded should be so based on an honest evaluation of their merit. The same is true when deciding what government regulations should exist for evaluating new drugs and chemicals. Perhaps here is the rub. In another context unrelated to medicine, Upton Sinclair observed back in 1935, "It is difficult to get a man to understand something when his salary depends upon his not understanding it." It will be better for all concerned if the merits of research methodologies are assessed honestly from within the boundaries of the scientific community (broadly conceived), than from without by those motivated by extra-scientific concerns and who care little for science!

CHAPTER 15

◆

IMPROVING PREDICTION: PRIMATE STUDIES

Facts do not cease to exist because they are ignored.
—Aldous Huxley

The promise of animal modeling was based on two assumptions. The first was that at the basal levels of the biological hierarchy of organization, humans and their animal *CAM*s were the same animal dressed up differently. The second assumption that we identified was that the best *CAM*s for human biomedical phenomena were those that were closest to us from a phylogenetic standpoint. In such *CAM*s, similarities are expected at levels in the hierarchy of organization above the molecular. This means looking at nonhuman primates in general, and chimpanzees in particular.

The lineage leading to modern chimpanzees diverged from the lineage leading to modern humans about seven million years ago (about the same amount of time separating deer from giraffes). It is certainly true that from an evolutionary standpoint, we expect there to be fewer differences between humans and chimpanzees than between humans and mice, or humans and yeast. But since humans and our closest phylogenetic relatives are complex, organized, interactive systems where small differences can be of great biomedical significance, it is far from clear what follows from this observation concerning the degrees of phylogenetic closeness between humans and mice or monkeys. We will examine three areas where nonhuman primates have been used to illuminate human biomedical phenomena.

We have already had occasion to comment upon the striking genetic similarities between humans and chimpanzees. Bringing out the significance of similarities and differences between humans and chimpanzees, Varki and Altheide have observed:

> The chimpanzee has also long been seen as a model for human diseases because of its close evolutionary relationship. This is indeed the case for a few disorders. Nevertheless, it is a striking paradox that chimpanzees are in fact not good models for many major human diseases/conditions (see Table 2 [Figure 15.1]). In retrospect, this should not be too surprising. After all, at least some major diseases of a species are likely related to (mal)adaptations during the recent evolutionary past of that species. Thus, comparisons with the chimpanzee genome could shed important light on the uniquely human pathogenic mechanisms of serious diseases. (Varki and Altheide 2005)

The point here is well taken. Comparative studies can, indeed, be very useful, bringing out as they do similarities and differences between species. Such comparative studies are of great interest to evolutionary biologists and are, indeed, a legitimate field of animal science. But comparative studies do not typically have as their endpoint the use of one species to predict results in another. This latter is a very different scientific activity, and one that can indeed be confounded by evidence of differences forthcoming from properly conducted comparative studies.

Though it is common for commentators to focus on the enormous similarities between humans and chimpanzees at the level of base-pair similarity (and even, in the light of recent genome studies, to try and measure the similarities quantitatively by "counting genes"—human and chimpanzees gene counts are very similar), the genetic reality is much more complex. Varki et al. comment:

> More recently it has emerged that many genes have undergone differential deletions or duplications in both humans and chimpanzees. The discovery of various non-coding RNAs and the increasing appreciation of the role of post-translational modifications and epigenetic factors add even more complexity when translating genotype to phenotype. Together, these findings have dashed the hope that it would be simple to determine the key genetic differences between humans and our closest evolutionary relatives, that is, the genomic aspects of 'what makes us human.' (Varki, Geschwind, and Eichler 2008)

Even at the genetic level of description, it has proven hard to see chimpanzees as humans dressed in ape suits!

Table 2. Differences between humans and apes in incidence or severity of medical conditions[a]

Medical condition	Humans	Great apes
Definite		
HIV progression to AIDS	Common	Very rare
Hepatitis B/C late complications	Moderate to severe	Mild
P. falciparum malaria	Susceptible	Resistant
Myocardial infarction	Common	Very rare
Endemic infectious retroviruses	Rare	Common
Influenza A symptomatology	Moderate to severe	Mild
Probable		
Menopause	Universal	Rare?
Alzheimer's disease pathology	Complete	No neurofibrillary tangles
Epithelial cancers	Common	Rare?
Atherosclerotic strokes	Common	Rare?
Hydatiform molar pregnancy	Common	Rare?
Possible		
Rheumatoid arthritis	Common	Rare?
Endometriosis	Common	Rare?
Toxemia of pregnancy	Common	Rare?
Early fetal wastage (aneuploidy)	Common	Rare?
Bronchial asthma	Common	Rare?
Autoimmune diseases	Common	Rare?
Major psychoses	Common	Rare?

[a]See Varki 2000; Olson and Varki 2003, and references therein. This list excludes disease states explained by anatomical differences, e.g., difficult labor, varicose veins, spine disorders, hemorrhoids, hernias, etc.

Figure 15.1. Human ape differences

Pursuing comparative genetic analysis, Perry et al. 2008 compared copy number variants (CNVs) between chimpanzees and humans:

> These analyses have identified specific genes, particularly those with inflammatory response functions, which may have faced exceptional natural selection pressures at the copy number level during human and chimpanzee evolution. The specific functional roles of these genes, such as *APOL1, APOL4, CARD18, IL1F7,* and *IL1F8* that are completely deleted in chimpanzees, are therefore of great interest. In humans, the *APOL1* gene (Apolipoprotein L-1) is involved in resistance to protozoan trypanosome parasites that cause sleeping sickness. Potential selective advantages of *APOL1* gene deletion for chimpanzees are not immediately apparent; however, we may be able to generate testable hypotheses once we understand better the other functions of this gene in humans. The interleukin-1 family member 7 (*IL1F7*) protein and *CARD18* both interact with caspase 1, which plays critical roles in innate immunity and inflammation. For example, *CARD18* binds to and inhibits the function of caspase 1, thereby inhibiting production of the interleukin-1-beta inflammatory cytokine. Therefore, this inflammatory pathway is likely regulated differently in chimpanzees than in humans. Interestingly, a mutation that inactivates the human *CASP12* gene has been driven to near fixation by positive selection, likely because loss-of-function of caspase 12 confers resistance to sepsis. Thus, it is particularly striking that gene-disrupting mutations affecting similar inflammatory response pathways have now been associated with potential signatures of positive selection in both humans and chimpanzees. (Perry et al. 2008)

Once again, comparisons are very illuminating and of great scientific interest. But they also highlight the existence of causal disanalogies between humans and chimpanzees—the very sorts of disanalogy that need to be considered when evaluating chimpanzees as predictive models of human biomedical phenomena.

Neuroscience

The great hope for primate studies is that they would help illuminate issues surrounding human cognition. This they have done through comparative studies of similarities and differences in brain structure and function. As before, it is often these revealed differences that confound hope for the construction of predictive primate models in this field of science, no matter how illuminating primate models may be for other purposes related to basic biological research in the field of neuroscience.

There are differences in the basic physiology and anatomy between humans and nonhuman primates, as one would expect from a study of evolution. In macaques, the most often used primates for brain research, the premotor cortex and the motor cortex are similar areas, but in humans the premotor is six times as large [(Kass and Huerta 1988) p606]. Twenty-nine percent of the human brain surface is the frontal area, but in baboons this area is 9.5%, and even less in Capuchin monkeys (9.2%) and marmosets (8.9%) [(Markowitch 1988) p102]. Studies on rhesus monkeys in the 1980s showed the human sensory cortex to be very different in the way cells are arranged (Kass and Huerta 1988). The thalamus is structurally different in humans from every other animal that has been studied [(Simmons 1988) p196]. Damage to a specific part of the supplementary motor system in humans causes loss of speech and global bilateral akinesia (complete loss of muscle response) while monkeys lose only minor function [(Hepp-Raymond 1988) p605]. Damage to the parietal lobes in humans causes apraxia (loss of the ability to make skilled movement with any accuracy) while monkeys suffer only temporary, slight loss of muscular function [(Hepp-Raymond 1988) p605]. Human visual systems take several years to develop after birth, but some monkeys, including marmosets and prosimians, mature in less than a year [(Kass and Huerta 1988) p327-391].

In this way, developmental differences between primate species, along with phenotypic endpoints, must be given due consideration when considering the claim that the monkey is man writ small, differing only in simple quantitative terms of the kind spoken of by those in the sway of the Bernardian research tradition which has done so much to shape the methods of predictive animal modeling, and expectations of enormous biomedical fruit to be forthcoming from such inquiries.

Human brains have a folded cerebral cortex (a gyrencephalic brain) whereas smaller primates, such as the marmoset, have a smooth cerebral cortex (a lissencephalic brain). Not only are there anatomical differences between a gyrencephalic and a lissencephalic brain but evidence suggests that there are functional differences, too. *Lower* and *higher* primates (ugly, pre-evolutionary terminology, which, though still common, dates back to a time when it was believed there was a *scale of nature* in terms of which organisms could be ranked as higher and lower, rather than as merely different), differ from one another by a number of structural features in their nervous systems and sense organs. The brains of lower primates are much smaller in relation to body size than those of the higher primates (this is something that can be crudely measured by considerations of the encephalization quotient (EQ)). The association areas, which govern the transfer of information between the different brain centres, differ in development between brains of higher and lower primates. The chimpanzee brain is about one quarter the size of the human brain and the macaque brain is around one quarter the size of the chimpanzee brain. Comparisons between the human brain and that of non-human primates are limited by the greater complexity of the human brain, due to its larger size, and exemplified by its capacity for language. Although there are counterparts of the macaque brain's structures in the human brain, their functions may have diverged over the course of evolution. Often, areas in the brain that appear to have a function in monkeys do not have the same role in humans.

Concerning the broad structural (architectural) details of human and primate brains, Hacia writes:

> Cerebrotypes, a species-by-species measure of brain size and architecture, have been examined in primate lineages. Although subject to interpretation, the cerebellum was found to occupy a constant fraction of the total brain volume in a diverse number of mammals. In primates, the telencephalon has grown considerably at the expense of the medulla, mesencephalon and diencephalons. This trend is more pronounced in humans and other hominoids than in lower primates. The neocortex component of the telencephalon shows the greatest expansion in hominoids whereas hippocampus, septum, schizocortex, piriform cortex and olfactory bulb components decreased. Relative to other primates, humans show increased intrahemispheric connectivity through the cerebral white matter, but lower interhemispheric connectivity through the corpus callosum and the anterior commissure. This appears to be a function of brain volume. Humans have larger brain volumes ($\sim$1300 cm^3) than other primates such as common chimpanzees ($\sim$340 cm^3), gorillas ($\sim$380 cm^3), and rhesus monkeys ($\sim$80 cm^3). It seems likely that the relative abundance of specific cell types will also be important in primate brain evolution. The pyramidal and nonpyramidal neurons, which comprise the neocortex, are conserved in diverse primate lineages. However, spindle neurons are present in humans, chimpanzees, gorillas and orangutans but not other primates. These neurons are most abundant in chimpanzees and humans. Interestingly, humans tend to have more spindle cells with higher cellular volumes than chimpanzees. Their localization in the anterior cingulate cortex has implicated them in communication, language comprehension and autonomic functions. Specific neurons in multiple individuals at different stages of development need to be meticulously examined to rigorously address brain evolution. However, major ethical, legal

and social issues concerning sample procurement are likely to prevent many such experiments in humans and hominoid apes. (Hacia 2001)

Enard et al. (Enard et al. 2002) compared the transcriptome in blood leukocytes, liver, and brain of humans, chimpanzees, orangutans, and macaques using microarrays, as well as protein expression patterns of humans and chimpanzees. They also studied three rodent species that are approximately as related to each other as are humans, chimpanzees, and orangutans. They identified species-specific gene expression patterns indicating that changes in protein and gene expression have been particularly pronounced in the human brain. They compared mRNA levels in brain and liver of humans, chimpanzees, and an orangutan. They examined approximately 12,000 human genes (see Figures 15.2 and 15.3). Concerning the data, Enard et al. comment:

> Our results show that that large numbers of quantitative changes in gene expression can be detected between closely related mammals. They furthermore suggest that such changes have been particularly pronounced during recent evolution of the human brain.
>
> The underlying reasons for such expression differences are likely to be manifold, for example, duplications and deletions of genes, promotor changes, changes in levels of transcription factors, and changes in cellular composition of tissues. (Enard et al. 2002)

Models of human disease are generally developed in nonhuman primates since they are subjects with behaviors and anatomical characteristics believed to be most similar to humans. But species differences play a role in the clinical expression as well as in the cellular specificity of disease. For example, striatal degeneration in humans is frequently associated with dyskinesia, whereas in nonhuman primates, striatal excitotoxic lesions alone are not sufficient to induce dyskinesia or chorea. Second, in addition to these species differences, the time course evolution of the nerve cell degeneration, which normally evolves over several years in neurodegenerative diseases in humans, is for practical reasons, being replaced over a much shorter period of time in animal models.

Researchers at the Salk Institute and the University of California have asked:

> What is known about the neuroanatomy of the human brain? Do we have a human cortical map corresponding to that for the

Comparison	Analyzed spots	Differences	
		Qualitative	Quantitative
Human–chimpanzee	538	41 (7.6%)	169 (31.4%)
M. musculus–M. spretus	8767	668 (7.6%)	656 (7.5%)

Table 1. Brain protein pattern differences between humans and chimpanzees as analyzed by 2D gel electrophoresis. Differences between humans and chimpanzees were scored if confirmed in three individual human-chimpanzee pairs and were analyzed in the same way as in a larger mouse study comparing M. musculus and M. spretus. Qualitative differences represent changes in electrophoretic mobility of spots, which likely result from amino acid substitutions, whereas quantitative differences reflect changes in the amount of protein.

Figure 15.2. Brain differences

> macaque? What does the human equivalent of the connectional map look like? The shameful answer is that we do not have such detailed maps because, for obvious reasons, most of the experimental methods used on the macaque brain cannot be used on humans … For other cortical regions, such as the language areas, we cannot use the macaque brain even as a rough guide as it probably lacks comparable regions. (Crick and Jones 1993)

For reasons largely unconnected with evolutionary biology, this comment is interesting for it is not the case that there is nothing to know about human brains on the basis of macaque studies. Rather, we are confronted with the old problem highlighted earlier in this book that even when animal models are fertile sources of hypotheses about the human condition, those hypotheses are just that until they are tested and validated in the subjects allegedly being modeled. Until that step is taken (and in this field of inquiry, barring fortuitous studies of humans as the result of treatment for disease or damage due to accidents, such evidence from deficit studies is not likely to be forthcoming), the evidential work of the putative predictive modeler has not begun. And this is true regardless of how scientifically useful the macaque data may be for numerous other scientific purposes.

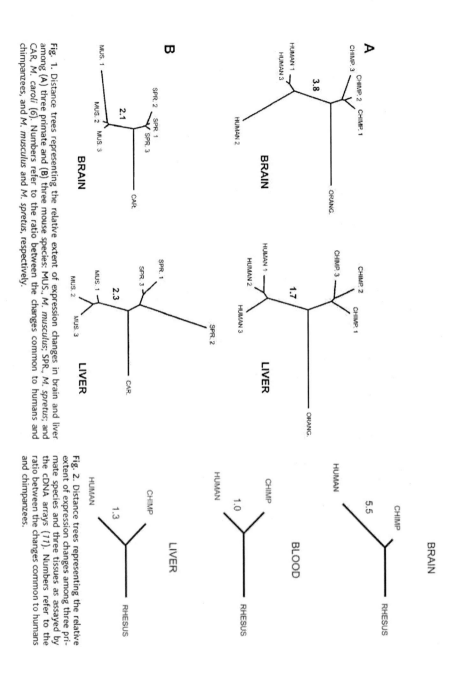

Fig. 1. Distance trees representing the relative extent of expression changes in brain and liver among (A) three primate and (B) three mouse species: MUS, *M. musculus*; SPR, *M. spretus*; and CAR, *M. caroli* (6). Numbers refer to the ratio between the changes common to humans and chimpanzees, and *M. musculus* and *M. spretus*, respectively.

Fig. 2. Distance trees representing the relative extent of expression changes among three primate species and three tissues as assayed by the cDNA arrays (11). Numbers refer to the ratio between the changes common to humans and chimpanzees.

Figure 15.3. Gene expression among mammals

Nonhuman primate brains and human brains *are* undeniably similar in structure, but that does not imply that structures in chimpanzee brains perform the same functions as similar structures in human brains. The chimpanzee brain is not a smaller, more primitive version of a human brain and whose differences from the human brain may be compensated for using quantitative scaling formulae; rather, it is a complex system with its own unique evolutionary history. There will be differential development of modules in humans and chimps that reflect the fact that the two species have taken distinct evolutionary trajectories, adapting and solving different cognitive problems in the course of evolution. Structural similarities in the primate visual system, for example, tell us very little about the subjective character of visual experience in chimpanzees, or indeed its functional integration with information forthcoming from other sensory modalities.

Courtney et al. studied working memory in humans and monkeys. Their comparative analysis brings out just what one would expect: the existence of similarities and differences. Of the differences uncovered the following get specific mention:

> First, there are three separate prefrontal regions associated with object working memory in humans, whereas only one has been identified in monkeys. Second, the frontal region associated with spatial working memory in humans occupies a more dorsal and posterior location relative to the homologous area in monkeys. However, as in the monkey, in which the area specialized for spatial working memory is located just anterior to the FEF, the same topological relationship exists in the human. Third, there is no evidence in monkeys for hemispheric lateralization of working-memory processes, whereas in humans the left hemisphere seems to dominate for analytical-based object representations, and the right for image-based object representations. (Courtney et al. 1998)

Comparative studies have also brought out the importance of differences in gene regulation in the brains of humans and other primates. Varki et al. have thus observed:

> Although microarray-based studies of inter-species gene expression are affected by many technical or methodological issues they revealed a few main themes. First, there is an apparent acceleration in the evolution of brain-enriched genes in humans versus chimpanzees, relative to other tissues. This finding has been interpreted to signify positive selection, but is also consistent with relaxation of

constraint. A second, related observation is that in the human lineage there are more increases in gene expression than decreases. This could indicate a general upregulation of brain energy metabolism in humans, a feature consistent with the expansion of the neocortex in the human lineage. Third, a neutral model has been proposed to account for much of the gene expression changes in the brain, because there is more variation in non-nervous-system enriched genes than in brain-enriched genes. (Varki, Geschwind, and Eichler 2008)

In our discussion of neuroscience, a theme has emerged: comparative studies have revealed much useful biological information about the similarities and differences between humans and their closest evolutionary relatives. Such comparative studies are not the target of the specific inquiries motivating the present volume. We are primarily interested in the use of nonhuman animals to predict human responses. Comparative studies (and we have provided here only a few examples to give the flavor of the enterprise) reveal some of the confounding factors, often rooted in evolution, that make the predictive enterprise more complex than many of its practitioners would like to admit. Another way to hammer this point home is to consider toxicology in the light of primate studies.

Toxicology

Nonhuman primates are frequently used as a nonrodent species to study potential new medications. Theoretically, because they are closest to us in evolutionary terms they should be better models than rats, mice and rabbits, and so on. However, there are many differences between humans and nonhuman primates in drug absorption, distribution, metabolism, excretion, and toxicity (ADMET). Here are some examples of the challenges the toxicologist encounters using nonhuman primates to predict human outcomes.

Gad (Gad 1990) observes that fenclofenac (an anti-inflammatory agent) was tested in two species of monkey (rhesus and patas) without observation of adverse effects. In humans, it caused acute cholestatic jaundice. Peculiarities sometimes occur with respect to the issue of phylogenetic distance, and these are also of interest, given our present concerns about the importance of relative phylogenetic closeness as something that might undergird the predictive modelers' strategies. The metabolism of the anti-cancer drug 5FU showed marked species differences. But of particular interest are matters of phylogeny. The relevant differences concerned the metabolism of 5FU by the enzyme dihydropyrimidine dehydrogenase. As observed by Sludden et al. (Sludden et al. 1998), rats and dogs were similar to humans, but more closely related species of monkey differed from humans. Sludden et al. even observed enzymatic metabolic differences between rhesus and cynomolgus monkeys. As the investigators noted at the time, this was odd given that these species of monkey belong to the same genus!

Tumor necrosis factor-related apoptosis-inducing ligand (TRAIL) kills some tumors in animals without harming the animal. Because of the affect on animals TRAIL was hailed as a miracle cure for cancer. As Kelley et al. noted:

> For pharmacokinetic studies, Apo2L/TRAIL was given as an i.v. bolus to mice (10 mg/kg), rats (10 mg/kg), cynomolgus monkeys (1, 5, and 50 mg/kg), and chimpanzees (1 and 5 mg/kg). Apo2L/TRAIL was rapidly eliminated from the serum of all species studied. Half-lives were ~3 to 5 min in rodents and ~23 to 31 min in nonhuman primates. Allometric scaling provided estimates of Apo2L/TRAIL kinetics in humans, suggesting that on a milligram per kilogram basis, doses significantly lower than those used in xenograft studies could be effective in humans. Apo2L/TRAIL clearance was highly correlated with glomerular filtration rate across species, indicating that the kidneys play a critical role in the elimination of this molecule. Safety evaluations in cynomolgus monkeys and chimpanzees revealed no abnormalities associated with Apo2L/TRAIL exposure. In conclusion, these studies have characterized the disposition of Apo2L/TRAIL in rodents and primates and provide information that will be used to predict the pharmacokinetics of Apo2L/TRAIL in humans. (Kelley et al. 2001)

Thus, in accord with the Bernardian paradigm, species differences appear to be of the quantitative kind that can be compensated for using scaling formulae, and prediction of human outcomes on the basis of animal data may proceed along these very lines.

Unfortunately, the drug was shown to cause liver toxicity in humans. Here the problems emerged as a serendipitous by-product of research on something else. As reported by Gura in *Science*:

> Strom did not set out to study TRAIL's role as a cancer drug. With graduate student Minji Jo and their colleagues, he was trying to find the molecular players that bring about liver cell death in diseases such as hepatitis and alcoholic cirrhosis. Previous work had shown that TRAIL kills cancer cells by triggering a form of cell suicide

called apoptosis, and the team wanted to find out whether TRAIL-induced apoptosis might also be involved in the premature death of diseased liver cells.

Much to their surprise, the researchers found that TRAIL readily induces apoptosis in cultured liver cells from both sick individuals and healthy liver donors. Because the protein had never before been shown to trigger death in animal liver cells, "we thought [the result] was a fluke," perhaps an artifact of the team's culture conditions, recalls Strom.

Further work ruled out that possibility. For example, the team found that TRAIL also induces apoptosis in cells anchored in liver slices. And, in accord with previous findings, the researchers found no effect of TRAIL on cultured human epithelial cells derived from liver tissue or on rat, mouse, or monkey liver cells. "The animal cells are resistant in the culture dish, in whole livers, and in the live animal," Strom sums up. "This was not a culture artifact." (Gura 2000)

Another example of nonhuman primates failing to accurately model humans occurred in 2002 when Elan Pharmaceuticals and Wyeth-Ayerst were forced to halt Phase II studies on their vaccine for Alzheimer's disease (called AN1792) after the discovery that 15 (now up to 18 (Frantz 2004)) patients (out of 360) had developed severe brain inflammation. Some used this human test failure as a reason for advocating tests of such drugs on nonhuman primates before beginning human trials. This case is particularly interesting because, unlike the thalidomide case discussed earlier in this book, we do not have to ask what *would* have happened *had* we, contrary to fact, done tests on non-human primates. Those who saw in the human failure of the drug a need to conduct tests in nonhuman primates didn't search the relevant literature carefully. Such tests had been done!

According to *The Scientist*, "When Elan researchers vaccinated transgenic mice that had developed AD-like pathology, plaques melted away. Two-and-a-half years of animal experiments yielded further encouraging results. The vaccine prevented and possibly reversed cognitive defects (Steinberg 2002)." In fact, Dale Schenk, the developer of the vaccine, and his team *did* use monkeys to test AN-1792 for its safety, as well as rabbits and guinea pigs. Schenk concluded that "the vaccine seems safe and is well tolerated . . . we found virtually no sign of any problems in the animals whatsoever (Marwick 2000)."

The 15 patients also showed symptoms of worsening AD. Human data was relevant here since the vaccine induced the expression of a peptide normally found in human brain cells, but which was absent in monkey brain cells. Many scientists said the vaccine would not work because it would set off a general immune response in the brain (Steinberg 2002). No such symptoms had been seen in the earlier animal studies. "We never saw a hint of this," Dr. Ivan Lieberburg, Elan's chief scientific and medical officer said. "It came as a total shock to Elan (Weiss March 2, 2002)."

Famously, monkey tests were not predictive for humans in the case of TGN1412 (a T cell agonist). Six volunteers who took TGN1412 were hospitalized in critical condition approximately one hour after ingesting the drug. TGN1412 was an anti-inflammatory agent that was supposed to treat conditions as like rheumatoid arthritis and multiple sclerosis but which instead caused multisystem organ failure. TGN1412 was tested in numerous animal species including rabbits and monkeys. As Hopkin reported at the time:

> There was no warning from animal tests, but last month the experimental antibody drug TGN1412 put six British men in intensive care. "We were shocked and surprised to see what happened in humans," Hünig [an immunologist at the University of Würzburg and researchers at TeGenero] told Nature. In preclinical trials, monkeys got a dose 500 times that given to the human volunteers, and the monkey CD28 receptor is identical to the human one, says Hünig. This means that the effects in the monkey trial should have been comparable.
>
> One possible source of the difference between the animal and human trials is that the 'tail' of the antibody molecule at the opposite end from the CD28-binding site may not be the same in humans and monkeys. Antibody tails can undergo a process called crosslinking, which amplifies an immune response by recruiting more immune cells or antibodies. Therapeutic antibodies are modified so that they have the same overall structure as a generic human antibody, but doing this may also have prevented the full extent of TGN1412's activity from showing up in animal tests. Hünig admits that this could have happened. (Hopkin 2006)

In fact there may be even more differences. Hansen et al.:

> Although TeGenero says that the amino-acid sequence of the critical portion of monkey CD28 (the C9D loop) is identical to that of human CD28, published amino-acid and DNA sequence data on rhesus CD28 indicate substantial genetic diversity within the spe-

cies. Differences of up to 4% (9 of 220 amino acids) between rhesus and human proteins have been found. (Hansen and Leslie 2006)

Once again, the devil is in the details of the differences, and not the broad similarities seen when one reads the literature "with a broad brush." The differences are not mere inconvenient anomalies to be swept under the carpet under the guise of *ceteris paribus* clauses. The differences are causally relevant—they are the disanalogies that undermine, in the present case, hopes for nonhuman primate *CAMs*. This invites the question as to why this should be the case.

The reason nonhuman primates (and other species) often fail to respond to drugs as humans do is, arguably because of speciation. But this then invites the question as to what it is about speciation that is relevant here. Merely invoking speciation as if it was a magic mantra explains nothing. That there are species differences in drug metabolizing activity is a fact that virtually all parties to these debates agree on. But we have seen here (as elsewhere in this book) that species differences do not admit of any simple organization, either from the standpoint of quantity (e.g., relative extents of competing reactions) or quality (e.g., unique characteristics of species-specific metabolic pathways).

We can say that some metabolic reactions are unique with respect to certain animal groups. The aromatization of quinic acid seems to be restricted to primates (or at any rate is only significant in primates). As we have seen earlier in this book, we know when it comes to studies of the metabolism of model substrates that cats seem incapable of significant conjugation with glucuronic acid (excreting phenol primarily through conjugation with sulfate), whereas the reverse is true of pigs. Humans and rats are capable of conjugation reactions with glucuronic acid as well as sulfate, though the relative extents of these reactions vary with species. The problem, as we have begun to see in this chapter, is that, while species differences with respect to the metabolism of xenobiotics are a metabolic reality, they are not readily organizable using measures of relative phylogenetic distance between species (either those rooted in traditional morphological and developmental thinking, or those recently forthcoming from a variety of molecular studies of proteins and genes).

While this is not the place to wax philosophical, it is interesting to note that one of the intellectual changes effected by the theory of evolution in Darwin's hands was a rejection of the so-called scale of nature (*scala naturae*). Traditional non-evolutionary biologists attempted to organize species in terms of the concepts of *higher* and *lower* (even today some biologists think in terms of higher and lower primates, and of primates and lower mammals, and so on). The question, of course is higher or lower relative to what? In medieval times an answer might have proceeded in terms of the natural order of God's creation (the Earth was the center of the created universe and in the hierarchy of life on Earth, humans were at the top). Whatever the theological merits of this proposal, untestable as it is, it does not have any traction by the standards of modern science.

Darwin himself rejected the ideas of higher and lower species, embracing instead the neutral idea of difference. Differences between species can be accepted while either rejecting or remaining neutral with respect to issues of nature's phylogenetic hierarchy. Darwin's achievement in the *Origin of Species* lay in the elucidation of an important mechanism—natural selection—whereby intraspecific differences could be amplified, in the course of evolutionary time into good and true species differences with no view to a phylogenetic hierarchy. However, the course of history is by no means straightforward and the significance of Darwin's ideas, even comparatively recently, has eluded even some of those who pay lip service to his theories.

To see what is going on here, reflect that pre-Darwinian biologists such as Lamarck (who had a theory of evolution, though not one recognizable by modern standards) thought the key to the issue of higher versus lower animals might lie in relative degrees of complexity. The idea is that humans are higher than other mammals because humans are more complex than they. As late as the final decade of the 20th century, some theorists were still trying to impose a natural order of complexity on species in terms of concepts such as *genome size*.

Humans, believed—just before the human genome project bore its fruit—to have more than 100,000 genes, were deemed more complex than animals with fewer genes. Attempts were even made to predict the number of cell types available to members of a given species using measures such as the square root of genome size (see Kauffman 1993). Thus, the square root of 100,000 is 317 and consistent with this is the observation of about 256 human cell types (perhaps there are more to be discovered or selection has found no role for some of the possible human cell types). The genome projects have revealed that humans, chimpanzees, dogs and mice have genomes that contain about the same number of genes (about 25,000) so species differences will not be accounted for by simple gene-counting measures. The matter is made all the more acute since orthologues of most human genes can be found in mice and dogs—including genetic systems for tails (expressed in mice and dogs but, barring accidents, mercifully silent in humans). It is also worth noting that the square root of 25,000 is 158, giving a prediction for the number of human cell types that falls well below observation.

The genome projects vindicate Darwin: monkeys, mice and men belong to different species, but they are not rankable in order of relative complexity. The idea that humans are more complex than other mammals has been remarkably persistent. And this is true in the animal model community when theorists engaged in basic biomedical research (as opposed to predictive animal modeling) claim that their animal models are good systems to study with respect to the heuristics of human biology because they are in some sense simpler systems. As we saw earlier in this book, one of the purposes that can be served by the use of models in science is the creation of simpler systems that mimic crucial behaviors observed in more complex, but less tractable systems of interest. For example, the Brusselator model of glycolysis constituted a gross simplification of a complex biochemical reality, but nevertheless captured important dynamical features observed in studies of glycolysis, including the existence of non-equilibrium steady states and the onset of oscillations.

Some animal modelers have had similar ideas concerning the use of nonhuman mammals as heuristic devices to prompt hypotheses about human biomedical phenomena. Forni et al. (1990) have reflected on the value of rodent cancer models using highly immunogenic tumors, which are very rare features of the human oncological condition, by observing:

> This is made possible by reducing the complex problem to simpler models in which some features of the situation they reflect are deliberately ignored. The use of highly immunogenic tumors in these studies is fully justified by the fact that currently we still have only a vague idea of the cell mechanisms involved in immune resistance to tumors. (Forni et al. 1990)

Such models are clearly intended to be *suggestive*, not *predictive* models, and as Forni et al. caution:

> Of course the use of models can be dangerous. Some biological functions that appear to be crucial to a simplified system may lose their importance or take on a different meaning in more complex systems. The fact that it is a model, not the real situation, that is being studied may be forgotten. (Forni et al. 1990)

In point of fact, immunogenic tumors in rodents are not simpler than human tumors; rather they are different (and possibly more amenable to study and to treat). It would be a mistake to assume that ease of generation, and amenability to study, are marks of fundamental biological simplicity, at least as that might be understood with respect to putative complexity-based phylogenetic ordering schemes.

The point of this excursion into issues of complexity and carcinogenicity is that it suggests, if only by analogy, a possible motivation for the use of nonhuman primates rather than other mammals, for the purposes of toxicological inquiry.

It is possible that at least some investigators believe that from a phylogenetic point of view, primate models of the human toxicological condition will fare better than more distant rodent models, because primate models are more complex than rodent systems and, moreover, exhibit similar kinds of complexity as the human systems they are intended to model. Perhaps something like this underlies the expectation that there is a natural phylogenetic scale, rooted in concepts of complexity, that justifies the use of nonhuman primate models. If so, as we have seen, such expectations have been falsified on many occasions. Following up on our discussion of transgenic models in the last chapter, it is also possible that some modelers believe that the insertion of human genes into nonhuman genomes makes nonhuman mammals, say rats and mice, not merely more complex than the native rodent, but also relevantly closer in complexity to humans. In essence, rodents so modified become creatures created in our own complex image. (These points about transgenic models are speculative and will not be pursued further here.)

What does a properly Darwinian account of species differences lead us to expect? First and foremost, we must live with the consequences at all levels in the biological hierarchy of organization (not to be confused and conflated with putative phylogenetic hierarchies) that species will differ from each other in large part as a reflection of the effects of selection as successive generations have adapted and insinuated themselves into a multiplicity of distinct niches in the economy of nature. Biochemical and metabolic adaptations (along with the accidental by-products thereof) and hence differences between species, will reflect the effects of selection as differently situated populations confront the problem of making a living in the face of different evolutionary challenges.

In this context, there will be no simple correlation (or indeed any correlation) between relative degrees of phylogenetic closeness and metabolic similarities with respect to the metabolism of xenobiotics. It is worth recalling at this juncture that organisms require detoxification mechanisms if they are to make a living *anywhere* in the house of nature. The extant adaptive mechanisms are legacies of long evolutionary processes of the kind that shape species themselves. The divergent responses of these adaptive detoxification mechanisms, by members of different species, to novel xenobiotic exposures of the kind of interest to drug companies and regulatory agencies will be hard if not impossible to predict.

However, even without a detailed understanding of evolution, (and eschewing grand but unwarranted phylogenetically-based metabolic ordering schemes), the mechanisms underlying species differences can be usefully discussed. The reason is simple. Whereas a general theory of species differences would examine questions surrounding the distant evolutionary origins of adaptive features of enzymatic detoxification mechanisms, natural selection acts on heritable variation between individuals from whatever cause proceeding. Adaptations are indeed the principle fruits of the operation of natural selection over successive generations. Of crucial relevance here are variation-generating stochastic events—events whose precise occurrence have nothing to do with degrees of phylogenetic relatedness. Important examples of stochastic events are mutations involving SNPs, CNVs and so on. These can be tracked, and their consequences studied, in the laboratory.

Thus, when Ulrich et al. (Ulrich et al. 2002) investigated the cellular expression of 9 cytochrome P450-isozymes (CYP1A1, CYP1A2, CYP2B6, CYP2C8, 9, 19, CYP2D1, CYP2E1, CYP3A1, CYP3A2, CYP3A4) and 3 glutathione S-transferase-isozymes (GST-p, GST-a, GST-l) in the pancreas of hamsters, mice, rats, rabbits, pigs, dogs and monkeys, and compared the results with the expression in the human pancreas, they found a wide variation in the distribution and cellular localization of the selected drug-metabolizing enzymes between the eight species on one hand and the pancreas of humans on the other hand. An exclusive expression of enzymes in the islet cells was found in the hamster (CYP2E1), mouse (CYP1A1, CYP1A2, GST-a, GST-l), rat (CYP2C8,9,19), rabbit (CYP1A2, CYP2B6, GST-p), and pig (CYP1A1). Although no polymorphism was found in the pancreas of animals, in human tissue four enzymes were missing in about 50% of the cases. They state:

> The differences in the distribution of these drug-metabolizing enzymes in the pancreas between the species call for caution when extrapolating experimental results to humans....In humans, a genetic polymorphism has been reported for CYP1A1, CYP2C9, GST-P1 (gene 1 of GSTp), GST-M1, GST-M3 (M1 and M3 represent 2 of the 5 GSTl genes), and GST-T1 (gene 1 of two GSTh genes). As shown in Tables 1–3 [Figures 15.4-6], seven of the enzymes showed differences in their expression in the human specimens. The differences could be related to several factors, including exposure to different substrates, nutrition and ethnic differences

TABLE 1.—The distribution of CYPs and GSTs in the ductal cells of the syrian hamster (SGH), nude mouse (NM), rat, rabbit, guinea pig, dog, monkey, and human.

CYP	SGH	NM	Rat	Rabbit	Pig	Dog	Monkey	Human
1A1	+	−	+	+	−	++	+	+
1A2	+	−	−	−	−	−	−	+
2B6	+	++	+	−	−	−	+	−/+[a]
2C8,9,19	+	++	+	−	−	−	+++	++[b]
2D1	+	+	+	++	+	+	+	++
2E1	−	+	+	+	−	−	−	+[c]
3A1	+	++	++	++	+	++	++	++
3A2	+	+	+	+	++	+	++	+
3A4	+	++	−	−	+	−	−	++
GST-π	++	+	+	−	+	++	+++	++
GST-α	+	−	++	−	+	++	++	++
GST-μ*	++	++	+	++	+	++	+++	++*,[d]

Staining intensities: −, none; +, weak; ++ moderate; +++ strong.
*Due to genetic polymorphism, GST-μ was expressed in 11/21 human specimens.
[a] No staining (9/21 samples), moderate staining (12/21).
[b] No staining (1/21), weak staining (13/21), moderate staining (7/21).
[c] No staining (3/21), weak staining (18/21).
[d] Weak staining (6/11), moderate staining (5/11).

Figure 15.4. CYPs and GSTs I.

TABLE 2.—The distribution of CYPs and GSTs in the acinar cells of the syrian hamster (SGH), nude mouse (NM), rat, rabbit, guinea pig, dog, monkey, and human.

CYP	SGH	NM	Rat	Rabbit	Pig	Dog	Monkey	Human
1A1	++	−	−	+	−	++	+	+
1A2	+	−	++	−	−	+++	+	+
2B6	+	++	−	−	+	++	++	−/+[a]
2C8,9,19	++	++	−	−	++	+	++	++[b]
2D1	++	+++	−	−	+	+	+	++[c]
2E1	−	+	+	+	−	−	−	−/+[d]
3A1	++	++	+	++	+	++	++	+
3A2	++	+	+	++	++	++	−	+
3A4	+	+	+	+	++	−	++	+
GST-π	−	−	−	−	−	−	−	−
GST-α	−	−	−	−	++	−	+	++[e]
GST-μ*	−	−	−	−	−	−	−	+*

Staining intensities: −, none; +, weak; ++ moderate; +++ strong.
*Due to genetic polymorphism, GST-μ was expressed in 10/21 human specimens.
[a] No staining (9/21 samples), weak staining (12/21).
[b] Weak staining (16/21), moderate staining (5/21).
[c] No staining (2/21), moderate staining (19/21).
[d] No staining (3/21), weak staining (18/21).
[e] Weak staining (7/21), moderate staining (12/21), strong staining (2/21).

Figure 15.5. CYPs and GSTs II.

TABLE 3.—The distribution of CYPs and GSTs in the islet cells of the syrian hamster (SGH), nude mouse (NM), rat, rabbit, guinea pig, dog, monkey, and human.

CYP	SGH	NM	Rat	Rabbit	Pig	Dog	Monkey	Human
1A1	++	+	++	+++[a]	+++	+	++	+/+++[b]
1A2	+++	+++[a]	+[a]	+++	−	−	−	+
2B6	+++[a]	+++	+++[a]	++[a]	+++[b]	−	+++[c]	−/+++[b,d]
2C8,9,19	+	+++	+++[a]	−	+++[b]	−	+	−/+++[b,e]
2D1	+	++	++	+	+	+	+	+++[f]
2E1	+++	++	+++	−	−	−	−	−/+[h]
3A1	+++[a]	++	+++	++	−	++[b]	+++	++[b]
3A2	+	+++[a]	+	+	+	+	+	++
3A4	+	++	+++	−	−	−	+	−/+++[k]
GST-π	+++	++[a]	+	+	−	++	++	−/+++[m]
GST-α	+++[a]	+	−	−	−	−	−	−/+[n]
GST-μ*	++	+	+	+++[a]	+[b]	−	++	++*

Staining intensities: −, none; +, weak; ++ moderate; +++ strong.

*Due to genetic polymorphism, GST-μ was expressed in 11/21 human specimens.

[a]Stronger staining in the periphery of the islet.

[b]Stronger staining of scattered cells in the islet.

[c]Stronger staining in the center of the islet.

[d]No staining (9/21), weak staining (12/21).

[e]No staining (2/21 samples), weak staining (16/21), strong staining (2/21).

[f]Moderate staining (8/21), strong staining (13/21).

[h]No staining (4/21), weak staining (17/21).

[k]Weak staining (12/21), moderate staining (4/21), strong staining (up to 25% of cells in 5/21).

[m]Strong staining of an average of 16% of the cells within the islets in all 21 samples.

[n]No staining (7/21 samples), moderate staining (up to 10% of cells in 14/21).

15.6. CYPs and GSTs III.

(e.g., more Asians than Caucasians have inactive alleles of CYP 2C19) . . . In contrast to humans, no interindividual differences existed between animals of the same strain.

Despite some similarities in the expression of the CYP enzymes between the species extreme care is needed when extrapolating the test results gathered from these animals to humans. Among the very closely related proteins there may be considerable catalytic differences. Even between the rodents, like rat and mouse, there is little comparison in the metabolic pathways for activation and detoxication of xenobiotics including carcinogens.

> Moreover, it appears that the metabolic capacity of the same tissue from different species varies considerably, as does the localization of the enzymes in different cells of the same tissue in the same species. (Ulrich et al. 2002)

Another example concerns indinavir. Nonhuman primates are supposedly the best models for drug toxicity and metabolism because of the homology of their drug-metabolizing enzymes. It is known that the nucleotide and amino acid sequences of P450 isoforms (CYP2D and 3A) in cynomolgus monkey and marmoset (*Callithrix jacchus*) are more than 90% similar to the human P450s. Chiba et al. studied the metabolism of indinavir, an HIV protease inhibitor, using liver microsomes from humans, cynomolgus monkeys (*Macaca fascicularis*), rhesus monkeys (*Macaca mulatta*), and chimpanzees (*Pan troglodytes*). (See Figure 15.7.) They found that in vitro metabolism of indinavir varied markedly between species:

> The overall rate of indinavir metabolism varied > 4-fold among primates (84 pmol/min/mg protein in cynomolgus monkey versus 20.4 pmol/min/mg protein in human) and followed the rank order: cynomolgus monkey > rhesus monkey > chimpanzee > human. The cis- (indan)hydroxylated metabolite of indinavir was formed only in cynomolgus and rhesus monkey livers, whereas trans-(indan)hydroxylation and N-dealkylation were observed as the major metabolites in all primates tested. Inhibition studies ... .indicated that a cytochromeP450 isoform of the CYP2D subfamily is involved in the formation of the unique cis-(indan) hydroxylated metabolite in monkey, whereas all other oxidative metabolites, including the trans-(indan)hydroxylated metabolite, are formed by CYP3A isoform(s)... (Chiba et al. 2000)

What is interesting here is that the investigators suspect that very small differences between monkeys and humans with respect to CYP2D isoforms are responsible for observed metabolic differences between monkey and man. Chiba et al. continue:

> The species differences in regio- and stereo-selective metabolism catalyzed by P450s have been studied for many compounds including digitoxin, mephenytoin, phenanthrene, progesterone, testosterone, tolbutamide and warfarin. It has been suggested that primates are better surrogates for human than rodent or dog in qualitatively predicting human *drug* metabolism. However, the present data

Comparative in vitro metabolism of indinavir in primates

Table 1 Hepatic microsomal metabolism of indinavir in primates[a].

Species[b]		Metabolite formation rate (pmol/min/mg)						
		M2 + M5[c]	M3	M4a	M4b	M6	M7	Total
Rhesus monkey[d]	mean	5.26	15.8	12.3	3.90	19.6	17.6	74.5
	SD	1.71	2.9	1.7	0.79	3.6	1.0	11.0
Cynomolgus monkey	mean	9.19	12.7	13.6	3.25	29.2	16.0	84.0
	SD	1.99	2.7	3.6	0.59	6.8	3.6	17.5
Chimpanzee	mean	5.62	2.94	4.83	0.80	12.5	nd[e]	26.7
	SD	1.53	0.89	0.92	0.10	2.5		2.6
Human	HL-1	7.82	7.09	5.94	2.5	16.0	nd	39.4
	HL-2	1.28	1.13	1.19	0.727	2.11	nd	6.44
	HL-3	1.06	0.763	0.744	0.202	1.61	nd	4.38
	HL-4	3.98	2.89	3.67	2.27	5.69	nd	18.5
	HL-5	2.56	0.959	1.46	0.478	3.65	nd	9.11
	HL-6	8.58	7.60	7.27	4.82	16.1	nd	44.4
	mean	4.21	3.41	3.38	1.83	7.53		20.4
	SD	3.27	3.15	2.73	1.75	6.75		17.4

[a] Metabolite formation rates were measured at 10 (monkeys) or 20 (chimpanzees and humans) min after the onset of incubation in the presence of NADPH and 4 mg/ml microsomal protein. Indinavir concentration was 10μM. Data are the mean of results from three individual microsomal preparations except human (n=6).
[b] All experimental animals were male.
[c] Formation rates for secondary metabolites (M2 and M5) were summed.
[d] Data were taken from Lin et al. (1996) for comparison.
[e] ND, Not detectable.

Figure 15.7. Indinavir

strongly indicate that the metabolism of indinavir in monkey is qualitatively different from that in human, whereas the indinavir metabolism in chimpanzee is similar to that in human . . .

The present results suggest that chimpanzee might be a good animal model in predicting drug metabolism in human. However, it was reported that the stereoselectivity of ABT-418 metabolism catalyzed by flavin-containing monooxygenase (FMO) in chimpanzee was different from rat, rabbit, dog and human. Therefore, the species differences and similarity in drug metabolism appears to depend on the enzyme system(s) involved in the metabolism of interest. (Chiba et al. 2000).

Once again, these sorts of results are not uncommon, but what we learn from them is that, notwithstanding well-established relationships of phylogenetic closeness, drug metabolizing capabilities vary from species to species with no obvious discernible pattern—and certainly not one capable of undergirding predictive modeling practices in any straightforward manner. In a way, what is bizarre about working through examples such as these is the disconnect between official theory (close phylogenetic kin good models make), and evidence (there seems to be neither rhyme nor reason behind patterns of drug metabolizing activity among closely related animals). That there is neither rhyme nor reason here is exactly what one might expect as the result of the differential occurrence of stochastic changes in the species lineages of interest, along with the effects of selection and other, non-selective evolutionary mechanisms.

If phylogenetic closeness does not seem to be of much help with respect to studies of drug metabolism, does it help with respect to modeling human disease? Here we examine HIV/AIDS and consider in some detail the lessons to be learned from experience with nonhuman primates in this context.

HIV/AIDS

Since appearing on the scene in the early 1980s, HIV has infected over 60 million humans worldwide. HIV/AIDS has been called, among other things the polio of our time and like polio no resources have been spared. Further, like polio, nonhuman primates have been used as a first choice when studying the disease. Billions have been spent in an attempt to find a cure and vaccine. According to the *LA Times* 2006: "Federal funding for HIV/AIDS research and treatment has grown from $200 million in 1985 to $21.7 billion in 2006 . . . . (Maugh II and Chong 2006)"

HIV/AIDS research is a good example of the implications of speciation resulting in accumulation of differences between evolutionarily closely related species.

Chimpanzees in AIDS research

A familiar pattern begins to emerge (as might be expected because of other known differences between human and chimpanzees with respect to pathology: chimpanzees do not develop cirrhosis from infection with hepatitis B or C, or suffer from rheumatoid arthritis, bronchial asthma, and type 1 diabetes malaria, and Alzheimer's disease). Stump and Vande-Woude have recently observed, drawing on the collective experience of primate investigators:

> Efforts to establish HIV1 infection in primates have proven largely unsuccessful. Early studies inoculating chimpanzees with plasma from HIV1-infected patients did not result in measurable pathogenic changes or AIDS induction in chimpanzees. (Stump and Vande-Woude 2007)

Probably no animal species has been more studied for the cause and effects of a disease as chimpanzees (and recently monkeys) have been for HIV/AIDS. But there are myriad differences in response to HIV between chimpanzees and humans:

- Chimpanzees do not develop any of the characteristic symptoms of AIDS, such as opportunistic infections or malignancies (Fiultz 1991; Fultz 1993; Koch and Ruprecht 1992).
- Chimpanzees develop only transient lymph node swelling (Fiultz 1991; Fultz 1993).
- Chimpanzees do not show HIV infection of brain tissue or macrophages, and HIV has not been found in chimpanzee cerebral spinal fluid or saliva (Gardner and Luciw 1989).
- The chimpanzee immune system mounts little antibody-mediated or cell-mediated responses to HIV-l (Ferrari et al. 1993).
- T4 to T8 lymphocyte ratios differ (Nara et al. 1989).
- HIV does not reproduce well in chimpanzees.
- Chimpanzees have higher baseline levels of T8 cells, a greater proliferative response, and a lower ratio of T4/T8 cells. Considering the fact that T4 cells are selectively attacked by HIV, this difference is not insignificant.
- Humans drop their antibody count prior to systemic illness; chimpanzees do not.
- Chimpanzees have HIV only in their blood cells, while humans also have the virus in their plasma also.
- Chimpanzees exhibit only a flu-like illness in response to being infected with the virus, while humans go on to full-blown AIDS.
- Humans develop opportunistic infections and cancers associated with HIV, which chimpanzees do not.

Margaret I. Johnston has pointed out:

> With few exceptions, HIV1 infection of chimpanzees is universally mild with no notable decline in CD4+ T-cell levels, immunosuppression or other signs of an AIDS-like illness. Rather, HIV-1 infec-

tion of chimpanzees results in detectable plasma HIV that decreases within 2–3 months of infection and becomes low to undetectable within a few years. The ability to detect or culture HIV after this initial time period is variable. (Johnston 2000)

Again we see that even close phylogenetic kin manifest profound differences in response to disease. Chimpanzees and bonobos appear to be less susceptible to many diseases. Some of the most well known differences between the human and chimp genomes are those in the genes encoding HLA types; there appear to be no shared alleles between human and chimp (Cooper et al. 1998).

HIV is a very simple virus; it has only 9 genes. But one thing that makes it interesting from an evolutionary point of view is the fact that it lacks the usual viral repair mechanisms and as a result, mutations are common. HIV has a mutation rate of approximately 1 percent per year. Stump and VandeWoude provide the following details concerning relevant comparative virology:

> Retroviruses were originally classified according to the morphology and position of the nucleocapsid core as determined by electron microscopy. Core characteristics together with location of virion assembly and budding were used to group viruses as A-, B-, C-, or D-type. The International Committee on Taxonomy of Viruses has replaced this classification system and now assigns viruses of the retroviridae family to the following genera: alpharetrovirus, betaretrovirus, gammaretrovirus, deltaretrovirus, epsilonretrovirus, lentivirus, and spumavirus. Members of the alpharetrovirus, betaretrovirus, and gammaretrovirus genera are considered simple retroviruses and encode only *gag, pro, pol*, and *env* genes. Members of the deltaretrovirus, epsilonretrovirus, lentivirus, and spumavirus genera are considered complex retroviruses and encode small regulatory proteins in addition to the genes of the simple retroviral groups. Organization of reading frames and type of tRNA necessary to prime the genome also are used to assign viral groupings. (Stump and VandeWoude 2007)

Three enzymes are vital for HIV's interaction with humans: HIV protease, reverse transcriptase and HIV integrase. HIV attacks human cells in at least four steps. First, the virus attaches itself to receptors located on the surface of the host cell, such as a CD4 cell and injects its RNA into the cell, where an enzyme converts it to DNA. Second, the DNA penetrates the cell's nucleus and uses the machinery of the nucleus to reproduce itself instead of making the human DNA. Third, to make copies of itself, HIV uses the protease enzyme to slice up the proteins into shorter strips, suitable for making new viruses. Finally, thousands of HIV containing capsules are released from the cell membrane, flooding the body with a new generation of the virus.

The virus used to infect the chimpanzee named Jerome, in Yerkes Primate Center (the only chimpanzee to come down with an AIDS-like illness) was different from the type that usually infects humans. After Jerome was infected, his blood was given to other chimpanzees that then dropped their CD4 counts; but in contrast to Jerome, they did not exhibit signs of illness. When HIV infects humans for the first time it binds to the CCR-5 receptor, then it develops a preference for the CXCR-4 receptor. The virus used to infect Jerome relied on the CXCR-4 receptor from the outset (Cohen 1999). These differences are significant.

Monkeys

Researchers discovered that it is possible to produce simian AIDS (SAIDS) in nonhuman primates by infecting one species of macaque with an SIV from a different species. This discovery led to the development of a large number of macaque AIDS models using SIVs in a number of monkey species. The monkey has become the primary nonhuman primate model used to study the virus SIV in hopes of learning something about how HIV affects humans. This is well and good in the context of basic research aimed at expanding our collective base of fundamental biological knowledge. It is much less clear whether the monkey model serves as a predictive model for medical interventions in humans.

Johnston, for example, has made the following observations about differences between SIV and HIV (the sorts of differences that matter when considering predictive models, but which may nevertheless be instructive in the context of basic biomedical research):

> Differences between HIV and SIV could prove important in vaccine evaluation. First, and perhaps foremost, SIV and HIV are distinct viruses. SIV and HIV envelope proteins, which are the key target of neutralizing antibodies, are considerably divergent. Antibodies directed against the envelope of SIV do not neutralize HIV and vice versa. Cytotoxic T lymphocytes (CTLs) specific for HIV do not recognize SIV-infected cells and vice versa. Thus, to utilize monkey models, an analog of the human HIV vaccine must be prepared. In terms of quality or efficacy, SIV analogs might or might

not be comparable to vaccine candidates optimized and manufactured for human trials.

Another difference is that SIV isolates use the CCR5 coreceptor for virus uptake into cells. In 40–50% of HIV-infected humans, HIV that uses CCR5 predominates early and throughout the asymptomatic phase of a typical HIV infection, but a shift of tropism to CXCR4 is observed as these humans progress to AIDS. This shift has not been reported in SIV-infected macaques. (Johnston 2000)

The CCR5 receptor is of particular importance as humans that have a mutation called the *CCR5-Delta32* (32 base pairs are missing from this version of the gene) produce a receptor that HIV cannot bind to. People with this mutation are resistant to HIV and develop AIDS more slowly if infected (Smith et al. 1997). The CCR5 receptor is of interest in other ways since it plays a role in HIV infection in humans and SIV infection in macaques. But here the issue of differences and similarities hinges less crucially on the similarities and differences between humans and macaques, and more crucially on evolved differences between HIV and SIV. This is to be expected for, in this case of host-parasite co-evolution, not only have the target organisms taken divergent evolutionary paths, the viruses have, too. Braun and Johnson comment:

> SIVmac and HIV share similar (though not identical) genetic organization (differing in the vpx and vpu genes), and are approximately 50% homologous based on nucleotide sequences. Both viruses are tropic for CD4+ T lymphocytes and macrophages and use CCR5 as a predominant coreceptor. The clinical course of SIVmac in rhesus macaques mirrors that of HIV infection in many respects, resulting in an initial burst of viral replication, followed by a partial control of viremia by host immune responses, followed by a more protracted course of progressive CD4+ T lymphocyte depletion, accompanied by increased susceptibility to opportunistic infections and malignancy (93). Although the outcome of SIVmac infection is variable and depends on the specific strain used, in general, macaques develop chronic levels of viremia of about 10^5 to 10^6 copies/ml (on average about 10-fold higher than most HIV-infected people) and progress to death in 1-2 years (in contrast to an average of 10 years in untreated HIV-infected people). The significant genetic diversity between HIV and SIV prevents efficacy testing of many HIV-specific genetic therapies in macaques. (Braun and Johnson 2006)

These points of difference have recently been echoed by VandeWoude and Apetrei:

> Although SIV infection in macaques presently represents the predominant animal model for the study of AIDS, consideration of several factors suggests that this model is not ideal. The clinical course of SIV infection in macaques is significantly more aggressive than that of HIV-1 infection in humans, with a significantly larger proportion of rapid progressors. This has hampered the identification of immune correlates of protection and resulted in vaccine study failures. (VandeWoude and Apetrei 2006)

These differences are not merely of academic importance. NIAID officials acknowledged at a summit they held in 2008 following the failure of a Merck AIDS vaccine in 2007 that the rhesus macaque system now used to test potential vaccines is not predictive and, in fact, has not been working out well for researchers (Kaiser 2008). The Merck vaccine failed to protect against HIV infection in humans despite doing so in monkeys. Commenting on this situation, Watkins et al. note:

> The adenovirus type 5 (Ad5)-based vaccine developed by Merck failed to either prevent HIV-1 infection or suppress viral load in subsequently infected subjects in the STEP human Phase 2b efficacy trial. Analogous vaccines had previously also failed in the simian immunodeficiency virus (SIV) challenge–rhesus macaque model. In contrast, vaccine protection studies that used challenge with a chimeric simian-human immunodeficiency virus (SHIV89.6P) in macaques did not predict the human trial results. Ad5 vector–based vaccines did not protect macaques from infection after SHIV89.6P challenge but did cause a substantial reduction in viral load and a preservation of CD4+ T cell counts after infection, findings that were not reproduced in the human trials. (Watkins et al. 2008)

Strangely, in the face of the evidence they just cited, Watkins et al. continue:

> *Although the SIV challenge model is incompletely validated, we propose that its expanded use can help facilitate the prioritization of candidate HIV-1 vaccines, ensuring that resources are focused on the most promising candidates. Vaccine designers must now develop T cell vaccine strategies that reduce viral load after heterologous challenge.* (Watkins et al. 2008) [Emphasis ours].

The issue here is more serious than one person seeing the proverbial glass half full while another sees it as half empty. One is left wondering what, if anything, could *completely invalidate* the allegedly predictive monkey model. Remember, the claim that monkey models predict human responses to vaccines (in the present case) is a scientific hypothesis. Hence it must be testable. For the hypothesis to receive a meaningful scientific test, we must know and state clearly in unambiguous terms (ideally, prior to testing) exactly what evidence, were it to be found would offer confirmation of the hypothesis in question, *and what evidence, were it to be found, would falsify it.* Hypotheses that are so plastic as to be able to accommodate anything that comes along by way of evidence (and the SIV challenge model appears to fall into this category), are not only not *incompletely validated* by actual contrary evidence (which most reasonable observers would consider to have refuted the hypothesis in question), they are not validatable *period* from a scientific point of view. Where the evidence loses its bite, as it evidently has here, there is no longer anything to be right or wrong about—but this means nothing more than that we are no longer doing science!

At the time of writing, field observations in Africa have uncovered what appears to be a strain of SIV that induces an AIDS-like illness in chimpanzees. In light of the forgoing, pending further research, it is unclear what the ultimate significance of these observations will be.

PART V

♦

CONCLUDING REMARKS

To be conscious that you are ignorant is a great step to knowledge.
—Benjamin Disraeli

Throughout this book we have been concerned with issues surrounding the claim that nonhuman animal models predict human biomedical outcomes. We have found that predictive modeling practices are highly problematic from a scientific point of view. There are problems arising from basic biological theory—in particular the theory of evolution (whose implications for present concerns are often either ignored or improperly understood—even by those who pay lip-service to these matters). There are problems arising from the particular research methods proposed by predictive modelers—especially insofar as these methods rest on an inadequate understanding of the nature and origin of well-documented species differences for the practice of their trade. Finally, thanks in large measure to predictive modelers themselves, there is an enormous body of evidence suggesting that there is indeed something rotten in the State of Denmark—or in this case the practice of predictive modeling.

We remind the reader once again that the target of our criticism of animal-based research is restricted to the practice of predictive modeling. We do not dispute that there are legitimate roles for animal test subjects in other kinds of experimental investigation—for example basic biological research aimed at increasing the sum total of human knowledge. Animal experiments in the context of basic research may enrich our knowledge of specific phenomena in mice, and, if painting is permitted with a broad enough brush, they may help delineate some of the important contours of mammalian biology, from which lessons about the Eukaryotes and even life itself might be forthcoming.

All this can be admitted by reasonable people, who may nevertheless have serious doubts about the informational value of experimental investigations on mice, for example, for specific issues in human biomedicine. Do experiments on nonhuman animals reveal specific predictions about the ways in which humans respond to xenobiotics, develop disease states, and, with proper medical intervention, survive or even recover from important diseases of interest? If they do, how have these predictions been tested (i.e., how was human data gathered? Were the predictions verified or falsified?) Here we must move from art painted with a broad brush to investigations of the devilish details of known differences and disanalogies between the species where these have implications for causal analogical reasoning. For too long researchers and their opponents (who often care little for science) have been allowed to get away with the view that critical thinking about animal-based research means you are either "with us or agin us." This binary choice of one of two polar opposites involves a false dilemma. It is time that the middle ground was sought through critical thinking about when and where animal experiments can be fruitful scientifically, and where, frankly, their fruits involve wishful thinking and promissory notes that seldom seem to come due.

It is now our turn to paint with a broad brush. The newly revived field of *systems biology* has influenced both authors of this volume. We contend that organisms—in sickness and in health—are complex interactive systems. Complexes of interactions internal to organisms and their complex interactions with the external environments in which they are embedded are, we think, of crucial importance to a proper understanding of the biology of organisms and their roles in biomedical research. Thus, a deep understanding of the implications of evolution will have to go beyond the study of molecular mechanisms, considered one by one with no focus on the broader biological contexts in which they occur, and into the domain of systems biology.

Systems biology in the present context arises from the realization that while understanding the molecular structure and organization of DNA and proteins is of great importance, it ignores those dimensions of function that reflect systemic molecular interactions. Kitano put it this way:

> System-level understanding, the approach advocated in systems biology, requires a shift in "what to look for" in biology. While an understanding of genes and proteins continues to be important, the focus is on understanding a system's structure and dynamics. Because a system is not just an assembly of genes and proteins, its

properties cannot be understood merely by drawing diagrams of their interconnections. Although such a diagram represents an important first step, it is analogous to a static roadmap, whereas what we really seek to know are traffic patterns, why such traffic patterns emerge, and how we can control them. (Kitano 2002)

More precisely, Leroy Hood explains the aim of systems biology as follows:

Systems biology is the ability to look at all the elements in a biological system—by elements, I mean genes, messenger RNA, proteins, protein interactions and so forth—and to measure their relationships to one another as the system functions in response to biological or genetic perturbations . . . What distinguishes systems biology from the more classical biology of the past 35 years or so, which looked at genes and proteins one at a time, is the attempt to look at all, or at least most, of the elements and their interrelationships. (Hood 2003)

Bringing out the role of theoretical and experimental modeling techniques, Ahn et al. observe:

Systems biology is an integrative approach that combines theoretical modeling and direct experimentation. Theoretical models provide insights into experimental observations, and experiments can provide data needed for model creation or can confirm or refute model findings. With this integrative approach, it becomes apparent that no single discipline is ideal to address systems biology. Scientists from molecular biology, computational science, engineering, physics, statistics, chemistry, and mathematics need to cooperate in order to explain how the biological whole materializes . . .

How is systems-level understanding achieved? The answer likely lies in the dynamic and changing nature of biological networks ...

Unlike the static depiction of many wiring network representations, both the molecular concentrations and enzyme activities are continually changing as a result of influences from other molecular substrates. The network is an interactive and dynamic web in which the properties of a single molecule are contingent on its relationship to other molecules and the activities of those other molecules within the network. Therefore, the behavior of the system arises from the active interactions of these biological components. To elicit the system-wide behavior, three factors need to be considered: (1) context, which values the inclusion of all compo-

nents partaking in a process; (2) time, which considers the changing characteristics of each component; and (3) space, which accounts for the topographic relationships between and among
components. (Ahn et al. 2006)

A Question of Network Structure

Having just seen systems biologists draw out attention to the importance
of networks, what, if anything, is known about global network structures?
According to the central dogma of molecular biology, the genome is the
ultimate repository for biological information, with information
processing being the business of the transcriptome, whereas the execution
of this processed information is the responsibility of the proteome and
metabolome. However, as Oltvai and Barabási observe:

> For example, although long-term information is stored almost ex-
> clusively in the genome, the proteome is crucial for short-term in-
> formation storage and transcription factor information retrieval is
> strongly influenced by the state of the metabolome. This integration
> of different organizational levels increasingly forces us to view cel-
> lular functions as distributed among groups of heterogeneous com-
> ponents that all interact within large networks. (Oltvai and Barabasi
> 2002)

This raises the issue of the nature of biological networks, and in particular
the issue of network architecture. From the discussion in Chapter 5 it became clear that pursuit of a system-level understanding of organisms will
involve at least 3 interrelated issues:

(a) System structures. Here, interest will be focused on the origins, development and nature of structures in the system. These are the structures
constitutive of the system at a given point in time.
(b) System dynamics. Here, interest will be focused on patterns of interactions among subsystems that give rise to temporal patterns in the behavior of the system (steady states, oscillations, quasi-periodic behaviors,
chaos, and so on).
(c) Control mechanisms. Here, the focus of attention is on regulatory
 mechanisms (e.g., positive and negative feedback loops), system-level
redundancies, and so on.

The basic structural issue in a network concerns the linkages between
nodes. Suppose you arrange some nodes representing molecules (sub-

strates and products in a biochemical network, perhaps) in two dimensions. These nodes can be linked together in various ways to represent reactions and pathways. As we saw in the discussion of biological networks in Chapter 5, in some networks the number of molecules (N) with a given number of connections (k) may fall off as a power law $N(k) \sim k^{-\delta}$. Since $N(k)$ does not have a characteristic peak value, it is said to be scale-free. Scale-free network are networks that exhibit a high degree of inhomogeneity. Most nodes in such networks have one or two links to other nodes, but a small number of nodes act as hubs with many connections to other system nodes. For the present, is there any evidence to support the claim that biochemical networks are scale-free networks?

As we saw in Chapter 5, Jeong et al. recently performed a comparative, bioinformatic metabolic analysis of the metabolic networks in 43 organisms drawn from all three domains of life. Despite variation with respect to constituents and pathways, the metabolic networks were found to be scale-free networks. It is too early to say what the properties are of other biological networks, for example, developmental networks, but this may be expected to be a matter to be illuminated as techniques of bioinformatics are extended to this and allied fields where networks are important. Questions of network architecture are currently the subject of on-going inquiries in evolutionary developmental biology (Davidson 2006).

Be this as it may, there is a clear medical implication of these observations derived from the new systems biology and its emphasis on network structure and network dynamics, and that is that there is a need to go beyond molecular structure in the analysis of disease. Sequencing technologies offer the promise of revealing to us where a patient's DNA may exhibit structural abnormalities. But knowing this tells us little about the phenotype that these abnormalities contribute to. The study of function and dysfunction will require a dynamical systems analysis of the consequences of changes in molecular structure. This will involve the study of the characteristics of network interactions.

In a recent study of the genetic interaction network in yeast, Tong et al. employed synthetic genetic array (SGA) analysis. They commented:

> In a typical SGA screen, a mutation in a query gene of interest is crossed into an array of viable gene deletion mutants to generate an output array of double mutants, which can then be scored for specific phenotypes. Synthetic lethal or sick interactions, in which the combination of mutations in two genes causes cell death or reduced fitness, respectively, are of particular interest because they can iden-

tify genes whose products buffer one another and impinge on the
same essential biological process. (Tong et al. 2004)

Tong et al. observe that the connectivity distribution of the array genes
they studied indicates that they were dealing with a scale-free network in
which there were many genes with few interactions and a few highly connected "hub genes." In view of their analysis of the yeast interaction network they observed:

> Highly connected "hub genes" are likely to be more important for
> fitness than less connected genes, because random mutations in or-
> ganisms lacking these genes would be more likely to be associated
> with a fitness defect. Indeed, hubs associated with conserved genes
> may be potential targets for anti-cancer drugs because cancer cells
> often carry a large mutation load and thus may be killed preferen-
> tially. (Tong et al. 2004)

Moreover, genetic interactions are already known to be important components of human diseases. For example, though cystic fibrosis is a typical example of a monogenic Mendelian disease, the severity of the disease
phenotype is known to be modulated by at least seven other alleles (Tong
et al. 2004). Furthermore, there are cases of digenic disease where single
mutations are asymptomatic, but where possession of two mutant alleles
causes disease. As Tong et al. observe:

> The synthetic of digenic interactions observed for a given gene can
> extend to multiple interacting partners. For example, Bardet-Biedl
> syndrome, a retinitis pigmentosa variant, results from combinations
> of mutant alleles in two genes from as many as six . . . Because
> asymptomatic mutations can accumulate in the population and
> probably have the potential to interact with a large number of dif-
> ferent genes, digenic effects may underlie many common diseases
> that are familial but not Mendelian in their inheritance. (Tong et al.
> 2004)

Moving now from global visions rooted in systems biology to the specific
concerns raised in this volume, we have seen that predictive animal modelers, explicitly or implicitly, rely on two assumptions when they have
tried to come to grips with the implications of the theory of evolution for
the practice of their trade. The first was that at the lowest (molecular) levels of the biological hierarchy of organization, humans and their animal
models were essentially "the same animal dressed up differently." We
have argued that the long reach of evolutionary causes and effects has,

contrary to this assumption, significant implications for biochemical processes and cellular processes. Some aspects of biochemistry and cell biology have obviously been conserved in the course of evolution, but some have not, with significant implications for the study of such diseases as cancer and such processes as drug metabolism. And even where basic components have been conserved across diverse lineages, they have often been subject to creative reuse in the course of evolution—reflecting systemic differences in the contexts in which these conserved biological units are employed to do their biological work.

Members of different species, thanks to the long reach of evolutionary processes, are differently complex. Even small differences between complex interactive systems can be of enormous significance. It is a mistake, then, to think of rodents as *humans writ small*, or indeed as *humans writ simple*. They are complex systems in their own right, with unique biological features that result from their having taken, in the course of evolutionary time, an evolutionary trajectory that differs from that taken by humans.

The second assumption was that, at any level of organization in the biological hierarchy, species differences become less marked for species exhibiting a relationship of phylogenetic closeness. But what exactly is a relationship of "phylogenetic closeness?" For some biomedical investigators this means humans and chimps (evolutionary divergence some 7 million years ago), for others it means humans and rodents (evolutionary divergence perhaps some 70 million years ago). We have argued in the last chapter that however one wishes to cash out relationships of phylogenetic closeness, there are biomedically significant disanalogies between humans and our closer, and indeed closest, extant relatives.

If the specifics of human biomedicine are the issue of interest, you must look at humans themselves, since what is needed is species-specific knowledge. This knowledge may be forthcoming from epidemiological studies and clinical observation, and it may be forthcoming at the cellular and molecular levels through the study of tissue cultures and cell cultures. It will not be forthcoming, in general, from the study of differently complex systems with different suites of adaptations at all levels in the biological hierarchy. What is needed is a truly Darwinian or evolutionary approach to medicine. There are, indeed, moral ways to get valuable human data. It is merely childish to argue that human knowledge can only be forthcoming from carefully controlled experimental studies of the kind investigators currently perform on nonhuman animals in their laboratories. Retrospective studies of human exposures to xenobiotics can be extremely valuable, as can prospective studies where those exposed volunteer, either freely and in an informed manner, or by lifestyle choice (e.g., smoking in the famous Framingham study on coronary heart disease).

In advocating the importance of human data, we are most assuredly not advocating a return to the horrors of Dr. Mengele! For those who have these concerns, you might reflect usefully that every time you take medication you are a participant in a (perhaps inadvertent) large-scale experiment by drug companies who make the medication in question. Side effects of low statistical frequency may only come to light through post-marketing surveillance, after there are large numbers of human exposures of the kind not practical in human tests for safety prior to marketing, and which are typically missed altogether by animal tests. You may also note that simply by living in a world shaped by the industrial revolution you are constantly exposed to noxious substances in the environment whose human effects, though often missed by animal studies, sometimes come to the fore in the light of careful epidemiological research. Finally, human diseases from the standpoint of etiology and therapy are often best studied in humans in the clinical context.

Though the disparate fields of biomedicine are inter-linked, all other things being equal, the relevant consultant for treatment of, say, prostate cancer is the oncologist, not the neurologist or the ophthalmologist. Medical specialties matter. In a similar vein, if the problem is rooted in human medicine, the relevant professionals are those with relevant training in human medicine. We know volumes about mouse-medicine, and if your mouse has cancer, it may indeed be curable (as we noted earlier) by a specialist in mouse-medicine, in ways that are typically inapplicable to human cancers. Given what we now know about the differences between human medicine and mouse-medicine, mouse-doctors typically have little to offer human patients (and to no one's great surprise, are not in great demand beyond the bounds of universities, research institutes, pharmaceutical companies and governmental regulatory agencies). Everyone except the putative mouse-doctor seems to understand this! As for the claim that it is only through, say, mouse-medicine that human medicine advances, it is high time that those making such claims in the public arena are called to give a proper explanation—one that goes beyond the realms of spin, public relations propaganda, and thirty-second sound bites.

Our position can be summarized as follows: Living complex systems belonging to different species, largely as a result of the operation of evolutionary mechanisms over long periods of time, manifest different responses to the same stimuli due to: (1) differences with respect to genes present; (2) differences with respect to mutations in the same gene (where one species has an ortholog of a gene found in another); (3) differences with respect to proteins and protein activity; (4) differences with respect to gene regulation; (5) differences in gene expression; (6) differences in protein-protein interactions; (7) differences in genetic networks; (8) differences with respect to organismal organization (humans and rats may be intact systems, but may be differently intact); (9) differences in environmental exposures; and last but not least; (10) differences with respect to evolutionary histories. These are some of the important reasons why members of one species often respond differently to drugs and toxins, and experience different diseases. Immense empirical evidence supports this position.

Predictive animal modelers may insist that animals, notwithstanding their causal disanalogies with humans, are still necessary because without animals researchers could not evaluate the drug or procedure in an *intact system*. We agree that life processes are interdependent; for example, the liver influences the heart, which in turn influences the brain, which in turn influences the kidneys, and so on. Thus, the response of an isolated heart cell to a medication does not confirm that the intact human heart will respond as predicted by the isolated heart cell. The liver may metabolize a drug to a new chemical that is toxic to the heart whereas the original chemical was not toxic.

We also concede that cell cultures, computer modeling, *in vitro* research, etc., cannot replace the living intact system of a *human being*. But while animal models may be *intact systems*, are they intact systems in ways that are causally relevant to *human intact systems*? Shifting the focus from genes, cells and tissues to intact animal systems does not evade the long reach of our concerns about causal disanalogy.

We realize that our claims are controversial, particularly among those whose livelihoods depend upon the use of animal models in predictive contexts, but our arguments are straightforward. If our arguments are unsound, they should be easy to refute. Here is how one could do this:

1. Explain why animals, when used as predictive models for the study of human disease and to test drugs, are not used as *CAMs*. (Remember, we fully accept that animal studies can yield fruitful insights in the context of basic biological research. If you want to know about rat biology, you must study rats. The issue here is whether you can study humans in ways that are predictively efficacious by studying rats.)

2. Show that animal models, when used as *CAMs*, are successful. Show that their positive predictive value and negative predictive values are high enough to be classified scientifically as predictive. This can be accomplished by comparing the results of drug toxicity studies in animals with studies in humans or by comparing the results of induced diseases in animals with the same disease in humans.

Concerning this last point, we have not been able to find such data contradicting our arguments; importantly, none of our critics have been able to present this data either. One hypothesis that explains this is that there is no such data. Either no one has compiled it, or it simply doesn't exist. We suspect that these are hypotheses worthy of further research. Until such data can be found, analyzed and interpreted we must tentatively conclude that the use of allegedly predictive animal models remains in vogue not for scientific reasons but for non-scientific reasons related to economic, legal, and social purposes. Those who have an interest in social policy being guided by science have good reason to demand that good science prevail and, thus, that society turn more of its attention to more fruitful methods of biomedical research.

Both authors of this volume have had the pleasure of being accused, to use Norman's fortuitous turn of phrase, of being on "the wrong side of the line of science orthodoxy" (Norman 2008), and critics interpret this to mean that we are in some sense *anti-science*. But there is no such thing as *science orthodoxy*. Orthodoxy pertains to religion not science. Science is a self-correcting and forward-looking activity rooted in human critical thinking skills. Questioning a very small area of science taken as a whole is not synonymous with questioning the scientific method or science in general. We mention this criticism merely to dismiss it.

Lenin famously asked, "What is to be done?" In the context of the present concerns, Horrobin answers:

> First, there must be a recognition that in the last analysis the human disease itself must be studied in human subjects. It is at least arguable that if we devoted as much effort to the human disease as we do to unvalidated models, then we might be much further forward in understanding. If we are to have any confidence our models are valid, then we must know at least as much about the diseases we investigate as the models we use. . .
>
> Furthermore, only when we have shown that our reductionist bottom-up studies are congruent with our studies of the human disease state in vivo will we be likely to achieve real progress. And in or-

der to do that we have to study human disease: this is where the system may be breaking down. Good clinical research is in decline and its practitioners are becoming demoralized. Sometimes I wonder whether the reluctance to put much more emphasis on clinical studies may involve a reluctance to find relatively simple answers which might obviate the need for much of our sophisticated research . . . (Horrobin 2003) (See Appendix 5 for full Horrobin article.)

APPENDICES

♦

INTRODUCTION

In these appendices we touch upon issues that are beyond the focus of the main body of our book but which some readers may nevertheless find interesting. We have deliberately avoided a discussion of the moral, political, economic and social issues raised by the practice of animal experimentation. These are topics, one way or the other, that have been dealt with by others who have professional interests in these matters. Earlier in this book, in Chapter 2, in the context of a discussion of history, we looked briefly at the distinction between writing history from an internalist perspective and an externalist position. The distinction corresponds, roughly, to looking at, say, the history of science from "inside" and from "outside."

Put slightly differently, we might look at a research practice, say animal-based research, and the institutions that support it, from inside. To do this is to look at the science from a scientific point of view. We might ask (as we have in the body of the book), given what we know about science, are these research practices good and sound practices as judged by the usual canons of science (which involve critical thinking and not simply a slavish acquiescence to the effect that since it is done by scientists it must, by definition, be good science)? In other words, an internalist perspective on the matters at hand looks at science from what may broadly be considered to be a scientific point of view.

The alternative is to take an externalist perspective and ask what social, economic and political factors, broadly conceived, impinge upon this particular branch of science and the institutions that support it. What are the extra-scientific factors (for example availability of funding) that influence behavior of practitioners and their critics. Generally speaking, though we think these matters are interesting and important, they have not been our primary focus of attention. We do not pretend for one minute that the distinction between internalist and externalist perspectives can always be sharply drawn, not do we pretend that they involve considerations locked away in watertight compartments, so that considerations of the one type can have no relevance for considerations of the other type. We do try to illustrate in what follows—even from the somewhat superficial treatment necessarily given for reason of space—that the matters at hand are both complex and difficult to resolve.

In Appendix 1, we consider the issues of the social utility of animal-based research (this includes basic as well as applied research). Animal modelers argue that there are enormous social benefits for human health and well-being forthcoming from their labors. Critics (from within science and from without) contend that these alleged social benefits are illusory largely on the basis of cost-benefit analysis. We note here (and we give no further consideration of them), that activists hostile to animal-based science not infrequently muddy these waters by abandoning their Enlightenment inheritance, with its emphasis on reason and evidence. None of the matters at hand can be properly understood if reason is abandoned and uncivil disobedience becomes the order of the day.

However, the light of reason has also been extinguished by lobbyists and public policy advocates, albeit in a more subtle manner. Claims have been and are being made to the public and their policy makers urging passive acceptance of the *status quo*. Those making the claims condemn as ignorant, irrational, and probably criminal, all who oppose in any way their view. Some critics of extant research practices may be unthinking Luddites, and some may be frankly dangerous. It doesn't follow that all of them are. When rational discussion is abandoned in favor shouting matches and thirty-second sound bites (and this is an issue that seems to contaminate all spheres of public debate, not just debates about animals in science), our basic, shared, democratic values are the ultimate price we all pay.

In Appendix 2 we examine cancer research to raise questions about the translation of experimental knowledge into clinical practice. In Appendix 3 we give examples of research proposals from the CRISP database to illustrate how current funding practices often require investigators in basic biological research to write promissory notes about human benefits in the future. In the nature of the case, such promises are usually unfulfilled. We believe that basic research, adding as it does to the sum total of human biological knowledge is inherently valuable, and not only because of anticipated human benefits. We offer no commentary on the proposals (readers can use the database to get more examples if they wish). We note here that those who hope for funding are often in the unenviable position of having to make claims about human benefits that have little likelihood of coming to pass.

APPENDIX I

◆

SOCIAL UTILITY OF ANIMAL MODELS IN BIOMEDICAL RESEARCH

It is often claimed that biomedical research is good science as judged by the standards of our current best views concerning what counts as good science (an internalist claim), and that such research is done for the sake of human health and well-being (an externalist claim). Obviously the results of an internalist inquiry into animal modeling practices do not logically determine answers to externalist issues. The point to be discussed in this appendix is somewhat different, for here we examine, among other matters, the role played by externalist claims concerning the societal benefits of biomedical research. While no reasonable person doubts the scientific importance of much (perhaps most) work done in the name of basic research, can such research support the claims of great societal benefits offered on its behalf to the public and their policy makers? While the principal concern of this volume has been with an internalist analysis of predictive animal modeling, the public policy debates often hinge on the presentation and analysis of competing externalist claims—either extolling the benefits of such research, or expressing skepticism about the alleged benefits.

The assessment of externalist claims concerning the social utility of biomedical research—let alone animal-based research—is a far more complex matter than is commonly supposed. Merely citing examples (for or against) settles very little: for each example, a counter example can usually be found. Popular "potted histories" of medicine are also of little help, as they are often written for the purposes of professional indoctrination, and/or consumption by the public and their policy makers (the former providing tax dollars, the latter deciding upon the disbursement of such monies).

When discussing externalist claims, context is often important. Externalist claims on behalf of basic research typically tend to be promissory in nature. For example, a research team might write a funding proposal claiming, among other things, that the results of the research may, at some point in the future, contribute to the discovery of a cure for some disease—say cancer. We will see below that the epistemological pathway from basic research to practical application is complex, and it may be hard to assess or quantify the actual role played by basic research in the formation of the beneficial end result. It is certainly not enough to say, for example, on an *ex post facto* basis, that heart transplants are now routine procedures at major medical centers, and that this is a direct result of Harvey's basic research on the circulation of the blood and the motion of the heart back in the early 17th century!

Externalist claims about applied research are somewhat easier to assess, if formulated with suitable precision. It is not enough here (or in the context of basic research, for that matter) to say all such research is valuable, since we can always learn from failed experiments. It is true that science advances as much from its failures as its successes, but this issue belongs firmly in the domain of internalist issues about the practice of good science. Perhaps good science has externalist benefits, but then again, perhaps it does not.

Another point: we quote sources in popular media outlets. While we recognize that these are not scholarly sources, they are important in a discussion of externalist issues relating to the way biomedical research is presented to the public at large. The general public and their policy makers typically do not read *The Journal of Experimental Biology*. They do read popular newspapers and magazines.

Contemporary biomedical science in the US and Europe has a deep methodological commitment to the use of animals as research subjects. The *American Medical Association* sees animal-based research as being essential to progress in human medicine [(American Medical Association 1992). *Sigma Xi*, the scientific research society, defends the use of animals in biomedical research by citing the enormous benefits:

> Results from work with animals have led to understanding mechanisms of bodily function in humans, with substantial and tangible applications to medicine and surgery (e.g., antibiotics, imaging technologies, coronary bypass surgery, anti-cancer therapies), public health (e.g., nutrition, agriculture, immunization, toxicology, and product safety) . . . As the Surgeon General has stated, research with animals has made possible most of the advances in medicine that we today take for granted. An end to animal research would mean an end to our best hope for finding treatments that still elude us. (Sigma Xi 1992)

Biomedical researchers argue forcefully that they need to be able to experiment on live animal systems. They also claim that the central (externalist) justification for doing this research (be it applied research or basic research) is the enormous societal benefit that flows from such research. Here it is claimed that animal-based research is important from the standpoint of both internalist and externalist considerations. Colin Blakemore, former head of the Medical Research Council in the United Kingdom:

> Antibiotics, insulin, vaccines for polio and cervical cancer, organ transplantation, HIV treatments, heart-bypass surgery—it reads like an A to Z of medical progress. But these major advances have something in common: they were all developed and tested using animals ... In fact, animal research has contributed to 70 per cent of Nobel prizes for physiology or medicine. Without it, we would—medically speaking—be stuck in the Dark Ages. (Blakemore 2008)

Botting and Morrison have stated in *Scientific American*: "In truth there are no basic differences between the physiology of laboratory animals and humans...we can not think of an area of medical research that does not owe many of its most important advances to animal experiments (Botting and Morrison 1997)." Dr. Wise Young, neuroscientist from Rutgers University stated: "There's never been a (medical) therapy developed without animals (LaFee 2005)." The Foundation for Biomedical Research stated in *Animal Research Fact vs. Myth*:

> Virtually all medical knowledge and treatment—certainly almost every medical breakthrough of the last century—has involved research with animals. There is a compelling reason for using animals in research. The reason is that we have no other choice ... There are no alternatives to animal research.

It might be countered, however, that the claims made by the Foundation for Biomedical Research, Drs Botting, Morrison, and Young, and Sigma Xi were merely public relations documents to defuse opposition to animal experimentation—we have certainly encountered this response. The trouble is that these organizations are not alone in the production of sweeping generalizations about the great benefits to humans of biomedical research practices using animals. The distinction between basic and applied research is a good place to begin an examination of these issues.

Basic Research and Applied Research

What is basic research? Confusion abounds, for as Arthur Kornberg argued in an editorial in *Science* in 1995:

We are urged: Do strategic basic research! Do targeted basic research! How can we make clear the oxymoronic nature of these terms? (Kornberg 1995)

J.J. Thomson, the discoverer of the electron, stated:

By research in pure science I mean research made without any idea of application to industrial matters but solely with the view of extending our knowledge of the Laws of Nature. (Lord Rayleigh 1942)

Following Grant et al. [(Grant, Green, and Mason 2003) p2] we will employ the distinction between *basic research* and *applied research* used by the Organization for Economic Co-Operation and Development (OECD). Basic research is experimental or theoretical work undertaken primarily to acquire new knowledge of the underlying foundation of phenomena and observable facts, without any particular application or use in view. By contrast, applied research consists also of investigations undertaken to acquire new knowledge, but it differs from basic research by being directed primarily to the achievement of particular aims and objectives. (In its discussion of R&D, the OECD also refers to activities characterized as experimental development, i.e., systematic work drawing on knowledge gained from research and practical experience directed either to the introduction of new materials, products, processes and systems, or to the improvement of such as are currently available.)

We do not pretend that it is always easy, or indeed possible to draw a clear line of demarcation between basic and applied research. There are, inevitably, gray areas. We will contend, however, that the predictive animal modeling of the kind we are interested in here belongs more or less in the domain of applied research—it is typically claimed to have more or less immediate and direct human clinical relevance. This promise of direct clinical relevance is part and parcel of the externalist attraction of such research, where extra-scientific pragmatic goals often trump internalist considerations of what constitutes good science. Claims about direct clinical relevance in this context can often be straightforwardly assessed. If, on the basis of animal studies, a drug is deemed safe for humans, this claim can be tested in the context of human trials and post-marketing surveillance. Clinical relevance, however, is also offered as a justification for the conduct of basic research, and here, as we shall see below, matters are far from straightforward. Examples of basic research proposals with claimed clinical relevance—

taken from the CRISP database (http://crisp.cit.nih.gov/) can be found in *Appendix 3*.

Such examples are not confined to grant applications. Gad writing in *Animal Models in Toxicology* 2007:

> Biomedical sciences' use of animals as models [is to] help understand and *predict* responses in humans, in toxicology and pharmacology . . . by and large animals have worked exceptionally well as *predictive* models for humans . . . Animals have been used as models for centuries to *predict* what chemicals and environmental factors would do to humans. . . . The use of animals as *predictors* of potential ill effects has grown since that time . . . If we correctly identify toxic agents (using animals and other *predictive* model systems) in advance of a product or agent being introduced into the marketplace or environment, generally it will not be introduced . . . The use of thalidomide, a sedative-hypnotic agent, led to some 10,000 deformed children being born in Europe. This in turn led directly to the 1962 revision of the Food, Drug and Cosmetic Act, requiring more stringent testing. *Current testing procedures (or even those at the time in the United States, where the drug was never approved for human use) would have identified the hazard and prevented this tragedy.* (Gad 2007) (Emphasis added.)

We have examined the actual lessons to be drawn from the thalidomide case earlier in this volume. They are by no means as simple as Gad suggests.

The above are not exceptions. There is a widespread and persistent theme in grant applications, the scientific literature, scientific representatives in the scientific press, and in the media in general, that results from animals translate more or less directly into human benefits. Michael F. Jacobson, executive director of the *Center for Science in the Public Interest* noted in 2008: "We must test animals to determine whether a substance causes cancer (CSPI 2008)." Similarly, Huff et al. observe: "Chemical carcinogenesis bioassays in animals have long been recognized and accepted as valid predictors of potential cancer hazards to humans (Huff, Jacobson, and Davis 2008)." In a similar vein, Fomchenko and Holland writing in *Clinical Cancer Research* 2006:

> GEMs [genetically engineered mice] closely recapitulate the human disease and are used to *predict* human response to a therapy, treatment or radiation schedule . . . GEMs that *faithfully recapitulate human brain tumors* and will likely result in high-quality clinical trials with sa-

tisfactory treatment outcomes and reduced drug toxicities. Additional use of GEMs to establish *causal links* between the presence of various genetic alterations and brain tumor initiation . . . (Fomchenko and Holland 2006) (Emphasis added.)

Science Daily recently reported Dec. 15, 2008:

Alzheimer's Research Using Animal Models Significantly Increases Understanding Of The Disease.

Very few species spontaneously develop the cognitive, behavioral and neuropathological symptoms of Alzheimer's disease (AD), yet AD research must progress at a more rapid pace than the rate of human aging. Therefore, in recent years, a variety of animal models have been created—from tiny invertebrates with life spans measurable in months to huge mammals that live several decades . . . .

"Because of the rare instances of spontaneous development of AD pathology in non-human species, animal models have been developed using various genetic, biochemical, or dietary manipulations to approximate full-blown symptoms of the disease," commented Dr. Woodruff-Pak. "The purpose of this Special Issue of the Journal of Alzheimer's Disease (JAD) is to provide an overview of the available animal models of AD and to highlight the power of these models in elucidating mechanisms and treatments. To bridge the wide gap between the molecular biology of AD and clinical therapeutics, it is essential to have valid non-human animal models to investigate disease mechanisms, test treatments, and evaluate preventative strategies and cures. While each animal model has limitations, the value of animal models for research on AD is immeasurable. Our progress in establishing a knowledge base about AD would be slowed, and in some cases prevented, without animal models." (Science Daily 2008)

The Australian paper *The Age* reported on October 31, 2008:

Melbourne scientists hope that a world-first combination drug therapy will revolutionise the treatment of cancer. Scientists at the Walter and Eliza Hall Institute of Medical Research have found that drug-resistant melanoma and colon cancer cells can be successfully treated using a combination of two new drugs. Using mice, scientists combined a cancer inhibitor—a drug that slows the growth of cancer cells—with a drug that harnesses the body's cell-killing machinery to induce the death of the cancer cells. (Milovanovic 2008)

HealthDay News December 4, 2008:

> A gene vital to lung development in newborn mice might be key to finding ways to handle respiratory problems in human infants, a new report says. Researchers at Cincinnati Children's Hospital Medical Center found that if they deleted the *Foxm1* gene from embryonic mice, lungs developed but did not mature fully or produce two critical proteins that line lung tissues and prevent them from collapsing. As a result, the mice died shortly after birth from respiratory distress. "Our findings demonstrate the *Foxm1* gene's central importance to lung maturation and surfactant production in mice," study senior investigator Dr. Vladimir Kalinichenko, a physician in the division of pulmonary biology at Cincinnati Children's, said in a news release issued by the hospital. "Ultimately, this information is important to newborn survival, as infants must breathe on their own at birth instead of getting oxygen from the mother's umbilical cord blood."
>
> In the study, Kalinichenko and his colleagues wrote that "identifying critical regulators of lung maturation, such as *Foxm1*, may provide novel strategies for diagnosis, prevention and treatment of respiratory distress syndrome (RDS) in preterm infants." The researchers are now looking for drugs that can activate *Foxm1* in hopes of developing drugs that treat diseases involving *Foxm1* deficiency, Kalinichenko said. Such a find could help treat premature babies with respiratory distress syndrome by inducing lung maturation and surfactant production. (HealthDay News December 4, 2008)

In a press release from St Jude's Children's Research Hospital, December 3, 2008 it was reported that:

> Using a harmless virus to insert a corrective gene into mouse blood cells, scientists at St. Jude Children's Research Hospital have alleviated sickle cell disease pathology. In their studies, the researchers found that the treated mice showed essentially no difference from normal mice. Although the scientists caution that applying the gene therapy to humans presents significant technical obstacles, they believe that the new therapy will become an important treatment for the disease. (St Jude December 3, 2008)

Biochemist William Reville wrote in the *Irish Times* November 6, 2008:

> Some anti-vivisectionists condemn biomedical animal research on the grounds that animals are so different from humans that any results obtained cannot be successfully extrapolated to humans. This

argument is wrong. Animals contract many of the same diseases as humans. For example, humans and dogs have 65 infectious diseases in common. The figure for cattle is 50, 42 for pigs, 35 for horses and 26 for fowl. We are susceptible to many of the same parasites, viruses and bacteria as animals and some of these can be transmitted between animals and people, eg rabies and malaria. Many non-infectious chronic human diseases, such as epilepsy, also afflict other species ... Many life-saving surgical procedures, including organ transplantation, heart-valve replacement, coronary artery bypass and open-heart surgery, have been developed using animal models first. Animal studies have also led to the development of drugs to treat epilepsy and certain forms of cancer. Animals are also considered essential for testing the safety of food additives, drugs, workplace chemicals and vaccines. (Reville 2008)

The above examples could be multiplied without effort. The above are neither few nor are they exceptions. They are the rule. The claims above go far beyond what could or should be claimed for basic science research. Indeed, the idea that animal results translate more or less directly to humans is made and promoted by research societies and researchers themselves both in private and in public.

Interestingly, the positions taken by lobbyists and other advocates of the interests of animal experimenters have critics within the scientific community, as well as among those outside who are friends of good science. Thus David F. Horrobin wrote the following in *Nature Reviews Drug Discovery*:

Congruence between *in vitro* and animal models of disease and the corresponding human condition is a fundamental assumption of much biomedical research, but it is one that is rarely critically assessed. In the absence of such critical assessment, the assumption of congruence may be invalid for most models. Much more open discussion of this issue is required if biomedical research is to be clinically productive.

Pharmaceutical research is failing in its ability to deliver new drugs. Furthermore, as much pharmaceutical research draws on the wider biomedical research community, the possibility is that biomedical research in general is also failing. There can be no doubt about the reality of the pharmaceutical failure. Almost all the major drug companies are generating inadequate numbers of new drugs and as a result their market values have fallen catastrophically. There is no relief in sight .... (Horrobin 2003)

So why does it persist? Horrobin continues:

> A wonderful metaphor of much modern medical and pharmaceutical research can be found in the book entitled The Glass Bead Game by Herman Hesse. In this story, the leaders of the real world conspire with the brightest of scholars to create a magical state within a state, the isolated world of Castalia. Castalia recruits the most thoughtful and scholarly youths, educates them wonderfully well, and persuades them that the highest achievement of the human mind is to play the almost infinitely complicated and subtle 'glass bead game', an intellectual Olympics which challenges and stretches the most exceptional. The world of the game is beautifully refined and internally self-consistent. The only problems are that Castalia makes almost no contact with the real world, and that playing the game makes no contribution to real world issues. The outcome, however, is acceptable to all sides. The real world sidelines its potentially troublesome intellectuals by seducing them to play in their heavily subsidised glass cage. The intellectuals avoid the need to test their brains on any real issues.

> When I look at the world of medical and pharmaceutical research it seems to me that we are well on the way to creating a Castalia which is entirely acceptable to the majority of its scientist-priests. They receive funding from the real world and are inducted into a complex organisation which, for those who know how to play the game, creates an ever-expanding universe of intellectual and social possibilities. This universe, like Hesse's Castalia, has few points of interaction with real medical problems: its inhabitants, like the original Castalians, are happy to keep it that way. The difference from Castalia is that, with medical research, the real world providing the funding does in the end expect the medical Castalia to provide a pay off. While Hesse's Castalians were confident of the absence of any need to justify their game, the medical Castalians are aware that their funding is to some degree conditional. So the latter constantly make announcements that exaggerate the importance for medicine of some trivial laboratory finding, and constantly assert that if only the public or the pharmaceutical industry will continue to supply money then success will be assured.

> I am increasingly fearful that much of the hyped promise is illusory. Most medical and pharmaceutical research is now of one of three broad types, which are informed by a fourth — the molecular biology of the genome. The other three types of research are cell culture, traditional animal models of disease (such as hypertension, arthritis, cancer or multiple sclerosis), and the new mouse models

involving various types of gene knockouts, knock-ins and their variants, which are currently threatening to overwhelm animal facilities worldwide. We are repeatedly told by the high priests of the medical Castalia that these amazing cascades of new knowledge will in the not-too-distant future lead to effective new individualized therapies for all our ills. (Horrobin 2003) (See Appendix 5 for full Horrobin article.)

In biomedical research today animals play roles in basic research and applied research. In the epistemological pathway from basic research to clinical application, basic research into some phenomenon may provide data, theoretical insights, or other information (e.g., experimental technique) that prompts the formation of hypotheses about clinical applications. These hypotheses may be formed by those conducting the basic research or by others who consume its informational fruits. Moving from the domain of promissory notes to something more concrete, however, takes us into the territory of applied research. Applied research aims to refine hypotheses concerning potential clinical application, and to test them for efficacy and safety. Such research almost always involves animal trials to see first if the fruits of basic research are likely to have the hypothesized clinical application, and secondly, to see if the proposed interventions are safe for humans. These trials are then followed by human trials to see if hypotheses about effectiveness and safety, gained from the animal trials, hold up in the human biomedical context. As Jann Hau puts it in the *Handbook of Laboratory Animal Science*, "The majority of laboratory animal models are developed and used to study the cause, nature, and cure of human disorders (Hau 2003)."

The process from beginning to end may take many years and may involve many groups of investigators. The epistemological pathway from basic research to applied research and on up to actual clinical application should not be thought of as a simple linear pathway in which basic research *determines* clinical application. Information (both experimental, technical, and theoretical) from many sources other than the initial basic research feeds into the epistemological pathway as applications are envisioned, tested for efficacy, tested for safety and ultimately applied in the clinical context. In this sense, basic research should be thought of as at most, an occasional contributing cause to the effect of ultimate clinical application. Such connections as there are between basic research and clinical application are, at best, indirect and involve a considerable element of chance.

Basic biomedical research does not necessarily have to involve animals—it might concern *in vitro* studies of cell or tissue cultures, for example, or it may involve the study of molecular processes. Basic research may also turn out to have no application whatsoever. Perhaps it does not prompt the formation of hypotheses concerning clinical application (for any number of reasons), perhaps the clinical hypotheses that it does prompt turn out to be erroneous—measures turn out to be ineffective and/or unsafe in animals or in humans. But typically, in bringing the fruits of basic research into the clinical context, applied research will almost inevitably involve the use of animal trials aimed at establishing both efficacy and safety of proposed interventions prior to the initiation of experiments on small numbers of humans. It is also worth noting that applied research may also proceed in the absence of basic research and/or human trials, as when animals are used in environmental toxicology to estimate human risk from exposure to substances in the environment.

The number of animals used today in scientific research is controversial since no exact measures are available. All parties agree that millions of animals are used in the US alone and consensus is that around 90% are mice and rats. Most sources agree that roughly half of all animals are used by industry (Home Office 2008; Hudson 2008; Taylor et al. 2008) (APHIS 2000; Mukerjee 2004; American Medical Association 1992; Langley 2006; Wikipedia 2008; HSUS 2008; APHIS 2008).

What is less controversial is the amount of money involved. Mouse sales alone amounted to over $200 million in 1999. Charles River Laboratories of Massachusetts sold animals to the tune of $140 million 1999. Experts estimate that Harlan Sprague Dawley of Indianapolis sold animals collectively worth over $60 million 1998 and Taconic had some $36 million in sales. TJL, a not-for-profit, taxpayer funded corporation sold $29 million worth of mice alone. Mice with specific genes missing cost from $100 to $15,000 (Pennisi 2000).

What percentage of NIH funding goes to animal-based research is largely unknown but can be estimated based on reports. Figure 2.1 is a chart from a report from the Committee on Models for Biomedical Research 1985 (Committee on Models for Biomedical Research 1985).

Mammals alone amounted to roughly 45% and if that number is combined with a fraction of the category that includes nonmammalian vertebrates (other) the number easily goes over 50%. Waltz in *Nature Medicine* recently observed:

> Making a single line of knockout mice can take up to a year and cost $100,000 . . . Making a new knockout line is not easy: it can take up to a year and cost up to $100,000, notes Muriel Davisson,

director of genetic resources at The Jackson Laboratory in Bar Harbor, Maine, one of the three NIH mouse banks. (Waltz 2005)

Primates are not extensively used in biomedical research because of high costs associated with the procurement and handling of such species, and the fact that natural populations are, in some cases close to extinction. Mammals such as dogs, cats and pigs have places in biomedical research, but it is the mouse who, to use Claude Bernard's turn of phrase, has become the *Job of physiology*.

Animals play roles in basic biomedical research aimed at uncovering fundamental biological knowledge, as well as applied research intended to have more or less immediate human clinical and/or medical relevance. The two contexts are not the same. Experiments on animals that expand the basic fund of biological knowledge may or may not be relevant to human medicine. Relevance cannot simply be assumed without evidence. After all, animal subjects, useful in the context of basic research, may nevertheless be causally disanalogous to the human target population that is the focus of applied research. Experiments, valuable in one context, may not be valuable in another. It doesn't follow from this observation that basic biological knowledge is irrelevant to human medicine, but it is at least worth asking whether we have reason to believe that it is likely to be directly or immediately relevant. If such knowledge is indirectly relevant, can we quantify the degree to which it is indirectly relevant? While a full analysis of the relationship between basic and applied research is beyond the scope of this volume, the following remarks are relevant in view of the fact that research has actually been conducted concerning the relationships between basic and applied research in the context of biomedicine.

In 1976, in the journal *Science*, Julius Comroe and Robert Dripps published a groundbreaking paper, "Scientific Basis for the Support of Biomedical Science" (Comroe and Dripps 1976), that purported to show that 41 per cent of all articles judged to be essential for later clinical advances in cardiovascular and pulmonary medicine and surgery, were not clinically oriented at the time they were conducted and that 62 per cent of key articles were the fruits of basic research. As noted by Grant et al.:

Since that analysis, support for basic research has increased in the G7 countries. In the UK, Research Council expenditure on basic research has increased from a low of £444 million (or 42 per cent of total civil R&D) in 1991/1992 to £769 million (or 61 per cent of total civil R&D). Although it would be difficult to argue that Com-

Table 4-1. Distribution of NIH Support of Extramural Research Among Humans, Laboratory Mammals, and Other Research Subjects, Expressed as Percentages of Total Dollars and of Total Projects and Subprojects[a]

Subject	Fiscal Year	Extramural Research Dollars, %	Total Projects and Subprojects, %
Humans	1977	27.5	32.4
	1978	26.8	31.2
	1979	26.8	29.2
	1980	25.0	28.9
	1981	23.8	29.7
	1982	23.2	31.5
	1983	22.9	32.2
Mammals	1977	43.5	41.9
	1978	44.0	42.5
	1979	44.9	43.8
	1980	45.0	44.2
	1981	47.3	44.1
	1982	48.1	43.5
	1983	47.9	42.7
Other[b]	1977	29.4	25.6
	1978	29.3	26.3
	1979	28.2	27.0
	1980	29.8	26.9
	1981	28.9	26.0
	1982	28.7	25.0
	1983	29.2	25.1

[a] Unpublished information provided by Division of Research Resources, National Institutes of Health.
[b] This category includes invertebrates, nonmammalian vertebrates, bacteria, viruses, mathematical and computer simulations, and other subjects.

Figure Appendix 1.1. Percentage of money to animal models.

roe and Dripps were directly responsible for a strategic shift (or drift) in the type of science supported by research funders, their arguments are often cited (albeit at times implicitly) in support of increased funding for basic biomedical research. [(Grant, Green, and Mason 2003) p 111]

Grant et al. go on to observe that, due to methodological flaws, the work by Comroe and Dripps, ". . . is not repeatable, reliable or valid" [(Grant, Green, and Mason 2003) p 111] and that it would probably not meet today's standards for peer review [(Grant, Green, and Mason 2003) p1]. As Farrar observed, among the methodological problems ". . . was a lack of clarity over whose opinions had been surveyed, how clinical advances were assessed and how a *key article* was defined (Farrar 2003)."

In the report by Grant et al., the focus was on five key advances in the context of neonatal intensive care. The study employed a bibliometric analysis of bibliographical databases in conjunction with the Science Citation Index (the data is thus open to public scrutiny and the results can be repeated by independent researchers). Grant et al. concluded that it takes about 17 years for basic research to have a clinical impact. More importantly:

> Using the revised bibliometric protocol, we have shown in this study that . . . between 2 per cent and 21 per cent of research was basic. This corroborates the findings of the clinical guidelines study that showed . . . only 8 per cent of research was basic. These two findings are at odds with Comroe and Dripps finding that 40 per cent of all research articles judged to be essential for later clinical advance were not clinically oriented at the time of the study, thus undermining the evidence base that has, in the past, supported the increased funding of basic research. [(Grant, Green, and Mason 2003) p 40]

Grant et al. do not question the claim that basic research contributes to clinical advances, nor do they question the role played by animals in the research process, but they do question the extent to which basic research contributes to key clinical advances. In fact the real situation may be less rosy than even Grant et al. concluded, for as pointed out by S. Jonathan Singer in *The Splendid Feast of Reason*:

> The majority of research scientists are merely competent and hardly innovative, and, as a result, much of the research that gets done and published is redundant and inconsequential. A few years ago, the Institute for Scientific Information (ISI), which has computerized scientific publications and the citations made to them by other scientists, revealed that nearly half of the scientific papers published in scientific journals were never cited in papers written by other scientists in the five years following publication. This revelation provoked a storm of fury in the scientific establishment. While the absence of citation on rare occasions may mean a study is so good that it is ignored by other scientists because it is well in advance of its time, the more general inference is that much published (let alone unpublished) scientific research is useless to any other scientist and is simply embalmed on library shelves. The ISI was attacked, however, for malicious distortion

and worse, in the best tradition of killing the messenger bearing bad tidings. Actually the situation is even worse than it appears, because the citation of a scientific paper in other publications is by no means proof of the excellence of the cited work. Generally, it attests merely to the work's immediate relevance, which, however, might be only marginal. [(Singer 2003) p173]

The issue Singer raises is not restricted to science. The current culture of "publish or perish" has encouraged the production of much useless research in a wide variety of non-scientific fields as well—with young academics seeking to find as many *LPUs* (least publishable units) as possible before going up for tenure!

Though Grant et al. did not challenge the clinical relevance of animal studies in the context of basic biological research, questions concerning the relevance of such studies have recently been raised by Pound et al. (Pound et al. 2004). These investigators examined the clinical relevance of basic animal-based research in six contexts (calcium channel blockers, e.g., nimodipine, for stroke; low level laser therapy for wound healing; fluid resuscitation for bleeding; thrombolysis for stroke; stress and coronary heart disease; and endothelin receptor blockade in heart failure). Pound et al.:

> The clinical trials of nimodipine and low level laser therapy were conducted concurrently with the animal studies, while clinical trials of fluid resuscitation, thrombolytic therapy, and endothelin receptor blockade went ahead despite evidence of harm from animal studies. This suggests that the animal data were regarded as irrelevant, calling into question why the studies were done in the first place and seriously undermining the principle that animal experiments are necessary to inform clinical medicine. (Pound et al. 2004)

Pound et al. then commented:

> Even if animal experiments provide valid results and sufficiently precise estimates of treatment effects to discount the effects of chance, the extent to which the results can reasonably be generalized to humans remains open to question. Perhaps it was because of this uncertainty that the data from animal studies were disregarded in the above cases. (Pound et al. 2004)

Others have noted the discrepancy between basic research findings and cures.

Dissenting voices exist in the popular press. The last week of April and first week of May 2003, Sharon Begley of *Newsweek* had two articles published in the *Wall Street Journal*. Both discussed the role of animals in biomedical research. She wrote: "Lab mice, after all, have responded quite well to an experimental Alzheimer's vaccine that blocked the formation of the amyloid plaques believed responsible for the disease. Lab rats with paralyzing spinal-cord injuries have walked again, albeit awkwardly, after treatments. And we've cured cancer in enough rodents to fill several New York City subway systems. For people, however, there is no cure for spinal-cord injury, Alzheimer's, Parkinson's disease, multiple sclerosis, cystic fibrosis, osteoporosis, brain and other cancers…the list goes on."

Begley suggests that the biomedical research industry has broken its contract with the taxpaying public. That contract being the taxpayers give the researchers billions of dollars to search for cures and treatments to diseases that afflict society and researchers are expected to produce said cures. Begley, writing this time in the *Wall Street Journal*:

> "Patients," says immunologist Ralph Steinman of Rockefeller University, New York, "have been too patient with basic research.".…Many of the brightest scientists have, therefore, plunged into the minutiae of roundworm genes and fruit-fly receptors, instead of human diseases. "Most of our best people work in lab animals, not people," says Dr. Steinman, who presents his case in a recent issue of the journal *Cerebrum*. "But this has not resulted in cures or even significantly helped most patients.".… "Human experiments are much more time-consuming and more difficult than animal studies," says Rockefeller's James Krueger, whose human research includes trying to correlate gene activity and changes in immune-system cells with the progression of psoriasis. "There are also funding issues. It's much easier to write a successful grant proposal for animal experiments. Animals are homogeneous, and let you say 'aha!' in a neat, clean experiment." Humans, in contrast, are genetically and behaviorally diverse, making it hard to tell whether some aspect of their disease reflects the disease alone, their DNA, how they live—or some messy permutation of all three. (Begley 2003)

The massive increase in funding that has resulted in the massive increase in the numbers of animals used has had little appreciable effect on human healthcare. Begley ends her first article by saying: "From 1998 to this year, the budget of the National Institutes of Health doubled. The 2004 budget request is $27.9 billion. Millions more in private money gushes into biomedical research. Despite those billions, it's the paralyzed rats that walk again. Solutions, anyone?"

In her second article Begley states: "About 30% of NIH's research budget supports clinical research on patients. The rest goes to basic science, from molecules in test tubes to tumors in lab mice." Begley's articles voice the concept we have been expressing for years. What is significant is the fact that the person saying this is a nationally renowned science journalist.

Similar concerns appear in scientific contexts, for as recently noted in a commentary in *Nature*:

> There is a growing disparity at the heart of biomedicine. In some ways, the field is experiencing a golden age: the quantity of basic research is shooting off the charts and budgets are far higher than they were two decades ago. Yet the impact of this research is growing at a much more modest rate: new cures and therapies are ever more expensive to develop and worryingly thin on the ground. (Translational research: getting the message across 2008)

Butler writing in the same issue of *Nature*:

> "NIH stands for the National Institutes of Health, not the National Institutes of Biomedical Research, or the National Institutes of Basic Biomedical Research." This jab, by molecular biologist Alan Schechter at the NIH, is a pointed one. The organization was formally established in the United States more than half a century ago to serve the nation's public health, and its mission now is to pursue fundamental knowledge and apply it "to reduce the burdens of illness and disability". So when employees at the agency have to check their nametag, some soul searching must be taking place.
>
> There is no question that the NIH excels in basic research. What researchers such as Schechter are asking is whether it has neglected the mandate to apply that knowledge. Outside the agency too there is a growing perception that the enormous resources being put into biomedical research, and the huge strides made in understanding disease mechanisms, are not resulting in commensurate gains in new treatments, diagnostics and prevention. "We are not seeing the breakthrough therapies that people can rightly expect," says Schechter, head of molecular biology and genetics at the National Institute of Diabetes and Digestive and Kidney Diseases in Bethesda, Maryland.
>
> Medical-research agencies worldwide are experiencing a similar awakening. Over the past 30 or so years, the ecosystems of ba-

sic and clinical research have diverged. The pharmaceutical industry, which for many years was expected to carry discoveries across
the divide, is now hard pushed to do so. The abyss left behind is
sometimes labelled the 'valley of death'—and neither basic researchers, busy with discoveries, nor physicians, busy with patients, are keen to venture there. "The clinical and basic scientists
don't really communicate," says Barbara Alving, director of the
NIH's National Center for Research Resources in Bethesda. (Butler 2008)

Heidi Ledford pointed out in the same issue of *Nature*:

In April this year, Nobel laureate Sydney Brenner brought the
crowd to its feet at the American Association for Cancer Research
meeting in San Diego, California. Brenner pioneered the use of the
nematode Caenorhabditis elegans as a simple model for studying
growth and development. But in his talk, he championed experiments on a more complicated creature: Homo sapiens. "We don't
have to look for model organisms anymore because we are the
model organism," he said.

Brenner is one of many scientists challenging the idea that
translational research is just about carrying results from bench to
bedside, arguing that the importance of reversing that polarity has
been overlooked. "I'm advocating it go the other way," Brenner
said. Bedside to bench means that clinical trials and patients' unex-
pected responses are valuable human experiments, and failed trials
can stimulate new hypotheses that may help refine the experiment
in its next iteration. (Ledford 2008)

Contopoulos-Ioannidis pointed out in *Science*:

Despite a major interest in translational research, development of
new, effective medical interventions is difficult. Of 101 very promising claims of new discoveries with clear clinical potential that were
made in major basic science journals between 1979 and 1983, only
five resulted in interventions with licensed clinical use by 2003 and
only one had extensive clinical use. Drug discovery faces major
challenges. Moreover, for several interventions supported by high-
profile clinical studies, subsequent evidence from larger and/or better studies contradicts their effectiveness or shows smaller benefits.
The problem seems to be even greater for nonrandomized studies.
(Contopoulos-Ioannidis et al. 2008)

Sharon Begley, again writing in *Newsweek* November 10, 2008:

> The nation's biomedical funding and training system are set up to do one thing, and they do it superlatively: make discoveries. That is what scientists dream of, that is what gets them published in leading journals (the coin of the realm in academia) and that is what gets them grants from the National Institutes of Health . . .

> These barriers to "translational" research (studies that move basic discoveries from bench to bedside) have become so daunting that scientists have a phrase for the chasm between a basic scientific discovery and a new treatment. "It's called the valley of death," says Greg Simon, president of FasterCures, a center set up by the (Michael) Milken Institute in 2003 to achieve what its name says. The valley of death is why many promising discoveries—genes linked to cancer and Parkinson's disease; biochemical pathways that ravage neurons in Lou Gehrig's disease—never move forward . . . (Begley 2008)

Where scarce funding dollars should be spent is not an inconsequential question. The federal government funds 80-85% of research at the top 20 research medical schools (Association of Medical Colleges 2009) with industry, state governments, and not-for-profits accounting for the remainder. Consider the $1.1 billion comparative research initiative funded by the US government. This initiative aims to find out which biomedical interventions work and for whom. Marc Berger, Vice President, Global Health Outcomes, Eli Lilly has observed:

> $1.1 billion is at best a beginning, or a down payment on the kind of research that we should be doing, because we spend far more trying to understand basic mechanisms of disease than we do trying to figure out what works and what doesn't work in the health care that we deliver. (Hughes 2009)

It has long been known that advances in basic research do not guarantee advances in patient care. What is interesting here is that there is evidently a debate of sorts about the societal benefits of biomedical research actually taking place within the scientific community. A treatment of these issues necessarily goes beyond the claims addressed in this volume. We leave it to others to pursue these matters. We can only observe here that the issues are far more complex than lobbyists for the research community have typically been willing to admit to both the public and their hapless policy makers.

APPENDIX 2

◆

CASE STUDY: CANCER RESEARCH

Some of the issues we have discussed above, in connection with claims to clinical relevance made by some of those engaged in basic research, can be illuminated through a consideration of the *war against cancer* launched by President Richard Nixon in the early 1970s. (This initiative culminated in the US with The National Cancer Act of 1971.) Since then billions of dollars have been invested in cancer research. Sharon Begley writing about the War on Cancer in *Newsweek* 2008:

> The meager progress has not been for lack of trying. Since 1971, the federal government, private foundations and companies have spent roughly $200 billion on the quest for cures. That money has bought us an estimated 1.5 million scientific papers, containing an extraordinary amount of knowledge about the basic biology of cancer. (Begley 2008)

There can be no doubt that our knowledge of the cell biology and molecular biology of cancer has grown exponentially since the early 1970s. Much of this enormous growth in knowledge is due to basic biological and biomedical research. This much is not in dispute. What, then, of the clinical fruits of such research—since it is potential clinical fruits that are used as lures to funding agencies, the general public and their policy makers? Certainly, achievements have been made with respect to cancers affecting young people. However, cancer is primarily a disease afflicting older people.

Here the picture is much less clear. We need to ask what sort of progress has been made in the war against cancer, and some prominent investigators think there is much less here than meets the eye. They argue, in effect, that the war against cancer is a war not won! As recently observed by Moss:

> In 1971 . . . we were assured that cancer would be cured by 1976. Since then this and other targets have come and gone, leaving the ultimate goal as distant as ever. Until recently the National Cancer Institute in Bethesda, Maryland stuck by the astonishing claim that

all suffering and death due to cancer would come to an end by
2015, and continues to quote a former director as saying that every
benchmark of the 1971 congressional mandate has been achieved.
(Moss 2006)

This is not to say that the accumulation of basic biological knowledge
since 1971 will ultimately show itself to have been pointless—perhaps it
will lead to great clinical results. It is a warning, however, against the making of incautious and misleading promissory notes to justify funding and
to maintain public support.

Back in 1997, Bailar and Gornik expressed skepticism about claims to
success, i.e., "mission accomplished"—in the war on cancer. They
pointed out:

> Age-adjusted mortality due to cancer in 1994 (200.9 per 100,000
> population) was 6.0 percent higher than the rate in 1970 (189.6 per
> 100,000). After decades of steady increases, the age-adjusted mortal-
> ity due to all malignant neoplasms plateaued, then deceased by 1.0
> percent from 1991 to 1994 . . . These trends reflect a combination
> of changes in death rates from specific types of cancer, with impor-
> tant declines due to reduced cigarette smoking and improved
> screening and a mixture of increases and decreases in the incidence
> of types of cancer not closely related to tobacco use. (Bailar and
> Gornik 1997)

Bailar and Gornik continue:

> The war against cancer is far from over. Observed changes in mor-
> tality due to cancer primarily reflect changing incidence or early de-
> tection. The effect of new treatments for cancer on mortality has
> been largely disappointing. The most promising approach to the
> control of cancer is a national commitment to prevention, with a
> concomitant rebalancing of the focus and funding of research (*ibid.*).

Concerning the externalist promissory notes that accompany claims made
in the war against cancer, Bailar and Gornik note:

> Will we at some future time do better in the war against cancer? The
> present optimism about new therapeutic approaches rooted in mo-
> lecular medicine may turn out to be justified, but the arguments are
> similar in tone and rhetoric to those of decades past about chemo-
> therapy, tumor virology, immunology, and other approaches. In our

view, prudence requires a skeptical view of the tacit assumption that marvelous new treatments for cancer are just waiting to be discovered (*ibid.*)

These remarks should serve as a caution against supposing that the epistemological pathway that leads from basic research to successful clinical application is a simple, straightforward one.

Notwithstanding the modest proposal from Bailar and Gornik, the cancer research community reacted with comments verging on the hysterical. In a news bulletin in *The Oncologist,* Samuel Broder, a former director of the National Cancer Institute was quoted as saying of Bailar and Gornik's claim that part of the reduction in cancer mortality was due to earlier diagnosis:

> That's insane. You don't treat somebody with a diagnosis. Do we have a rule that its unfair to fight the tumor? You have to give it a head start, or it wouldn't be fair to diagnose it early? As though early diagnosis is not part of treatment. When you say it doesn't count because it's due to earlier diagnosis, well, who brought you the earlier diagnosis? Your friendly National Cancer Program. This is really an attack on basic science under the guise of being an attack on treatment. (News Bulletin 1997)

The *Straw Man fallacy* here is obvious. The attack is *not* on basic science, but on the externalist rationales offered to justify support in the court of public opinion.

Where, then, do we stand a decade after the great debate of 1997? The following comments from Moss seem to be in order:

> We are told that while a cancer diagnosis 35 years ago was inevitably a death sentence, many cancers today are curable. Yet it has been known for 100 years that cancers are generally curable if they can be removed while still in their early stages. When somebody dies from cancer it is usually because it has spread from one site in the body to another, yet over the past 35 years the death rate from most of these metastatic cancers has remained largely unchanged. (Moss 2006)

Moreover, there are the usual concerns about statistical artifacts when looking at the data. In the 1970s, five-year survival rates were of the order of 50%. Today 66.6% of those diagnosed will survive 5 years. The problem here concerns *lead time bias*, not improved treatment. Moss explains:

Thanks to widespread screening, people are often now diagnosed with cancer earlier in the course of their disease than they would have been in the past. However, the natural history of the disease has not changed at all: the time of death is typically the same as it would have been had the disease been diagnosed later. (Moss 2006)

In 2002 the number of cancer deaths in the U.S. was 557,272. In 2003, it was 556,902, an absolute decline of 370 deaths. This apparently led Andrew von Eschenbach, then director of the National Cancer Institute to observe, "It proves our expectation of continued progress against cancer is well-founded" (quoted in (Moss 2006) p. 19). As Moss sarcastically observed, at this rate, cancer will be eliminated from the U.S. by the year 3508. An article by Leaf in *Fortune* magazine on 2004 entitled, "Why We're Losing the War on Cancer" laid much of the blame for our failure in combating cancer on using animals as models for humans:

The cancer community has published an extraordinary 150,855 experimental studies on mice, according to a search of the PubMed database. Guess how many of them have led to treatments for cancer? Very, very few. In fact, if you want to understand where the War on Cancer has gone wrong, the mouse is a pretty good place to start. Says Weinberg: "A fundamental problem which remains to be solved in the whole cancer research effort, in terms of therapies, is that the preclinical [animal] models of human cancer, in large part, stink" . . .

Even more depressing is the very real possibility that reliance on this flawed model has caused researchers to pass over drugs that would work in humans. After all, if so many promising drugs that clobbered mouse cancers failed in man, the reverse is also likely: More than a few of the hundreds of thousands of compounds discarded over the past 20 years might have been truly effective agents. Roy Herbst, who divides his time between bench and bedside at M.D. Anderson and who has run big trials on Iressa and other targeted therapies for lung cancer, is sure that happens often. "It's something that bothers me a lot," he says. (Leaf 2004)

Alexander Kamb of Novartis Institutes for BioMedical Research wrote in *Nature Reviews Drug Discovery* 2005:

But as a rule, investigational drugs do not enter the clinic without some rationale and supporting preclinical evidence of efficacy. Given that many of these investigational anticancer drugs eventually fail, *the animal models on which clinical trials are predicated must at best be limited in power, and at worst wildly inaccurate.* Either the models are too simplistic or they possess hidden complexities that hamper the collection of reliable data — grist for the mill of prediction. (Emphasis added.) (Kamb 2005)

An editorial in *Nature Reviews Drug Discovery* 2006 referred to the *Fortune* article:

This lack of therapeutic progress has prompted much debate on where the war on cancer has gone wrong, as highlighted by an article in *Fortune* two years ago entitled "Why we're losing the war on cancer —and how to win it". *Among the key culprits identified in this article were standard mouse models of cancer, which are clearly far from optimal — around 90% of potential drugs that show promise in these models subsequently fail in clinical trials. More candidate anticancer drugs fail in Phase II trials — when efficacy in humans is generally first assessed — than in any other major therapeutic area, emphasizing how poorly efficacy in current preclinical models predicts that in humans.* (Emphasis added.) (Editorial 2006)

Begley continues in *Newsweek* 2008:

A widely discussed 2004 article in Fortune magazine ("Why We're Losing the War on Cancer") laid the blame for this at the little pawed feet of lab mice and rats, and indeed there is a lot to criticize about animal studies. The basic approach, beginning in the 1970s, was to grow human cancer cells in a lab dish, transplant them into a mouse whose immune system had been tweaked to not reject them, throw experimental drugs at them and see what happened. Unfortunately, few of the successes in mice are relevant to people . . . "Far more than anything else," says Robert Weinberg of MIT, the lack of good animal models "has become the rate-limiting step in cancer research." (Begley 2008)

Deshaies echoes Begley when discussing drug discovery in general and cancer treatments specifically:

Drug discovery is hard. Even harder is drug discovery aimed at biological systems that haven't previously been tested as therapeutic targets. The extraordinary challenge is brought home with sobering clarity by numbers: in 2008, only one new compound was approved by the US Food and Drug Administration (FDA) for treating cancer. In the same year, the US pharmaceutical industry alone spent about $65 billion in the pursuit of new medicines, much of which went towards anticancer research—so the apparently slow progress in discovering new cancer medicines is certainly not due to lack of investment. (Deshaies 2009)

The point of this excursion into the field of cancer medicine is to sound a cautious note when claims are made about the immediate and direct relevance of basic scientific research (which we have already acknowledged has greatly contributed to the sum of human knowledge).

APPENDIX 3

♦

GRANT PROPOSALS FOR BASIC BIOMEDICAL RESEARCH

Here are some examples of basic research proposals with claimed clinical relevance—taken from the CRISP database (http://crisp.cit.nih.gov/). The institutional pressures in the struggle for research funding are such that researchers must often make promissory notes about the potential human relevance of their research. Consistent with our earlier distinction between internalist and externalist considerations in the conduct of science, is the observation here that the proposed science may be good (an internalist claim)—and we do not dispute the worthiness of the proposals below—while the implied human benefits (an externalist claim) may be *pro forma*, unrealistic or speculative at best. It is often forgotten in these debates that good science is not necessarily socially relevant science. Biomedical researchers do not have the mathematicians' luxury of toasting their discipline by saying, "To pure mathematics, may it never be any use to anyone," but slavish service to social relevance is likely to stifle creativity, and forced lip service to such relevance is just plain wrong!

Grant Number: 1R03EY016425-01
Project Title: *A new model for corneal disease*
Abstract: Corneal diseases such as infections and injury are major causes of blindness with alterations in both epithelium and stroma leading to the loss of corneal transparency. Co-factors of LIM (Clims; also called Ldb and Nli), which regulate transcription of target genes by associating with DNA-binding proteins, are highly expressed in epithelia. To study their function in epithelial cells, we expressed a dominant negative Clim under control of the keratin 14 promoter in mice (K14-DN-Clim mice)… *This new mouse model may be important for understanding the molecular pathways that prevent keratinization of corneal epithelium and vacularization of corneal stroma. In addition, characterization of the K14-DN-Clim mice may provide insights into the pathogenesis of corneal causes of blindness and provide a potential model to test therapeutic approaches.* (Emphasis added.)

Grant Number: 5R01NS035439-08
Project Title: *Febrile Seizure Model: Neuronal Injury and Mechanisms*
Abstract: This proposal focuses on the mechanisms and consequences of febrile seizures, the most prevalent seizure type in young children. An im-

mature *rat model* of prolonged febrile seizures, those associated with potential development of limbic epilepsy, has been characterized, and has already shed considerable light on the neuroanatomical basis of these seizures and on their functional consequences... *The proposed studies should provide novel and important insight into the remarkable age-and seizure-specific effects of prolonged experimental febrile seizures* on the developing hippocampus, changes leading to enhanced excitability long-term.... (Emphasis added.)

Grant Number: 5R24RR015088-03
Project Title: *New vertebrate model organism cDNA libraries*
Abstract: A variety of mammalian and nonmammalian *animal models* are currently used in research designed to ultimately improve human health...In recent years, genomic tools have become increasingly important for the molecular genetic analysis of important biological questions. However, aside from the mouse, there are rather few genomic resources available for important vertebrate model organisms such as the frogs Xenopus laevis and X. tropicalis and the zebrafish Danio rerio... These libraries will be made freely available to the research community at modest cost and will provide an important resource required to jump-start research in Xenopus tropicalis. In addition, they will be provided to the ongoing model organism EST project directed by Dr. Steve Johnson at Washington University of St. Louis. It is expected that the availability of these libraries and the data derived from them will serve to facilitate progress in the adoption of Xenopus tropicalis *as an important vertebrate model system.* (Emphasis added.)

Grant Number: 5R01AG012694-10
Project Title: *Canine as an animal model of human aging*
Abstract: During the past five years a canine model of aging has been evaluated, with emphasis on categorization of cognitive decline and links with neuropathology... The present proposal seeks first to extend our evaluation of the canine model and establish the cognitive processes that are particularly sensitive to aging, using both cross sectional and longitudinal strategies. (Emphasis added.)

Grant Number: 5R01MH057483-06
Project Title: *Dopamine deficit and schizophrenia*
Abstract: ... In the monkey, we have found that subchronic exposure to PCP induces a decrease in dopamine function in the prefrontal cortex (PFC), which persists for more than a month. Demonstrates, neurochemical and anatomical specificity. This PCP-induced PFC dopamine deficiency correlates with cognitive impairments in the monkey, which resemble those occurring in schizophrenia .... *These data will aid in the development of novel strategies for ameliorating the neurochemical and behavioral deficits in this potential animal model, and in the cognitive dysfunctions associated with schizophrenia and other psychiatric disorders.* (Emphasis added.)

Grant Number: 1R01NS050156-01A2
Project Title: *Focal Dopamine Indicated in Dyskinesias in MPTP Monkeys*
Abstract: The broad aim of this proposal is to test the hypothesis that L-dopa-induced dyskinesias (LID) in Parkinson's disease (PD) arise at least in part from non-uniform dopaminergic denervation of the striatum, whereby islands of dopaminergic activity (hotspots) are created within the most severely affected part of the striatum, the post-commissural putamen . . . This research program is important from three perspectives. First, it promises to establish an in vivo model of LID in which neural correlates of LID can be investigated in a controlled setting. This work will provide insight on the mechanisms behind L-dopa induced dyskinesias and *will provide a basis for using gene therapy approaches to treat patients with Parkinson's disease.* (Emphasis added.)

Grant Number: 1U01AA014829-01
Project Title: *Testing fasd therapeutic agents: neonatal rodent models*
Abstract: The long-term goals of this component are to use rodent models of binge alcohol exposure during the 3rd trimester equivalent to *screen and identify molecular agents that may be effective in preventing prenatal alcohol-induced brain damage and neurodevelopmental disorders* . . . . A key advantage of integrated approaches across the Consortium is that as promising candidate molecular agents emerge, these animal models can provide in vivo tests of their therapeutic effectiveness. (Emphasis added.)

Grant Number: 5U54CA119367-020006
Project Title: *Mouse Cancer Models for Integrated Tissue/Serum Proteomics and Molecular Imaging*
Abstract: Nanotechnology has the potential to significantly impact the development of small animal models of cancer including models to test new antineoplastic therapies. In this project we will develop mouse tumor xenograft models that will allow us to test if we can combine both tissue/serum nanosensor based proteomic analysis and molecular imaging with targeted fluorescent quantum dots to predict and monitor treatment response with specific therapies . . . . The significance of this work is that it should help set the foundation for using ex vivo nanosensors in clinical trials, to allow development of novel molecular imaging probes for clinical trials, and to improve drug testing in small animal cancer models. This should lead to marked improvement in predicting and monitoring response to therapy in cancer patients.

Grant Number: 1R41AI071451-01
Project Title: *Novel Model to Predict Safety and Efficacy of Microbicides*
Abstract: Genital herpes is 1 of the most prevalent sexually transmitted infections (STI) worldwide and is associated with substantial morbidity . . . A major limitation in the development of this novel class of drugs is the ab-

sence of a small animal model to predict safety and effectiveness. Preliminary studies indicate that the cotton rat will fill this niche.

Grant Number: 1U01AA013497-01
Project Title: *Ethanol Driven Neuroadaptation/Cholinergic Interneurons*
Abstract: . . . We seek to characterize ethanol induced IEG expression, dynamic trafficking of key DA/GLU synaptic components and synaptic rewiring within shell nucleus accumbens cholinergic neuronal networks in animal models of excessive ethanol consumption. We will also seek to determine the alterations of key DA/GLU components in relation to ethanol mice genetic models and ethanol exposure. The findings from this work should contribute to a better understanding of the neuronal mechanisms that cause or predict excessive ethanol consumption and toward the development of improved behavioral and pharmacological prevention treatments for alcoholism and alcohol abuse.

Grant Number: 2R01DA001442-28A1
Project Title: *The Behavioral Pharmacology of Phencyclidine*
Abstract: . . . The presence of PCP-like side effects and abuse liability will seriously limit the therapeutic usefulness of glutamate antagonists, so knowing how to predict them from animal studies is essential.

Grant Number: 1Z01AI000734-07
Project Title: *Experimental Tb Chemotherapeutics: Animal Models And Mode of Action*
Abstract: . . . The project relies on the development of advanced animal models for predicting drug efficacy under real world conditions.

Grant Number: 5K08CA084044-03
Project Title: *Use of the mouse for prostate cancer gene discovery*
Abstract: . . . Progress is further impeded by the lack of a prostate cancer animal model where the contribution of individual genetic lesions to the neoplastic phenotype can be explored. An understanding of the genetic changes underlying prostatic cancer would have a major impact on our ability to 1) identify individuals with a hereditary predisposition to prostate cancer, 2) predict which tumors will progress, and in the long term 3) design rational therapeutic modalities. We propose to develop the mouse as a model system for the identification and study of genes involved in prostate cancer.

Grant Number: 3P01CA023099-22S10022
Project Title: *Pharmacokinetics/pharmacodynamics of anticancer drugs in childhood solid tumors*
Abstract: Treatment failure in children with malignant solid tumors may be directly related to variability in anticancer drug disposition. However, children with cancer usually receive identical drug dosages, based on body size,

without accounting for pharmacokinetic variability. Clinical pharmacokinetic-pharmacodynamic investigations in children are ofter descriptive, while animal models provide greater flexibility in designing preclinical hypothesis-testing studies. This project focuses on the application of pharmacokinetic/dynamic methodologies to characterize concentration-effect relationships of anticancer drugs in animal model systems, and to develop rational drug therapy regimens for the treatment of children with solid tumors, with particular emphasis on camptothecin analogs. The specific aims of this project are: 1) to establish parmacokinetic [sic] -pharmacodynamic relationships of camptothecin anticancer drugs in human tumor xenograft models; 2) to determine the concentration profile of camptothecins in the extracellular fluid (ECF) of xenografts; 3) to elucidate the underlying mechanisms responsible for altered drug disposition in tumor-bearing mice compared to nontumor-bearing mice; 4) to develop a pharmacokinetic model to predict disposition of camptothecins in the CSF; and 5) to identify the optimal schedules and pharmacokinetic exposures producing antitumor activity for other new agents. These aims will be addressed primarily with the murine pediatric tumor xenograft model and in nonhuman primates to characterize cerebrospinal fluid disposition of drugs. Models of drug distribution into extracellular fluid (ECF) will be developed and validated using microdialysis methods. Within this program, the concentrations, exposure times, and intervals between exposures that optimize cytotoxicity will be characterized in vitro. In this project, these studies will be extended in vivo, using unique xenograft models of childhood solid tumors, to evaluate systemic exposure, tissue exposure, and tumor responses, and to construct pharmacodynamic models for clinical use. Studies of ECF and plasma drug concentrations will determine whether rates or duration of drug administration alter distribution into tumor tissues. Laboratory studies will be undertaken to identify altered drug metabolism pathways in mice bearing xenographs, compared to normal mice. Translation of the findings from this project will be incorporated into clinical studies to develop optimal dosing regimens and to verify the relatonships [sic] identified in the animal models.

Grant Number: 3P50CA058185-08S30008
Project Title: *Prevention models*
Abstract: . . . Retinoids and antiestrogens such as tamoxifen have also been proposed as chemopreventive agents in breast cancer. Preliminary results in one animal model of mammary carcinogenesis demonstrate for the first time that 9-cis retinoic acid, the natural ligand for the RXR class of retinoid receptors, can significantly decrease tumor incidence and tumor burden. Importantly, in combination with tamoxifen, 9-cis retinoic acid significantly extends tumor latency, further decreases tumor incidence and number, and increases the number of tumor free animals. An important goal of the proposed work will be To [sic] use a number of other animal models of mam-

mary carcinogenesis to confirm the chemopreventive effects of 9-cis retinoic acid and synergy with tamoxifen. We plan to optimize chemoprevention strategies in these animal models prior to initiating combined retinoid and tamoxifen chemoprevention trials in humans.

Grant Number: 1R43HL083555-01
Project Title: *Gene-targeted miniature swine models of atherosclerosis*
Abstract: . . . These genetically modified miniature swine will be expanded through breeding and then sold to meet a growing need of medical device and pharmaceutical companies for uniform animal models of human pathologies that can help predict the outcome of human therapeutic interventions.

Grant Number: 1R21NS062367-01A1
Project Title: *Understanding Mechanisms of Fetal Hypoxic Brain Injury Resulting in Cerebral Palsy*
Abstract: . . . This study is designed to use magnetic resonance imaging to improve our understanding of the fetal brain injury resulting in cerebral palsy and other motor deficits. Based on animal model data, this understanding has the potential to identify human fetuses-at-risk, early diagnose fetal brain injury and implement preventive or therapeutic strategies, ultimately reducing incidents and disabilities of cerebral palsy.

APPENDIX 4
♦
OLSON ARTICLE

Regulatory Toxicology and Pharmacology 32, 56–67 (2000)
doi:10.1006/rtph.2000.1399, available online at http://www.idealibrary.com on IDE**L®

Concordance of the Toxicity of Pharmaceuticals in Humans and in Animals

Harry Olson,[1] Graham Betton,[2] Denise Robinson,[3] Karluss Thomas,[3] Alastair Monro,[1] Gerald Kolaja,[4] Patrick Lilly,[5] James Sanders,[6] Glenn Sipes,[7] William Bracken,[8] Michael Dorato,[9] Koen Van Deun,[10] Peter Smith,[11] Bruce Berger,[12] and Allen Heller[13]

[1]Pfizer Inc., Groton, Connecticut; [2]AstraZeneca Pharmaceuticals, Macclesfield, England; [3]ILSI-HESI, Washington, DC, 20036; [4]Pharmacia & UpJohn, Kalamazoo, Michigan; [5]Boehringer Ingelheim Pharmaceuticals, Ridgefield, Connecticut; [6]Rhone-Poulenc Rorer, Collegeville, Pennsylvania; [7]University of Arizona, Tucson, Arizona; [8]Abbott Laboratories, Abbott Park, Illinois; [9]Eli Lilly and Co., Greenfield, Indiana; [10]Janssen Research Foundation, Beerse, Belgium; [11]Monsanto-Searle Laboratories, Skokie, Illinois; [12]Sanofi-Synthelabo, Inc., Malvern, Pennsylvania; and [13]Bayer Corporation, West Haven, Connecticut

Received January 22, 2000

This report summarizes the results of a multinational pharmaceutical company survey and the outcome of an International Life Sciences Institute (ILSI) Workshop (April 1999), which served to better understand concordance of the toxicity of pharmaceuticals observed in humans with that observed in experimental animals. The Workshop included representatives from academia, the multinational pharmaceutical industry, and international regulatory scientists. The main aim of this project was to examine the strengths and weaknesses of animal studies to predict human toxicity (HT). The database was developed from a survey which covered only those compounds where HTs were identified during clinical development of new pharmaceuticals, determining whether animal toxicity studies identified concordant target organ toxicities in humans. Data collected included codified compounds, therapeutic category, the HT organ system affected, and the species and duration of studies in which the corresponding HT was either first identified or not observed. This survey includes input from 12 pharmaceutical companies with data compiled from 150 compounds with 221 HT events reported. Multiple HTs were reported in 47 cases. The results showed the true positive HT concordance rate of 71% for rodent and nonrodent species, with nonrodents alone being predictive for 63% of HTs and rodents alone for 43%. The highest incidence of overall concordance was seen in hematological, gastrointestinal, and cardiovascular HTs, and the least was seen in cutaneous HT. Where animal models, in one or more species, identified concordant HT, 94% were first observed in studies of 1 month or less in duration. These survey results support the value of in vivo toxicology studies to predict for many significant HTs associated with pharmaceuticals and have helped to identify HT categories that may benefit from improved methods. © 2000 Academic Press

INTRODUCTION

A vitally important theme in toxicology is the search for and the assessment of in vitro and in vivo models that are predictive for adverse effects in humans exposed to chemicals. The conduct of toxicology studies in laboratory animals is driven by experience, historical precedence, and governmental requirements, and the results of these studies usually, and reasonably, lead to restrictions on the use, or method of use, of the chemicals concerned. Such a process must be based on the assumption that the current choice of animal models and the design of the studies are truly predictive of human hazard. The reliability of this assumption has far-reaching repercussions in terms of the potential for inappropriate use of animals and the unnecessary deprivation of, or restrictions in the use of, valuable chemicals including pharmaceuticals. Identification of any weaknesses in the assumption could lead to revisions of existing regulations and stimulate the search for better methods for the safety evaluation of chemicals in the future.

There have been relatively few attempts to methodically assess the correlation between the toxicity caused by chemicals in animals and in humans. This is not surprising, given that the toxicity of many chemicals observed in humans is after accidental exposure, the quantitative details of which in terms of duration and intensity are often not known. Chemicals, which are components of the diet, either macro- or micro-, are more susceptible to evaluation of their toxicity in animals and in humans, provided that the means to carry out epidemiological studies are available. However, a rich source of relevant information is pharmaceutical chemicals. For these, the human exposure is controlled and measured accurately. In addition, clinical studies of drugs employ systematic clinical examinations and

56

tests of organ function, aimed specifically at the detection of adverse effects.

There are few published analyses of comparative animal–human toxicity data on pharmaceuticals, with progress presumably inhibited by the perceived confidential nature of such data. The first brave foray into this area was by Litchfield when he analyzed an array of toxicities in rats, dogs, and humans for six diverse drugs being developed by his company (Litchfield, 1962). He reported that toxicities that occurred in rats only were rarely observed in humans and those in dogs only occurred slightly more frequently in humans, while those that occurred in both rats and dogs showed about a 70% concordance with humans.

Cytotoxic anticancer agents, by their very nature, tend to cause much toxicity in humans and several groups have examined the extent to which toxicities seen in humans can be predicted from animal data for these drugs (Owens, 1962; Schein et al., 1970; Rozencweig et al., 1981). Generally, these drugs caused qualitatively similar toxicity in animals and in humans, with data from dogs predicting gastrointestinal toxicity in humans particularly well and data from dogs and monkeys grossly overpredicting hepatic and renal toxicity. Rozencweig et al. (1981) warned that the predictability of such data is highly dependent on the prevalence of the particular human toxicity, with rare toxicities being essentially unpredictable from animal data.

Clinical toxicity data for more diverse types of drugs have also been the subject of several workshops and overviews (Lawrence et al., 1984; Fletcher, 1987; Lumley and Walker, 1990; Parkinson et al., 1994). In one small series in which the toxicity in clinical trials led to the termination of drug development, it was found that in 16/24 (67%) cases the toxicity was not predicted in animals (Lumley, 1990). In another analysis, 39/91 (43%) clinical toxicities (from 64 marketed drugs) were not predicted from animal studies (Igarashi, 1994). This latter publication forms part of the largest data set known to us, that of the Japanese Pharmaceutical Manufacturers Association (JPMA, 1994). This was derived from the literature (as distinct from questionnaire-derived data) and refers to data from 139 drugs approved in Japan from 1987 to 1991. The animal toxicity data are drawn from 468 repeated-dose studies, mainly in rats and dogs but with a few studies in mice and monkeys. No indication was given about the importance of the clinical toxicity, e.g., whether it was trivial or whether it led to a restriction in the use of the drugs. There were few correlations across species with, overall, the best predictivity being for cardiovascular events, and the poorest for cutaneous and hypersensitivity phenomena. Despite its relatively high incidence in all species, hepatobiliary toxicity in humans was surprisingly poorly predicted from animal studies. The JPMA also conducted an enterprising review of the

extent to which general pharmacology studies were useful in predicting adverse effects in humans (Igarashi et al., 1995). A total of 141 drugs were reviewed and the analysis showed considerable value in tests of spontaneous locomotor activity in mice, gastrointestinal transit time in mice, gastric secretion in rats, and urinary retention and sodium excretion in rats.

Two reviews addressed those drug cases where the clinical toxicity was so severe as to lead to withdrawal from marketing in the approximate period 1960–1990 (Heywood, 1990; Spriet-Pourra and Auriche, 1994). In one report only 4 of 24 cases were predictable from animal data; in the other report, only 6 of 114 clinical toxicities had animal correlates. Such a poor correspondence is not surprising, given that these late-onset phenomena once on the market are usually idiosyncratic in nature, i.e., of very low incidence, not dose-related, and apparently not related to the pharmacology of the compound.

A knowledge of pharmacology in various species, including humans, tells us that species can differ markedly in their response to pharmacological agents. Indeed, it has been reported that 29% of withdrawals of drugs from development are attributable to an inappropriate (e.g., lack of efficacy or selectivity) pharmacological response (Prentis et al., 1988). Against that background, one would expect diverse species responses to many toxic stimuli. Several reviews (Oser, 1981; Calabrese, 1984, 1987; Garratini, 1985; Zbinden, 1993) summarize and discuss the many differences in anatomy, physiology, or biochemistry between laboratory animals and humans and these can provide useful points of reference for anticipating whether a particular chemical is likely to show a similar response in an animal species and in humans. In addition to the fundamental differences between species in biological response, Zbinden (1991) cautioned against too great an expectation from animal toxicology studies for a host of reasons inherent in the designs of such studies and of clinical trials (see Table 1).

In a symposium that addressed the question of the relevance of animal toxicology studies for humans, several contributors urged the pharmaceutical industry to collaborate and pool its data on toxicity of drugs in development with a view to drawing broadly based conclusions (Brimblecombe, 1990; McLean, 1990). In this paper we report the first product of just such a collaboration. Twelve companies provided codified data to the International Life Sciences Institute (ILSI) who coordinated the compilation and analysis of the data. The primary objective was to examine how well toxicities seen in preclinical animal studies would predict actual human toxicities for a number of specific target organs using a database of existing information. To a lesser degree, the symposium also sought to better understand the duration of dosing required in animals

OLSON ET AL.

TABLE 1
Some Differences between Animals and Humans
Critical to Prediction of Toxicity

	Animals	Man
Subjects		
Number	Large groups	Individuals
Age	Young adult	All ages
State of health	Healthy	Usually sick
Genetic background	Homogeneous	Heterogeneous
Doses		
Magnitude	Therapeutic to toxic	Therapeutic
		Therapeutic
Schedule	Usually once daily	optimum
Circumstances		
Housing	Uniform, optimal	Variable
Nutrition	Uniform, optimal	Variable
Concomitant		
therapy	Never	Frequent
Diagnostic procedures		
Verbal contact	None	Intensive
Physical exam	Limited	Extensive
Clinical lab	Limited, standardized	Individualized
Timing	Predetermined	Individualized
Autopsy	Always	Exceptional
Histopathology	Extensive	Exceptional

to reveal the same toxicity in man where the same toxicity was seen in animals and man.

APPROACH TO DATA COLLECTION AND METHODS OF ANALYSIS

Although a considerable effort was made to collect data that would enable a direct comparison of animal and human toxicity, it was recognized from the outset that the data could not answer completely the question of how well animal studies predict overall the responses of humans. To achieve this would require information on all four boxes in Fig. 1, and this was not practicable at this stage. The magnitude of the data collection effort that this would require was considered impractical at this stage. The present analysis is a first step, in which data have been collected pertaining only to the left column of Fig. 1: true positives and false negatives. By definition, therefore the database only contains compounds studied in humans (and not on those that never reached humans because they were considered too toxic in animals or were withdrawn for reasons unrelated to toxicity). Despite this limitation, it was deemed useful to proceed in the expectation that any conclusions that emerged would address some of the key questions and focus attention on some of the strengths and weaknesses of animal studies.

The approach chosen was to collect and collate toxicities in humans (HTs) associated with exposure to pharmaceuticals. Since most data resided in the archives of the larger companies with long histories of drug development, these companies were approached

to participate. The companies were asked to tabulate HTs that had occurred during drug development. No restriction was placed on the time frame from which the data were drawn, except that it was requested that participants provide data from compounds in clinical development over a consecutive period of years, to avoid any bias from selected data sets.

A working party of clinicians from participating companies developed criteria for "significant" HTs to be included in the analysis. For inclusion a HT (a) had to be responsible for termination of development, (b) had to have resulted in a limitation of the dosage, (c) had to have required drug level monitoring and perhaps dose adjustment, or (d) had to have restricted the target patient population. The HT threshold of severity could be modulated by the compound's therapeutic class (e.g., anticancer vs anti-inflammatory drugs). In this way, the myriad of lesser "side effects" that always accompany new drug development but are not sufficient to restrict development were excluded. The judgments of the contributing industrial clinicians were final as to the validity of including a compound. The clinical trial phase when the HT was first detected and whether HT was considered to be pharmacology-related were recorded. HTs were categorized by organ system and detailed symptoms according to standard nomenclature (COSTART, National Technical Information Service, 1999).

In a subsequent step the company toxicologist examined the reports of toxicology studies in animals for the compounds that met the criteria of HT. Data examined included clinical signs, physiological measurements, hematology and clinical chemistry assays, and histopathology evaluations from rodent and nonrodent toxicology or safety pharmacology studies. A toxicity correlation was considered to be positive if the same target organ was involved in humans and in animals in the judgment of the company clinicians and the toxicologists. In the case of a positive correlation, the data provided were the rodent and nonrodent species tested and the duration of exposure (often the same as the duration of the study) at which the toxicity in question was first observed. In the event of lack of correlation, the data provided included all species tested, the longest duration of dosing in each species tested, whether dose-limiting toxicity was achieved, and whether there

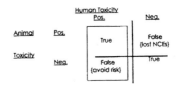

FIG. 1. Predictivity of animal toxicity data.

was a qualitative similarity in the patterns of metabolism in humans and in animals that were available.

All data were provided to the ILSI staff who entered them in codified fashion to maintain confidentiality. A central database of 221 HT examples from 150 compounds was developed for analysis. Three academic experts (Dr. G. Sipes of University of Arizona, Dr. R. Bain of George Washington University, and Dr. R. Abernethy of Georgetown University) advised on the study conduct of the analyses and interpretation of the data.

An ILSI Workshop was held in Virginia in April 1999, involving academic scientists, government regulators, and industry scientists, to review the data. At the Workshop, the participants were divided into groups to discuss the six principal types of HT reported: hepatic, neurological, cardiovascular, hematological, gastrointestinal, and hypersensitivity. These groups were asked to address the same series of generic questions on the data for their respective toxicities. Finally, a panel of experts was convened to assess the value and utility of the database and to make recommendations about the continuation of the project in the future.

RESULTS (PART 1)

Results are presented in two parts: first, overall analyses of the total database; and second, the answers of the breakout groups to the generic questions.

Overall Analysis of Findings (Total Database)

Preliminary results of analysis of the incomplete database have been reported previously (Olson, 1998).

Distribution of Human Toxicities by Therapeutic Class

Overall, a total of some 221 separate cases of compounds associated with significant human toxicity were recorded. A total of 150 compounds contributed to this series with multiple HTs being recorded in 47 cases. The distribution of the therapeutic class of compounds studied is shown in Table 2.

For each of the compounds in the database, information was collected regarding the route of administration. The routes of administration employed in humans were 168 HTs by oral, 52 intravenous, 7 by inhalation, and 2 dermal, with two routes of administration being used in the case of 9 HTs.

The therapeutic classes showed significant variations in their COSTART organ-system-associated HT profile as shown in Table 3. Detailed analyses of signs and symptoms within COSTART groupings are addressed in the following sections.

The rate of project termination for various HTs was highest for (in order) urogenital, cutaneous, hepatic,

TABLE 2
Distribution of Compounds by Therapeutic Class

Therapeutic class	No. of compounds
Anticancer	14
Anti-infection	21
Anti-inflammatory	15
Antiviral	8
Cardiovascular	16
Endocrine	10
Gastrointestinal	9
Hematology	1
Immunology	2
Impotence	2
Metabolism	5
Neurologic	31
Renal	2
Respiratory	13
Trauma	1
Total	150

and cardiovascular HTs and, by therapeutic class, highest for anti-inflammatory, antiviral, endocrine, and respiratory therapeutic classes (Table 3).

Relationship to Dosing Duration and Clinical Trial Phase

The time of first onset of HTs according to the clinical trial phase was analyzed according to HT class. Overall, over half of HTs were first manifest in Phase I trials. HTs seen after single-dose administration to man numbered 62 cases with 158 cases seen following multiple doses (remainder unspecified). Classes of HTs detected with frequency in Phases II and III were cutaneous and hepatic types (Table 4).

The survey also recorded the frequency of development project termination. In those instances where the HT led to project termination, 39% were terminated in Phase I, 43% were terminated in Phase II, and 10% were terminated in Phase III.

Only four HTs were considered to be idiosyncratic in nature, two cases of rash (one in Phase I, one in Phase II) and two cases of thrombocytopenia in Phase II.

Pharmacologic Basis of Human Toxicities

The characterization of HTs as being related to the primary pharmacological activity of the drug is given according to HT class in Fig. 2.

The overall distribution of pharmacological HTs according to clinical trial phase was 35% in Phase I, 39% in Phase II, and 43% in Phase III.

Concordance by Animal Models

Concordance by one or more species: Overall and by HT. Overall, the true positive concordance rate (sensitivity) was 70% for one or more preclinical animal

60   OLSON ET AL.

TABLE 3
Frequency of Human Toxicities According to Therapeutic Class and Percentage of Terminations

Therapeutic class	BCH	CUT	HEP/LFT	CV/ECG	END	NRL	HEM	GI	MSK	REPRO	URN	OTH	TOTAL	%TERMN
Anticancer	1	1	2	1	1	5	6	3	0	0	2	3	25	20
Anti-infection	0	3	6	2	0	6	2	10	3	0	1	5	38	37
Anti-inflammatory	0	4	2	2	0	6	0	6	0	0	2	0	22	55
Antiviral	0	1	3	0	0	1	1	2	0	0	3	0	11	54
Cardiovascular	0	0	1	11	0	4	1	1	0	0	0	0	18	39
Endocrine	0	2	3	2	2	1	0	2	0	0	0	0	12	50
Gastrointestinal	0	0	2	2	0	4	0	3	0	0	0	3	14	36
Hematology	0	0	0	0	0	0	1	0	0	0	0	0	1	0
Immunology	0	0	0	0	0	1	0	1	0	0	0	0	2	0
Impotence	0	0	0	2	0	0	0	0	0	0	0	2	4	0
Metabolism	0	0	1	1	0	1	0	2	0	0	0	0	5	20
Neurologic	0	2	5	8	0	18	0	11	1	0	1	2	48	33
Renal	0	0	0	2	0	0	0	0	0	0	0	0	2	100
Respiratory	0	1	6	2	2	1	0	0	0	1	1	3	17	47
Trauma	0	0	0	1	0	1	0	0	0	0	0	0	2	0
Total	1	14	31	36	5	49	11	41	4	1	10	18	221	37
% Terminated by HT	0	64	55	47	40	35	27	10	25	100	70	6		

Note. BCH, biochemical; CUT, cutaneous; END, endocrine; GI, gastrointestinal; HEM, hematologic; HEP/LFT, hepatobiliary and liver function test abnormalities; MSK, musculoskeletal; NRL, neurological; REPRO, reproductive; URN, urinary; OTH, other.

model species (either in safety pharmacology or in safety toxicology) showing target organ toxicity in the same organ system as the HT (Fig. 3). For the remaining 30% of HT there was no relationship between toxicities seen in animals and those observed in humans. Concordance was seen in 63% of nonrodent studies (primarily the dog) and 43% of rodent studies (primarily the rat). There was considerable overlap in toxicology with 36% of HTs being concordance with two species (i.e., a rodent and a nonrodent) with concordance by only one species occurring in the nonrodent (27% of HTs) and the rodent (7%).

The ratio of positive concordance versus nonconcordance by rodent and nonrodent species is shown in Fig. 4. The total incidence of usage of each species (concordance and nonconcordance) was incorporated and showed the nonrodent species of dog and primate to have a higher frequency of positive concordance than did rodents.

Concordance varied significantly according to the human target organ system affected as shown in Fig. 5. The best concordance was for hematological, gastrointestinal, and cardiovascular toxicities and the least

was for cutaneous toxicity. The proportional contribution of nonrodent versus rodent models to concordance for given types of HTs is discussed below for the main organ system HTs observed. There were marked differences in the relative contribution of nonrodent and rodent toxicology species to concordance depending on the HT category (Table 5).

Analysis of overall prediction rates according to clinical trial phase when the HT was first observed showed a slightly higher frequency in Phase I onset HTs (75%) compared to Phase II (58%) and Phase III (52%).

Concordance by therapeutic class. Since certain therapeutic classes were prone to expression of given types of HTs, e.g., hematotoxicity with anticancer agents, the variation in concordance rates according to HT class could indirectly influence the concordance rates for certain therapeutic classes. This is reflected in Fig. 6.

Time to first appearance of concordant animal toxicity. Where the animal model(s) were successful in predicting for a given HT, the survey requested the earliest time at which the relevant animal toxicity was

TABLE 4
Distribution of Clinical Trial Phase Time of First Onset by HT Class

Therapeutic class: Phase	BCH	CUT	CV/ECG	END	GI	HEP/LFT	HEM	MSK	REPRO	NRL	URN	OTH	TOTAL	TERMN	%
I	1	8	23	1	30	14	7	0	1	29	6	15	135	52	39
II	0	6	10	4	7	13	4	2	0	14	2	3	65	28	43
III	0	0	3	0	4	4	0	2	0	6	2	0	21	2	10

APPENDICES

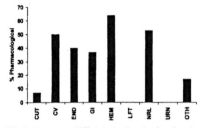

FIG. 2. Percentage of HTs judged to be related to the primary pharmacology.

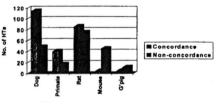

FIG. 4. Concordance rates versus species.

observed. Overall, 94% of animal target organ toxicities correlated with HTs were first observed in studies less than or equal to 1 month in duration (Fig. 7). A large proportion of animal toxicities was observed following single dose administration; 25% of these observations were from safety pharmacology rather than toxicity studies.

False Negative Prediction by Animal Models

Duration of animal studies where HT was not concordant (false negatives). Given the high rate of detection by animals of HTs in studies of 1 month duration or less, it is possible that failure to detect HTs may have resulted from insufficient duration of exposure in animal models. Analysis of the longest duration of studies conducted in animals shows that overall 73% of HT cases had been studied in animals for 2 months or more (Fig. 8).

Since 61% of HTs first occurred during Phase I clinical trials when chronic toxicology programs may not have been completed, the duration of animal testing on this subset was also analyzed: 45% of these Phase I onset cases had one or more toxicology species tested for 2 months or longer.

Achievement of limiting toxicity in animals in false negative cases. Since interspecies differences in exposure (toxicokinetic data were not collected in the sur-

vey) may have resulted in failure to demonstrate relevant target organ toxicity, the proportion of animal studies achieving dose limiting toxicity [up to maximum tolerated dose (MTD) in some cases] was analyzed. For cases where HTs were not predicted, 91% of rodent and 90% of nonrodent toxicology studies were judged to have been performed at limiting doses. Hence, insufficient exposure of animals to drug alone could not account generally for the 30% false negative rate.

Correlation of animal metabolite profile with man in false negative cases. Using a qualitative judgment of whether the main human metabolites were present in one or more animal toxicology species (data available in 29 of 63 false negative cases), animal metabolism profiles were considered to correlate with that of man in 86% of cases. Therefore, metabolic differences between animals and man alone probably do not explain the false negative cases. Taking both concordant and nonconcordant cases overall where comparative metabolism data were available, there was a 89% animal: human metabolite correlation rate.

Correlation of animal pharmacological responsiveness in false negative cases. Since approximately 40% of HTs were evaluated as being pharmacology-related, pharmacological unresponsiveness in the animal species could result in false negative prediction. Taking false negative prediction cases, the animal species used as models were, in one or more species, considered to be

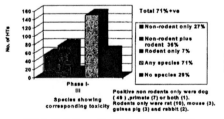

FIG. 3. Concordance of human toxicity from animals.

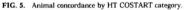

FIG. 5. Animal concordance by HT COSTART category.

　　　　　　　　　　　　　　　　OLSON ET AL.

TABLE 5
Animal Species Concordant Number of Cases for Frequent HT COSTART Systems

COSTART:	CUT	CV/ECG	GI	HEM	HEP/LFT	NRL
Animal correlates						
Yes	5	29	35	10	17	31
No	9	5	6	1	14	13
Nonrodent only	1	16	16	0	7	11
Rod and nonrodent	1	12	18	10	8	17
Rodent only	3	1	1	0	2	3

pharmacologically responsive in 63% of cases. Taking only those false negative cases of pharmacology-related HT prediction, 85% of animal models were pharmacologically responsive. Pharmacological unresponsiveness of the animal models therefore cannot alone account for the false negative rate. Taking both the true positive and the false negative cases together, 69% of animal species used were pharmacologically responsive with 87% being responsive for pharmacology-related HTs overall.

RESULTS (PART 2)

This section is a compilation of the responses of the six breakout groups at the ILSI Health and Environmental Science Institute's (HESI) Workshop (April 1999) which focused on the six principal types of HT. Each breakout group addressed the four main questions, listed below. These results include group comments which are similar; specific remarks from individual breakout groups are so noted.

Q.1. Evaluate the Database Generally in the Context of the Breakout Group's Specific Endpoints and Comment on: (i) Any Animal–Human Toxicity Correlations That Can Be Made; (ii) Whether the HT Was Related to the Therapeutic Class or the Known Primary Pharmacology of the Compound

Most groups were critical of the database to the extent that it often lacked specific detail regarding the

exact nature of the HT, including its incidence in patients, severity, and time to onset. Uncertainties in nomenclature, application of the COSTART terminology, and whether the HT was perceived to be problems. The *liver* group would like to have known whether the HT was hepatocellular or biliary cell injury, jaundice, or fulminant liver failure. The *hematology* group would like to have had more information on the type of cells affected and to have known if neutropenia existed alongside other hematotoxicity. The *cardiovascular* group would have been interested to know whether hypotension was accompanied by tachycardia.

The tentative conclusions regarding the incidence of HT and of termination by clinical phase (Results (Part 1), "Relationship to Dosing Duration and Clinical Trial Phase") must be tempered with caution. The HT data supplied to the database did not distinguish between the severity of the HT and its incidence. Thus a HT first observed in Phase III may have been the result either of a longer drug exposure or of a greater number of patients being in the trial than in earlier clinical phases. For example, termination may have been caused by 1% incidence of a severe HT or a 50% incidence of a mild HT.

Several groups noted limitations (already alluded to, see Table 1) inherent in such a human–animal comparison exercise. Thus, failure to distinguish between toxic signs and symptoms, and especially nausea and many of the subjective neurologic HTs (e.g., headache, hallucinations), emphasized the need to limit expectations for prediction of such HTs. Interpretation was

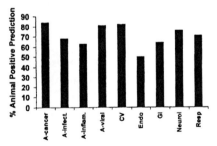

FIG. 6.　Preclinical concordance for HTs by therapeutic class.

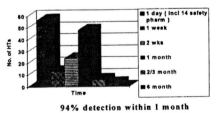

94% detection within 1 month

FIG. 7.　Time to first detection of relevant toxicity in animals.

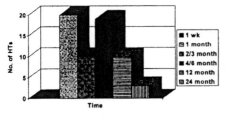

FIG. 8. Longest duration of nonconcordant toxicity studies ($N =$ 73).

also impeded by uncertainty over what specific evidence of toxicity in animals was considered to have predicted each HT. In general, it had to be assumed that any disturbance of the relevant organ system in animals was judged to be a positive correlation. For example, diarrhea in animals might correspond in humans to abdominal pain or nausea; excitation in animals to, say, dizziness in humans; and so on.

The *hypersensitivity* group observed that tests for such activity were rarely included in preclinical toxicology packages and thus correlations would not be expected with the data routinely obtained from animals. Also, while most HTs in this category were cutaneous but there was a notable data gap with few compounds applied via the dermal route. Also a proven immunological basis for the symptoms of rash was not evident in all cases.

The discrepancies between the incidence of each HT and the frequency with which that HT led to termination in clinical trials were interesting. It was recognized by several groups that the decision to terminate or continue in clinical trial will, of necessity, depend on the therapeutic ratio and the seriousness of the condition being treated and also the stage of clinical development and investment with few Phase III HTs resulting in project termination.

Liver toxicity was only the fourth most frequent HT (Table 3), yet it led to the second highest termination rate. There was also less concordance between animal and human toxicity with regard to liver function, despite liver toxicity being common in such studies. There was no relation between liver HTs and therapeutic class. This poor correlation prompted the question of whether a partial explanation might lie in the assumption that the same biomarkers were as appropriate in humans as in animals. The first signs of unwanted effects on the hepatobiliary system in humans are usually rises in circulating aminotransferase enzyme levels. The JPMA study (Igarashi, 1994) reported that these enzymes are relatively insensitive markers of liver toxicity in animals, but that the correlation with human toxicity was much better if one incorporated histopathology data from animals (also included in the present survey).

Neurologic HTs were the most common category (22%) and they occurred disproportionately more frequently for neurologic drugs. It was noted that whereas the correlation with animal studies was independent of the therapeutic class of drug, these HTs led to termination of only 17% of neurologic drugs but 45% of nonneurologic drugs. Such HTs are apparently more acceptable in neurologic drugs, this obviously depending to some extent on the concurrent therapeutic benefit. Correlations with findings in animals were much better for nonrodents than for rodents. The cases in rodents were all peripheral neuropathies in anticancer agents.

Cardiovascular HTs had a high rate of concordance (80%) with animal studies. Most were caused by cardiovascular drugs (Table 2) and interpretable in terms of their primary pharmacological activity. The main categories were: rhythm changes (tachycardia, bradycardia), ECG abnormalities, hypotension, and vasodilatation. While data from rodents were indicative of a potential for hypotension, rodents contributed no information that was not available from nonrodents although the rabbit detected an ECG HT case. By contrast, 28/36 of the HTs were observed in nonrodents (25 in dogs, 6 in monkeys).

Hematologic HTs correlated in high degree (91%) with animal findings, with rodents and nonrodents both being responsive. Termination due to this HT was relatively low (27%). This HT was strongly associated with anticancer and anti-infective agents, with only two compounds coming from other therapeutic classes.

Gastrointestinal HTs overall correlated well (85%) with animal findings, especially in nonrodents. This HT was associated primarily with anticancer, anti-infective, and anti-inflammatory agents, all known to provoke such toxicity by well-understood pharmacological mechanisms. Gastrointestinal HTs were the second most common category, yet had the lowest rate of termination. One might say that these HTs are apparently often just regarded as "nuisance" effects.

Hypersensitivity HTs (essentially various cutaneous reactions) were too few to allow conclusions about relationship to therapeutic class or pharmacological mode of action, in particular whether the HTs were immunological or not in character. Nevertheless, the high rate of termination (64%), the highest in the database, highlighted the need to improve preclinical testing methodology in this area (*vide infra*).

Q.2. What Duration of Dosing in Animals Was Sufficient to Reveal the Toxicity That Corresponded to a HT?

Several groups pointed out that the minimum times given in Fig. 7 may well be overestimates for many toxicities, i.e., those dependent on time of sampling for biochemical or hematological tests, on time of functional investigation, or on time of termination for histopathological examination. Even so, the data in Fig. 7 are striking in that 57 (38%) of the relevant toxicities were observed on Day 1, these being primarily cardiovascular (16), gastrointestinal (15), and neurological (12) phenomena. Liver toxicity was never reported after a single dose. Most of the single dose observations for cardiovascular events came from specific safety pharmacology studies; the others came mainly from clinical observations on the first day of a multidose study.

The nine HTs that required over 1 month to manifest themselves in animals showed no pattern. They comprised one or two agents from each of the HT categories.

Q.3. Which HTs Were Not Detected by the Animal Studies?

The *neurologic* group inevitably found many examples of nondetection (or perhaps "not known") because of the need to communicate symptoms such as headache, dizziness, etc. The only objective sign not predicted (seizure) was associated with a difference in *metabolism* between humans and the animals tested.

The *cardiovascular* group found a case of myocardial infarction not predicted from animal studies, perhaps because of a lack of an animal counterpart. Two cases of hypotension were not observed in animals: in one the MTD was not achieved in animal studies; in the other a clear difference in animal/human metabolism was noted.

The *hematologic* HT that sometimes escaped detection was thrombocytopenia: two cases, even after 6-month studies in animals. The panel suggested that this toxicity could have been detected by nonstandard methodology.

The only *gastrointestinal* HT that did not correlate with animal studies was nausea. This is not surprising, given its subjective nature and the uncertainty regarding whether it should be classified here or as a neurologic HT.

No cutaneous (classified here as *hypersensitivity*) HTs were observed using the standard animal studies conducted, although phototoxicity did correlate well with the response of guinea pigs in special tests.

Q.4. Identify Novel and/or Available Technologies That Could Be Included in Existing Animal Studies to Address Deficiencies in Identifying Target Toxicity, and Identify Testing Strategies— Including Additional Studies—That Could Be Implemented to Improve for Screening Compounds into Development

The major recommendations (not in any particular order) were the following:

a. It was suspected that some of the poor correlations may have been due to apparent design deficiencies in animal studies. Thus, the hematology breakout group recommended the use of toxicokinetic and tissue distribution studies to lead to better timing of blood sampling and function tests. Safety pharmacology studies should be subject to the same considerations, as recommended previously by Jorkasky (1998).

It was recognized that interspecies differences in pharmacokinetics are unlikely to underlie many of the noncorrelations in this database, given the manifestly different pharmacological responses of species and the fact that almost all the animal studies were carried out at toxicity-limited dose levels. Nevertheless, the quantitative aspect cannot be completely disregarded. There must be certain pharmacological mechanisms that allow approximate allometric relationships in dose–responses across species. If established, this would give a perspective on toxicity observed in animals only at enormous multiples of the expected clinical dose and would provide guidance to a clinician planning to increase the dose level of a drug in a clinical trial.

b. The *cardiovascular* and *neurologic* groups, cognizant of the importance of safety pharmacology studies for these HTs, urged that these studies (including a functional observational battery) be extended, refined, and better integrated with general toxicology studies. Others have emphasized the critical contribution of general pharmacology studies to safety in the conduct of early clinical trials (Igarashi *et al.*, 1995; Jorkasky, 1998; Williams, 1990). The limitations of conducting pharmacological measurements in animals under the restrained conditions of a normal toxicology study were acknowledged. Safety pharmacology studies should align with toxicology studies in terms of choice of species and dosage regimen as appropriate.

The value of supporting mechanistic studies, often *in vitro*, was mentioned, but these will not usually be conducted broadly as part of a preclinical screen; however, certain compound classes may trigger particular screens. The *cardiovascular* group drew attention to the use of blood pressure measurement by telemetry, Purkinkje fiber preparations, electrophysiology of myocardial cells, etc. The *hematology* group mentioned flow cytometry and bone marrow culture. While these and other similar experimental tools have unquestionable value for exploration of the mechanism of toxicity, they have not hitherto found application in the prediction of HT. Companies were asked, when a HT had an animal correlate, if it was derived from a standard or a nonstandard toxicology study; only 13/148 cases were from nonstandard studies.

c. The choice of species might also be the subject of more thoughtful consideration. Often studies are conducted, in the dog and the rat, without an open-minded consideration of whether an alternative species might be better in terms of pharmacodynamics, physiology, biochemistry, metabolism, etc. Guinea pigs are obviously well-established in testing for *hypersensitivity;* the *neurological* group drew attention to the utility of specially trained primates, though again presumably not as a routine preclinical screen. The *liver* and *gastrointestinal* groups suggested that animal disease models could be put to better use. The human diseases being treated may increase patient susceptibility to a, e.g., through increased gastrointestinal permeability which would not be reflected in animals with normal gastrointestinal reserve.

d. Several groups urged the more imaginative use of biomarkers. These could perhaps reveal, on the one hand, hitherto undetected changes in animals and, on the other, earlier detection of HTs. The biomarkers could be different in animals and in humans. Examples included troponin T and CK-MB in the *cardiovascular* area and α2-antitrypsin in the *gastrointestinal* area. Along the same lines, the *hypersensitivity* group mentioned the underestimated value of lymph node assays and that systematic weighing of immune/lymphoid organs in animal studies often provides a first indication of a disturbance to the immune system. The *hematologic* group drew attention to the potential for wider exploitation of newer technique such as flow cytometry and bone marrow culture.

e. The *hypersensitivity* group pointed out that since the present design of animal preclinical toxicology studies has little scope for assessment of immunological endpoints, the poor human–animal correlation for this HT category could only be rectified by the routine addition of tests or models aimed at the detection of systemic or cutaneous hypersensitivity and, where relevant, phototoxicity, in relevant species. This group also acknowledged that whereas phototoxicity testing in guinea pigs is fairly reliable as an indicator of human hazard, other test systems for hypersensitivity are far from satisfactory and there is need for urgent research in this area.

f. One way to diminish dependence on the animal to human extrapolation process is to work directly with human tissue. The *liver* group noted the potential value of human liver slices and other *ex vivo* or *in vitro* preparations to obtain of information on metabolic transformation of test compounds and, conversely, of potential effects of the test compound on the liver.

g. Several groups speculated, without being specific, on the possible future use of molecular biological techniques, such as gene expression profiling proteomics, the use of gene chips, etc. The main value may be in identifying in advance individuals with intrinsic susceptibility to various HTs.

DISCUSSION

This study did not attempt to assess the predictability of preclinical experimental data to humans. What it evaluated was the concordance between adverse findings in clinical data with data which had been generated in experimental animals (preclinical toxicology).

This HESI Workshop and collaborative project is unique (to our knowledge) in magnitude of the database and scope of project. This is an initial step to develop a quantitative understanding of concordance of animal target organ toxicology and manifest HT associated with pharmaceutical development. The intent of this project at its inception was to relate the value of preclinical testing models and methods to identify important HT, which by definition is "relevant" toxicity. This approach provides useful perspective for the types of HT evaluated. It is recognized to be limited by not being able to fully explore all aspects of "predictivity" of HT (see Fig. 1).

No restrictions were placed on the time period from which qualifying data sets could be submitted. Indeed, the inclusive years of data collection for the full database are unknown. Factors influencing the individual data submissions include the refinement of protocols in recent years, unavailability of certain types of data and endpoints from earlier studies, and the development of GLP protocols which might impact data quality. Other than limiting the availability of certain types of information such as data on metabolites and toxicokinetics, it is unknown whether the unevenness of study designs over time might have other influences on the outcome of the database analysis.

A significant message from this survey is that the two HTs with the poorest correlation with animal studies (liver and hypersensitivity/cutaneous reactions) were also the two HTs that led most often to termination of clinical development. This highlights the need for progress in these two areas. The way forward must surely be to increase investigations of mechanism; each occurrence of any unpredicted HT (not just liver or skin) should be followed by investigation of the mechanism involved which, in turn, should lead to a search for a nonclinical predictive model.

The results of this survey and the workshop breakout group discussions have identified several key findings and have also revealed several areas for additional evaluation to pursue in a future project. The main finding of this study is the true positive concordance rate of over 71% for comparable target organs in animal toxicity studies for identified HTs. In addition the survey supports the utility and relevance of studies of up to 1 month in duration with target organ toxicity alerts seen in over 90% of cases. Prior to ICH recommendations on parity of preclinical versus clinical dosing duration, many companies performed 1-month studies before entry into Phase I.

OLSON ET AL.

TABLE 6
Distribution of Therapeutic Classes (%)

	This database	World scene	
		In development Dec. 1997[a]	Marketed 1989–1998[b]
Anticancer	9	16	8
Anti-infective	14	9	16
Cardiovascular	11	15	21
Endocrine	7	10	8
Gastrointestinal	6	3	3
Immunological	1	7[c]	13[c]
Neurological	21	24	16
Respiratory	9	6	4
Other	22	9	11

[a] *Source:* Ashton *et al.* (1998) (data from 42 companies).
[b] From CMR (1999).
[c] Includes anti-inflammatory therapies.

Since 39% of the HTs described in this study were first observed in short-term duration Phase I clinical studies, including 28% seen following single-dose administration to man, it may be important for a future survey to determine the exposure ratio (duration and therapeutic index) between animal study findings and HT occurrence. This may help to explain why compounds were progressed in the clinic despite preclinical evidence of potential toxicity. Additionally, in a future prospective survey it would be useful to identify the duration of animal studies required to identify all HTs of a specific type and especially the time to onset of those HTs observed in Phase II and III clinical trials. Of course there remains important value to the conduct of longer duration preclinical studies including changes in NOELs over time, progression of target organ toxicities with chronic administration, and evaluation of processes leading to carcinogenesis. These aspects were not covered in the current survey.

A remaining need, and shortcoming of this survey—as pointed out in the Introduction—is that the design did not include "false positive" and "true negative" outcomes to determine the discriminating value of prospective preclinical toxicity biomarker signals to predict HTs. A more complete evaluation of this predictivity aspect will be an important part of a future prospective survey.

A question raised at the Workshop concerned to what extent this database was representative of the range of drug types under development and marketed worldwide. If it were not, caution would be called for in drawing general conclusions about the reliability of animal models. A comparison of the distribution of therapeutic classes in the present database with that of drugs in development and marketed in recent years is shown in Table 6. This shows that relative to drugs in development, the HESI database is somewhat overrepresented by anti-infective agents and underrepresented by anti-

cancer agents; in comparison with marketed drugs, there is some overrepresentation of neurologic and respiratory classes and some underrepresentation of cardiovascular and immunological classes. Although unlikely, it cannot be excluded that this may be due, in part, to the varied vintage of the compounds submitted. None of the discrepancies would be sufficient to seriously distort the overall conclusions of this survey.

The Workshop concluded that the project had been of real value in bringing together scientists from the pharmaceutical industry, academia, and government regulatory agencies to discuss the strengths and weaknesses of the current toxicology strategies. It was agreed that it was desirable that the project should proceed to a second phase, essentially to broaden and extend the present database to ideally add measures of true negative and false positive rates for preclinical toxicology. Lessons learned in this first phase, discussed above in response to Q.1, would be applied to make the next data collection more informative and open to additional questioning. Additionally, in a future exercise one would be assured that the data submitted were representative of current pharmaceutical research activity. Other companies wishing to participate are invited to contact Karluss Thomas, one of the authors of this paper.

ACKNOWLEDGMENTS

We recognize the contributions of the cochairs of the six breakout groups to the HESI Workshop in providing organization and leadership at the workshop sessions. They are Dr. Louis Cantilena, Armed Forces Institute of Pathology; Dr. Gabre-Mariam Mesfin, Pharmacia & Upjohn; Dr. Ralph Heywood, Independent Consultant; Dr. Fred Radzialowski, Monsanto/Searle Laboratories; Dr. Anthony Dayan, University of London; Dr. Nasir Khan, Monsanto/Searle Laboratories; Dr. Neal Shear, Sunnybrook Health Science Center; Dr. Joseph Vos, National Institute of Public Health and the Environment (The Netherlands); Dr. Francois Ballet, Rhone Poulenc Rorer; Dr. Paul Watkins, University of Michigan Medical School; Dr. Marc Bonnefoi, Rhone-Poulenc Rorer; and Dr. Pamela Cyrus, Bayer Corporation. Summary reports of each breakout session topic are available upon request from Karluss Thomas (HESI). Acknowledgments are also extended to the management of all sponsor companies for providing input into the database used in this survey. Participating companies included Abbott Laboratories, Bayer Corporation, Boehringer-Ingelheim Pharmaceuticals Inc., Eli Lilly & Company, Janssen Research Foundation, Monsanto/Searle Laboratories, Novartis Pharmaceutical Corporation, Pfizer Pharmaceuticals, Pharmacia & Upjohn Inc., Rhone-Poulenc Rorer Inc., Sanofi Pharmaceuticals Inc., and Zeneca Pharmaceuticals. Additionally, acknowledgments also are extended to all HESI Workshop participants for providing input and critique of the database and the survey instrument, to improve the planned Phase 2 survey.

REFERENCES

Ashton, G. A., Lobo, L. I., Griffiths, S. A., and McAuslane, N. (1998). Activities of the international pharmaceutical industry in 1997: Pharmaceutical investment and output. *Centre Med. Res. CMR* 98–104R.

Brimblecombe, R. (1990). The importance of retrospective comparisons. In *CMR Workshop—Animal Toxicity Studies: Their Rele-*

vance for Man (C. E. Lumley and S. W. Walker, Eds.), pp. 15–19. Quay, Lancaster, UK.

Calabrese, E. J. (1984). Suitability of animal models for predictive toxicology: Theoretical and practical considerations. Drug Metab. Rev. 15, 505–523.

Calabrese, E. J. (1987). Principles of Animal Extrapolation. Wiley, New York.

CMR (1999). CMR News, Vol. 17. Centre for Medicines Research International, Epsom, Surrey, UK.

Fletcher, A. P. (1987). Drug safety tests and subsequent clinical experience. J. R. Soc. Med. 71, 693–696.

Garratini, S. (1985). Toxic effects of chemicals: difficulties in extrapolating data from animals to man. Annu. Rev. Toxicol. Pharmacol. 16, 1–29.

Heywoood, R. (1990). Clinical toxicity—Could it have been predicted? Post-marketing experience. In Animal Toxicity Studies: Their Relevance for Man (C. E. Lumley and S. W. Walker, Eds.), pp. 57–67. Quay, Lancaster, UK.

Igarashi, T. (1994). The duration of toxicity studies required to support repeated dosing in clinical investigation—A toxicologists opinion. In CMR Workshop: The Timing of Toxicological Studies to Support Clinical Trials (C. Parkinson, N. McAuslane, C. Lumley, and S. R. Walker, Eds.), pp. 67–74. Kluwer, Boston/UK.

Igarashi,T., Nakane, S., and Kitagawa, T. (1995). Predictability of clinical adverse reactions of drugs by general pharmacology studies. J. Toxicol. Sci. 20, 77–92.

Jorkasky, D. K. (1998). What does the clinician want to know from the toxicologist? Toxicol. Lett. 102–103, 539–543.

JPMA (1994). Seiyakukyo data No. 65. Translation of a report from the Japanese Pharmaceutical Manufacturers Association, provided by Dr. T. Igarashi.

Lawrence, D. R., McLean, A. E. M., and Weatheral, M. (1984). Safety Testing of New Drugs. Academic Press, New York.

Litchfield, J. T. (1962). Evaluation of the safety of new drugs by means of tests in animals. Clin. Pharmacol. Ther. 3, 665–672.

Lumley, C. (1990). Clinical toxicity: Could it have been prevented? Premarketing experience. In CMR Workshop—Animal Toxicity Studies: Their Relevance for Man (C. E. Lumley, and S. W. Walker, Eds.), pp. 49–56. Quay, Lancaster, UK.

Lumley, C. E., and Walker, S. R. (Eds.) (1990). CMR Workshop—Animal Toxicity Studies: Their Relevance for Man. Quay, Lancaster, UK.

McLean, A. (1990). In CMR Workshop—Animal Toxicity Studies: Their Relevance for Man (C. E. Lumley and S. W. Walker, Eds.), pp. 79–82. Quay, Lancaster, UK.

National Technical Information Service (1999). COSTART: Coding symbols for Thesaurus of Adverse Reaction Items, Version 5. National Technical Information Service, Springfield, VA.

Olson, H., Betton, G., Stritar, J., and Robinson, D. (1998). The predictivity of the toxicity of pharmaceuticals in humans from animal data—An interim assessment. Toxicol. Lett. 102–103, 535–538.

Oser, B. L. (1981). The rat as a model for human toxicological evaluation. J. Toxicol. Environ. Health 8, 521–642.

Owens, A. H. (1962). Predicting anticancer drug effects in man from laboratory animal studies. J. Chron. Dis. 15, 223–228.

Parkinson, C., McAuslane, N., Lumley, C., and Walker, S. R. (Eds.) (1994). CMR Workshop: The Timing of Toxicological Studies to Support Clinical Trials. Kluwer, Boston/UK.

Prentis, R. A., et al. (1988). Pharmaceutical innovation by the seven UK-owned pharmaceutical companies. Br. J. Clin. Pharmacol. 25, 387–396.

Rozencweig, M., et al. (1981). Animal toxicology for early clinical trials with anticancer agents. Cancer Clin. Trials 4, 21–28.

Schein, et al. (1970). The evaluation of anticancer drugs in dogs and monkeys for the prediction of qualitative toxicities in man. Clin. Pharmacol. Ther. 11, 3–40.

Spriet-Pourra, C., and Auriche, M. (Eds.) (1994). SCRIP Reports. PJB, New York.

Williams, P. D. (1990). The role of pharmacological profiling in safety assessment. Regul. Toxicol. Pharmacol. 12, 238–252.

Zbinden, G. (1991). Predictive value of animal studies in toxicology. Regul. Tox. Pharm. 14, 167–177.

Zbinden, G. (1993). The concept of multi-species testing in industrial toxicology. Regul. Toxicol. Pharmacol. 17, 85–94.

PERSPECTIVES

OPINION ◉

Modern biomedical research: an internally self-consistent universe with little contact with medical reality?

David F. Horrobin

Congruence between *in vitro* and animal models of disease and the corresponding human condition is a fundamental assumption of much biomedical research, but it is one that is rarely critically assessed. In the absence of such critical assessment, the assumption of congruence may be invalid for most models. Much more open discussion of this issue is required if biomedical research is to be clinically productive.

Pharmaceutical research is failing in its ability to deliver new drugs. Furthermore, as much pharmaceutical research draws on the wider biomedical research community, the possibility is that biomedical research in general is also failing. There can be no doubt about the reality of the pharmaceutical failure. Almost all the major drug companies are generating inadequate numbers of new drugs and as a result their market values have fallen catastrophically. There is no relief in sight. The US FDA and the European Medicines Evaluation Agency (EMEA) have recently announced that 2002 applications for new drug approvals are well below the depressingly low levels of 2001. What is going on?

The failure is certainly not due to lack of investment, as the major drug companies are spending well over $30 billion per year on R&D. Two other explanations are common. One is that all the easy problems have been solved: what is left is much more difficult. The other is that the new technologies will

bear fruit in time given enough further investment. There is, of course, no way to critically evaluate these two alternatives: all we can do is hope.

A fourth possibility does deserve critical evaluation, but attracts surprisingly little open discussion. This is that biomedical science, and hence pharmaceutical science, has taken a wrong turn in its relationship to human disease. This discussion paper raises some issues which have been swept under the carpet by the pharmaceutical and medical research communities.

The glass bead game

A wonderful metaphor of much modern medical and pharmaceutical research can be found in the book entitled *The Glass Bead Game* by Herman Hesse[1] (FIG. 1). In this story, the leaders of the real world conspire with the brightest of scholars to create a magical state within a state, the isolated world of Castalia. Castalia recruits the most thoughtful and scholarly youths, educates them wonderfully well, and persuades them that the highest achievement of the human mind is to play the almost infinitely complicated and subtle 'glass bead game', an intellectual Olympics which challenges and stretches the most exceptional. The world of the game is beautifully refined, and internally self-consistent. The only problems are that Castalia makes almost no contact with the real world, and that playing the game makes no contribution to real world issues. The outcome, however, is

acceptable to all sides. The real world sidelines its potentially troublesome intellectuals by seducing them to play in their heavily subsidised glass cage. The intellectuals avoid the need to test their brains on any real issues.

When I look at the world of medical and pharmaceutical research it seems to me that we are well on the way to creating a Castalia which is entirely acceptable to the majority of its scientist-priests. They receive funding from the real world and are inducted into a complex organisation which, for those who know how to play the game, creates an ever-expanding universe of intellectual and social possibilities. This universe, like Hesse's Castalia, has few points of interaction with real medical problems: its inhabitants, like the original Castalians, are happy to keep it that way. The difference from Castalia is that, with medical research, the real world providing the funding does in the end expect the medical Castalia to provide a pay off. While Hesse's Castalians were confident of the absence of any need to justify their game, the medical Castalians are aware that their funding is to some degree conditional. So the latter constantly make announcements that exaggerate the importance for medicine of some trivial laboratory finding, and constantly assert that if only the public or the pharmaceutical industry will continue to supply money then success will be assured.

I am increasingly fearful that much of the hyped promise is illusory. Most medical and pharmaceutical research is now of one of three broad types, which are informed by a fourth — the molecular biology of the genome. The other three types of research are cell culture, traditional animal models of disease (such as hypertension, arthritis, cancer or multiple sclerosis), and the new mouse models involving various types of gene knockouts, knock-ins and their variants, which are currently threatening to overwhelm animal facilities worldwide. We are repeatedly told by the high priests of the medical Castalia that these amazing cascades of new knowledge will in the not-too-distant

PERSPECTIVES

Hermann Hesse
The Glass Bead Game

'The crowning glory of Hesse's long life and one of the truly important books of the century, in any language' *The Times*

Figure 1 | **Hesse's *The Glass Bead Game*.** The cover of a modern edition of *The Glass Bead Game*, also known as *Magister Ludi*, originally published as *Das Glasperlenspiel* in 1943, which describes the fantasy world of Castalia.

future lead to effective new individualized therapies for all our ills.

There is no doubt about the cascade of new information. The fundamental issue, however, is whether that new information is in any way congruent with the real world of medical illness? I fear that it is not and that we are creating a modern glass bead game that bears as little relation to real medicine as did Hesse's Castalia to the reality of the surrounding world. The arguments in favour of this view are potentially long and complex, but can be summarized briefly by addresssing three key questions.

Cell culture and in vivo *function.* Does the functioning of cells in culture bear a sufficiently strong relationship to the functioning of cells in an organ *in vivo* such that conclusions drawn from the former are useful in predicting behaviour of the latter?

An important distinction must be made between what might be called the anatomical biochemistry of the cell and its functional biochemistry. It is reasonably safe to say that if a particular biochemical step is present *in vitro*, then that particular biochemical step is also likely to be present in at least some form *in vivo*. We can therefore construct a network of all possible biochemical events *in vivo* by examining all possible biochemical events *in vitro*.

But what the *in vitro* system cannot do is construct a functional and valid *in vivo* biochemistry. And that is potentially a fatal flaw. For in most human diseases it is the functional biochemistry and not the anatomical biochemistry which goes wrong. When we ask cell culture to inform us about *in vivo* cell function, in most cases we ask too much. Some examples of the lack of congruence between cell culture and other *in vitro* models and the *in vivo* systems they attempt to model are given in BOX 1.

The list of the differences between cell culture and *in vivo* systems provided in BOX 1 is a brief and partial one, and could be considerably expanded. Given the potential lack of relevance of *in vitro* research, one might have expected rigorous theoretical discussion and experimental exploration of the problems of the issue of congruence of *in vitro* and *in vivo* studies. But there is almost nothing: certainly there is no general sense among medical researchers using *in vitro* systems that their work involves so many untested and unjustified assumptions that its congruence with any useful *in vivo* world must be in serious doubt. And if that congruence is unproven, there must exist the risk that not only might the *in vitro* work be useless, it may be actively misleading.

The Castalians are perhaps concerned that *in vitro* studies which form a very large part of their work are open to criticism. However, their unwillingness to discuss the paucity of evidence that *in vitro* studies can mimic either the complexities or realities of the *in vivo* world has the effect of acting in favour of the anti-animal research lobby. The Castalians are not as clearly spoken as they should be when the anti-animal research lobby advocates switching from animal to *in vitro* studies. The idea that *in vitro* systems — or even computer models — might be able to reflect acurately and reliably the complex *in vivo* systems of an intact human being involves a leap of faith. Such a leap, without critical and sustained discussion of congruence, should have no place in rational science.

Animal models and human disease. Does the use of animal models of disease take us any closer to understanding human disease? With rare exceptions, the answer to this question is likely to be negative. The reasoning is simple. An animal model of disease can be said to be congruent with the human disease only when three conditions have been met: we fully understand the animal model, we fully understand the human disease and we have examined the two cases and found them to be substantially congruent in all important respects.

These conditions have not been fully fulfilled for any human disease, although perhaps the closest examples come from research into infections and endocrine-deficiency diseases. Even in infectious-disease research, the animal model is often very different from the supposed human disease because of differences in the immune response. We have largely forgotten that when we apply the term 'guinea pigs', rather than 'rats' or 'mice', to experimental human subjects, we do so because in the early days of animal research the guinea pig was the only common animal which reacted to some infections in a near-human way.

All the other animal models — including those of inflammation, vascular disease, nervous system diseases and so on — represent nothing more than an extraordinary, and in most cases irrational, leap of faith. We have a human disease, and we have an animal model which in some vague and almost certainly superficial way reflects the human disease. We operate on the unjustified assumption that the two are congruent, and then we spend vast amounts of money trying to investigate the animal model, often without bothering to test our assumptions by constantly referring back to the original disease in humans.

These unexplored assumptions are the fundamental flaws in any animal model scenario. The animal rights campaigners are justified in pointing out that there is little rationale for using animal models which frequently simply draw attention and funds away from the careful investigation of the human condition. The Castalian establishment is wrong in not drawing attention to the unjustified assumption of congruence in most cases of animal experimentation on disease models.

Genetically modified mice and human disease. Will genetically modified mice lead to better understanding of human disease? The only appropriate answer at this stage is perhaps or perhaps not. But the omens are not good and the confidence of so many in the Castalian establishment seems to me to be entirely misplaced. Why am I so sceptical?

First, most human disease is highly unlikely to be due to a single abnormal gene. It may well be that the consequences of catastrophic failure of a single gene can be partly understood with the assistance of appropriate genetically modified mouse models. But such diseases are for the most part rare and tend, in any case, to be reasonably well understood from direct human studies.

Second, consistent phenotypes are rarely obtained by modification of the same gene even in mice[4]. The disruption of a gene in one

PERSPECTIVES

strain of mice may be lethal, whereas disruption of exactly the same gene in another strain of mice may have no detectable phenotypic effect. If this is true of the impact on one gene of the rest of the mouse genome, how much more is it likely to be true of the impact of the rest of the genes in the human genome?

Third, the great majority of human diseases that affect large numbers of the population are likely to be the result of the interaction of several different genes. If one mouse gene is so difficult to understand in a mouse context, and if the genome of a different inbred strain of mouse has so much impact on the consequences of that single gene's expression, how unlikely is it that genetically modified mice are going to provide insights into complex gene interactions in the non-interbred human species? At the least one must conclude that most predictions of near term human benefit are not only overblown but are actually fraudulent.

Castalia or not?

So for three heavily funded and heavily hyped techniques of medical research — which are also now fundamental techniques on which many drug discovery efforts are based — there are grave doubts about the congruence between their underlying scientific assumptions and reality. The charge that we may be building a vast and internally consistent medical research game that has lost touch with patients is a serious one, and deserves serious discussion. For if it is only partly true — and it may be more than that — then two things will follow. First, the promised medical pay offs will not be delivered[5,6]. Second, the research enterprise will fall into such disrepute that there could be a dramatic loss of public support for biomedical and pharmaceutical research. The long-term survival of the medical research effort requires a much more self-critical attitude.

What can be done to reduce the risk of isolated self-consistency? First, there must be a recognition that in the last analysis the human disease itself must be studied in human subjects. It is at least arguable that if we devoted as much effort to the human disease as we do to unvalidated models, then we might be much further forward in understanding[5,7,8]. If we are to have any confidence our models are valid, then we must know at least as much about the diseases we investigate as the models we use.

"Good clinical research is in decline and its practitioners are becoming demoralized."

Second, although we know little about the congruence between particular animal models and particular human diseases, we do know that with respect to many aspects of their function there is indeed substantial congruence between animal physiology and biochemistry and human physiology and biochemistry. When critically examined, much of the strongest evidence for the importance of animal experiments to human disease comes from efforts to understand *normal* animal physiology and biochemistry. The insulin story, the importance of cardiovascular function for cardiac surgery and renal function for the management of kidney failure are examples in which the main value of the animal models lay in the understanding they generated about normal human physiology and its responses to stressors. There we can indeed be reasonably confident that animal and human studies will be substantially congruent.

It is therefore arguable that the biggest failure of modern medical and pharmaceutical research lies in its near abandonment of attempts to learn more about normal function in intact normal animals. Whole-animal studies teach two lessons that many medical researchers have forgotten. First, whole organisms are extremely complex. Although the apparent clarity provided by simpler systems can often be useful, it almost inevitably also introduces distortions. The simpler systems must always be subject to reality checks by constant referrals back to the way the whole organism works.

The second major lesson that comes from whole-animal studies is that kinetic understanding is crucial. In order to understand diabetes, it is not necessary merely to understand the anatomical layout of the biochemical pathways but also how they interact kinetically. Without that functional kinetic knowledge, which in the end can come only from *in vivo* animal studies, then no sense at all can be made of the information from, for example, cell culture studies or applied to human physiology

One of the distressing aspects of modern genomics and molecular biological studies is that they are almost entirely kinetic-free zones. Their practitioners are merely sketching out anatomically what pathways might be possible, and are not describing functionally what pathways actually do take place *in vivo*. Only when we get to that state of functional knowledge will the medical benefits begin to come through. The lack of insight into the real world is as touching — and misguided — as that of the Renaissance anatomists who were equally confident that once we understood gross structure we would also understand function. Five hundred years later those anatomists might well be surprised not at how much progress we have made but at how little. Our anatomy has moved from the level of the organ down to that of the genome but anatomy is in essence the level at which our genomic studies presently are. Only when we effectively study functional interactions in a kinetic way will we begin to understand what we are doing.

Furthermore, only when we have shown that our reductionist bottom-up studies are congruent with our studies of the human disease state *in vivo* will we be likely to achieve real progress. And in order to do that we have to study human disease: this is where the

Box 1 | **Examples of lack of congruence between *in vitro* and *in vivo* models**

- The anatomical constraints and the cellular populations present in culture and *in vivo* are different. There is no circulation *in vitro*.
- The types and rates of nutrient and oxygen supply, and carbon dioxide and metabolite removal, are different.
- The restraints on cell multiplication are different.
- The endocrine environment is different, both in terms of the amounts and patterns of hormones present and their kinetic changes.
- The antibiotic environment is different: *in vivo* cells are not normally bathed in penicillin, streptomycin and other antibiotics, but there has been no systematic evaluation of the effects of any of these exogenous agents on metabolism.
- The lipid environment is different. The phospholipid composition of cells in culture is quite different from the phospholipid composition of the parent *in vivo* cells[2]. As phospholipid composition determines the quaternary structure and therefore function of a high proportion of a cell's proteins, and also determines signal transduction responses to most protein changes, it is likely that the functions of proteins *in vitro* will be, for the most part, somewhat different from the functions of those same proteins *in vivo*[3].
- Even when appropriate constituents are present in culture fluid, their concentrations may be dramatically different from anything seen *in vivo*.

PERSPECTIVES

system may be breaking down. Good clinical research is in decline and its practitioners are becoming demoralized[9]. Sometimes I wonder whether the reluctance to put much more emphasis on clinical studies may involve a reluctance to find relatively simple answers which might obviate the need for much of our sophisticated research and so undercut the need for Castalia!

For example, I am puzzled as to why most people other than professional geneticists seem uninterested in the high levels of non-concordance among identical twins for common diseases. And even the geneticists are more interested in the concordance than the non-concordance. For almost every common disease, whether it be inflammatory, malignant, degenerative, psychiatric or any other type, when one identical twin is affected the other twin is *not* affected from 40–90% of the time. Therefore some factor or combination of factors in the environment must have prevented or switched off

the disease process in the non-affected twin[5]. Whatever factors are involved, they have occurred during a normal life and so do not require high-tech genetic or pharmaceutical manipulation. The corollary is that it cannot be that difficult to switch off genetic disease processes if only we can understand them better. Is it really too much to think that a direct assault on human disease by studying humans might be at least as productive as the massive investment in investigation of unvalidated animal and *in vitro* models? At least what we find in humans will be both real and relevant. It will also be necessary as a reality check on what may be nothing more than a self-consistent but ultimately irrelevant Castalian game.

Laxdale Ltd, Kings Park House, Laurelhill Business Park, Stirling FK7 9JQ, Scotland.
doi:10.1038/nrd1012

1. Hesse, H. *Magister Ludi* (Bantam Press, New York, 1970) Originally published as *Das Glasperlenspiel*, Fretz and Wasmuth Verlag, Zurich, 1943.
2. Reynolds, L. M., Dalton, C. F. & Reynolds, G. P. Phospholipid fatty acids and neurotoxicity in human neuroblastoma SH-SY5Y cells. *Neurosci. Lett.* **309**, 193–196 (2001).
3. Witt, M. R. & Nielsen, M. Characterisation of the influence of unsaturated free fatty acids on brain GABA/benzodiazepine receptor binding *in vitro*. *J. Neurochem.* **62**, 1432–1439 (1994).
4. Pearson, H. Surviving a knock-out blow. *Nature* **415**, 8–9 (2002).
5. Horrobin, D. F. Innovation in the pharmaceutical industry. *J. R. Soc. Medicine* **93**, 341–345 (2000).
6. Horrobin D. F. Realism in drug discovery — could Cassandra be right? *Nature Biotechnol.* **19**, 1099–1100 (2001).
7. Persson, C. G., Erjefalt, J. S., Uller, L., Andersson, M. & Greiff, L. Unbalanced research. *Trends Pharmacol. Sci.* **22**, 538–541 (2001).
8. Rees J. Post-genome integrative biology: so that's what they call clinical science. *Clin. Med.* **1**, 393–400 (2001).
9. Rosenberg, L. Physician-scientists — endangered and essential. *Science* **283**, 331–332 (1999).

⟨⟩ Online links

FURTHER INFORMATION
US Food and Drug Administration:
http://www.fda.gov/
European Medicines Evaluation Agency:
http://www.emea.eu.int/
Encyclopedia of Life Sciences
Human disease: mouse models
Access to this interactive links box is free online.

REFERENCES

Agarwal, S., and N. Moorchung. 2005. Modifier genes and oligogenic disease. *J Nippon Med Sch* 72 (6):326-34.

Ahn, A. C., M. Tewari, C. S. Poon, and R. S. Phillips. 2006. The limits of reductionism in medicine: could systems biology offer an alternative? *PLoS Med* 3 (6):e208.

Aithal, G. P., C. P. Day, P. J. Kesteven, and A. K. Daly. 1999. Association of polymorphisms in the cytochrome P450 CYP2C9 with warfarin dose requirement and risk of bleeding complications. *Lancet* 353 (9154):717-9.

Alberts, Bruce, Alexander Johnson, Julian Lewis, Martin Raff, Keith Roberts, and Peter Walter. 2002. *Molecular Biology of the Cell.* 4th ed: Garland.

Alon, U. 2003. Biological networks: the tinkerer as an engineer. *Science* 301 (5641):1866-7.

Altman, LK. 1998. *Who Goes First? The Story of Self-Experimentation in Medicine.* University of California Press.

American Medical Association. 1992. *White Paper on Animal Research.* American Medical Association.

Anisimov, V. N., S. V. Ukraintseva, and A. I. Yashin. 2005. Cancer in rodents: does it tell us about cancer in humans? *Nat Rev Cancer* 5 (10):807-19.

Anisimov, VN. 1987. *Carcinogenesis and Aging.* Vol. 1 and 2. Boca Rotan: CRC Press.

———. 2003. *Molecular and Physiological Mechanisms of Aging.* St Petersburg: Nauka.

APHIS. 2000. RATS/MICE/ and BIRDS DATABASE: RESEARCHERS, BREEDERS, TRANSPORTERS, AND EXHIBITERS. A Database Prepared by the Federal Research Division, Library of Congress under an Interagency Agreement with the United States Department of Agriculture's Animal Plant Health Inspection Service, edited by USDA. Washington, DC.

———. 2008. *Animal Welfare* 2008 [cited December 21 2008]. Available from http://www.aphis.usda.gov/animal_welfare/index.shtml.

Association of Medical Colleges. 2009. *The Medical School Profile System.* Association of Medical Colleges 2009 [cited March 13 2009]. Available from http://www.aamc.org/data/msps/start.htm.

Babloyantz, A. 1986. *Molecules, Dynamics, and Life.* New York: Wiley.

Backhouse, Roger E. 2002. *The Ordinary Business of Life: A History of Economics from the Ancient World to the Twenty-First Century.* Princeton University Press.

Bahls, Christine, Jonathan Weitzman, and Richard Gallagher. 2003. Biology's Models. *The Scientist* 17 (S1):2.

Bailar, J. C., 3rd, and H. L. Gornik. 1997. Cancer undefeated. *N Engl J Med* 336 (22):1569-74.

Bailey, Jarrod, Andrew Knight, and Jonathan Balcombe. 2005. The future of teratology research is in vitro. *Biogenic Amines* 19 (2):97-145.

Begley, Sharon. 2003. Financial Obstacles Help Keep Doctors From Patient Research *Wall Street Journal,* May 2.

———. 2008. We fought cancer... and cancer won. *Newsweek,* September 6.

————. 2008. Where Are the Cures? *Newsweek*, November 10.

Behringer, R. R. 2007. Human-animal chimeras in biomedical research. *Cell Stem Cell* 1 (3):259-62.

Beniashvili, Dzhemal Sh. 1994. *Experimental Tumors in Monkeys*. CRC Press.

Berlocher, S H. 1998. Can sympatric speciation be proven from biogeographic and phylogenetic evidence? In *Endless Forms: Species and Speciation*, edited by D. J. Howard and S. H. Berlocher. New York: Oxford University Press.

Bernard, Claude. 1949. *An Introduction to the Study of Experimental Medicine*. Henry Schuman Inc.

————. 1973. *An Introduction to the Study of Experimental Medicine*. Translated by H. Green. New York: Dover.

Blakemore, Colin. 2008. Should we experiment on animals? Yes. *Telegraph*, October 28, 2008.

Blume, E. 1992. Scientists, activists discuss DES issues. *J Natl Cancer Inst* 84 (12):925-6.

Boguski, M. S. 2002. Comparative genomics: the mouse that roared. *Nature* 420 (6915):515-6.

Botting, J. H., and A. R. Morrison. 1997. Animal research is vital to medicine. *Sci Am* 276 (2):83-5.

Brady, Catherine A. 2008/9. Of Mice and Men: the potential of high-resolution human immune cell assays to aid the pre-clinical to clinical transition of drug development projects. *Drug Discovery World* (Winter):74-78.

Braithwaite, Richard Bevan. 1953. *Scientific explanation : a study of the function of theory, probability and law in science*. Cambridge: Cambridge University Press.

Braun, S. E., and R. P. Johnson. 2006. Setting the stage for bench-to-bedside movement of anti-HIV RNA inhibitors-gene therapy for AIDS in macaques. *Front Biosci* 11:838-51.

Bray, D. 2003. Genomics. Molecular prodigality. *Science* 299 (5610):1189-90.

————. 2003. Molecular networks: the top-down view. *Science* 301 (5641):1864-5.

Brigandt, I. 2003. Homology in comparative, molecular, and evolutionary developmental biology: the radiation of a concept. *J Exp Zoolog B Mol Dev Evol* 299 (1):9-17.

Bugge, T. H., K. W. Kombrinck, M. J. Flick, C. C. Daugherty, M. J. Danton, and J. L. Degen. 1996. Loss of fibrinogen rescues mice from the pleiotropic effects of plasminogen deficiency. *Cell* 87 (4):709-19.

Burggren, W W, and W E Bemis. 1990. Studying Physiological Evolution: Paradigms and Pitfalls. In *Evolutionary Innovations*, edited by M. H. Nitecki. Chicago: University of Chicago Press.

Burns, R. S., C. C. Chiueh, S. P. Markey, M. H. Ebert, D. M. Jacobowitz, and I. J. Kopin. 1983. A primate model of parkinsonism: selective destruction of dopaminergic neurons in the pars compacta of the substantia nigra by N-methyl-4-phenyl-1,2,3,6-tetrahydropyridine. *Proc Natl Acad Sci U S A* 80 (14):4546-50.

Butler, D. 2008. Translational research: crossing the valley of death. *Nature* 453 (7197):840-2.

Bynum, W F, E J Browne, and Roy Porter. 1981. *Dictionary of the History of Science*. Princeton University Press.

Cairns-Smith, A G. 1986. *Seven Clues to the Origin of Life: A Scientific Detective Story.* Cambridge University Press.

Calabrese, E. J. 1984. Suitability of animal models for predictive toxicology: theoretical and practical considerations. *Drug Metab Rev* 15 (3):505-23.

Calabrese, Edward J. 1991. *Principles of Animal Extrapolation.* CRC Press.

Caldwell, J. 1980. Comparative Aspects of Detoxification in Mammals. In *Enzymatic Basis of Detoxification,* edited by W. Jakoby. New York: Academic Press.

Caldwell, J. 1992. Problems and opportunities in toxicity testing arising from species differences in xenobiotic metabolism. *Toxicol Lett* 64-65 Spec No:651-9.

Caraco, Y., J. Sheller, and A. J. Wood. 1996. Pharmacogenetic determination of the effects of codeine and prediction of drug interactions. *J Pharmacol Exp Ther* 278 (3):1165-74.

Carroll, Sean, Jennifer Grenier, and Scott Weatherbee. 2004. *From DNA to Diversity: Molecular Genetics and the Evolution of Animal Design.* Wiley-Blackwell.

Chauret, N., A. Gauthier, J. Martin, and D. A. Nicoll-Griffith. 1997. In vitro comparison of cytochrome P450-mediated metabolic activities in human, dog, cat, and horse. *Drug Metab Dispos* 25 (10):1130-6.

Check, E. 2005. Human genome: patchwork people. *Nature* 437 (7062):1084-6.

Cheung, Connie, and Frank J. Gonzalez. 2008. Humanized Mouse Lines and Their Application for Prediction of Human Drug Metabolism and Toxicological Risk Assessment. *J Pharmacol Exp Ther* 327 (2):288-299.

Chiba, M., J. A. Nishime, W. Neway, Y. Lin, and J. H. Lin. 2000. Comparative in vitro metabolism of indinavir in primates--a unique stereoselective hydroxylation in monkey. *Xenobiotica* 30 (2):117-29.

Claude, N. 2007. Are non-clinical studies predictive of adverse events in humans? *Ann Pharm Fr* 65 (5):292-7.

Clayson, DB. 1980. The carcinogenic action of drugs in man and animals. In *Human Epidemiology and Animal Laboratory Correlations in Chemical Carcinogenesis* edited by F. Coulston and P. Shubick: Ablex Pub.

Clemmensen, J, and S Hjalgrim-Jensen. 1980. On the absence of carcinogenicity to man of phenobarbital. In *Human Epidemiology and Animal Laboratory Correlations in Chemical Carcinogenesis,* edited by F. Coulston and S. Shubick: Alex Pub.

Clendening, Logan. 1960. *Source Book of Medical History.* Dover.

CNN. 2008. *FDA panel recommends continued use of controversial diabetes drug.* CNN 1999 [cited 10-18 2008]. Available from http://www.cnn.com/HEALTH/9903/26/rezulin.review.02/index.html.

Coe, J. E., K. G. Ishak, J. M. Ward, and M. J. Ross. 1992. Tamoxifen prevents induction of hepatic neoplasia by zeranol, an estrogenic food contaminant. *Proc Natl Acad Sci U S A* 89 (3):1085-9.

Cohen, I Bernard. 1985. *The Birth of a New Physics.* WW Norton and Co.

Cohen, J. 1999. AIDS vaccine. Chimps and lethal strain a bad mix. *Science* 286 (5444):1454-5.

Coleman, William. 1978. *Biology in the Nineteenth Century: Problems of Form, Function and Transformation (Cambridge Studies in the History of Science):* Cambridge University Press.

Collins, J. M. 2001. Inter-species differences in drug properties. *Chem Biol Interact* 134 (3):237-42.

Committee on Models for Biomedical Research, Board on Basic Biology. 1985. *Committee on Models for Biomedical Research. Board on Basic Biology. Commission on Life Science. National Research Council. Models for Biomedical Research: A New Perspective.* Washington, DC: National Academy Press.

Committee on Toxicity Testing and Assessment of Environmental Agents, National Research Council. 2007. *Toxicity Testing in the 21st Century: A Vision and a Strategy.* Washington DC: National Academy of Science.

Comroe, J. H., Jr., and R. D. Dripps. 1976. Scientific basis for the support of biomedical science. *Science* 192 (4235):105-11.

Conn, PM, and JV Parker. 2007. Letter. Animal research wars. *Skeptic* 13 (4):18-19.

Contopoulos-Ioannidis, D. G., G. A. Alexiou, T. C. Gouvias, and J. P. Ioannidis. 2008. Medicine. Life cycle of translational research for medical interventions. *Science* 321 (5894):1298-9.

Cooper, D. W., M. Carpenter, P. Mowbray, W. R. Desira, D. M. Ryall, and M. S. Kokri. 2002. Fetal and maternal effects of phenylephrine and ephedrine during spinal anesthesia for cesarean delivery. *Anesthesiology* 97 (6):1582-90.

Cooper, S., E. J. Adams, R. S. Wells, C. M. Walker, and P. Parham. 1998. A major histocompatibility complex class I allele shared by two species of chimpanzee. *Immunogenetics* 47 (3):212-7.

Coulston, F. 1980. Final Discussion. In *Human Epidemiology and Animal Laboratory Correlations in Chemical Carcinogenesis*, edited by F. Coulston and P. Shubick: Ablex.

Council on Scientific Affairs. 1981. Carcinogen regulation. . *JAMA* 246 (3):253-6.

Courtney, S. M., L. Petit, J. V. Haxby, and L. G. Ungerleider. 1998. The role of prefrontal cortex in working memory: examining the contents of consciousness. *Philos Trans R Soc Lond B Biol Sci* 353 (1377):1819-28.

Couzin, J. 2007. Cancer research. Probing the roots of race and cancer. *Science* 315 (5812):592-4.

Crick, F., and E. Jones. 1993. Backwardness of human neuroanatomy. *Nature* 361 (6408):109-10.

Crombie, A C. 1959. *Medieval and Early Modern Science*: Doubleday and Anchor.

CSPI. *Longer Tests on Lab Animals Urged for Potential Carcinogens.* CSPI 2008 [cited November 17. Available from http://www.cspinet.org/new/200811172.html.

Cziko, Gary. 1997. *Without Miracles: Universal Selection Theory and the Second Darwinian Revolution*: MIT Press.

Dally, A. 1998. Thalidomide: was the tragedy preventable? *Lancet* 351 (9110):1197-9.

Darwin, Charles. 1970 [1859]. *Origin of the Species*. Edited by P. Appleman. New York: W W Norton.

Davidson, E H. 2001. *Genomic Regulatory Systems: Development and Evolution.* San Diego: Academic Press.

Davidson, Eric H. 2006. *The Regulatory Genome: Gene Regulatory Networks in Development and Evolution.* Academic Press.

Davis, G. C., A. C. Williams, S. P. Markey, M. H. Ebert, E. D. Caine, C. M. Reichert, and I. J. Kopin. 1979. Chronic Parkinsonism secondary to intravenous injection of meperidine analogues. *Psychiatry Res* 1 (3):249-54.

Davis, M. M. 2008. A prescription for human immunology. *Immunity* 29 (6):835-8.

Dennis, C. 2006. Cancer: off by a whisker. *Nature* 442 (7104):739-41.

Deshaies, Raymond J. 2009. Drug discovery: Fresh target for cancer therapy. *Nature* 458 (7239):709-710.

Devlin, Keith. 1981. *Sets, Functions, and Logic: An Introduction to Abstract Mathematics.* Chapman & Hall.

Di Carlo, F. J. 1984. Carcinogenesis bioassay data: correlation by species and sex. *Drug Metab Rev* 15 (3):409-13.

Dieckmann, W. J., M. E. Davis, L. M. Rynkiewicz, and R. E. Pottinger. 1953. Does the administration of diethylstilbestrol during pregnancy have therapeutic value? *Am J Obstet Gynecol* 66 (5):1062-81.

Dilman, V. M., and V. N. Anisimov. 1980. Effect of treatment with phenformin, diphenylhydantoin or L-dopa on life span and tumour incidence in C3H/Sn mice. *Gerontology* 26 (5):241-6.

Dipple, K. M., J. K. Phelan, and E. R. McCabe. 2001. Consequences of complexity within biological networks: robustness and health, or vulnerability and disease. *Mol Genet Metab* 74 (1-2):45-50.

Donehower, L. A., M. Harvey, B. L. Slagle, M. J. McArthur, C. A. Montgomery, Jr., J. S. Butel, and A. Bradley. 1992. Mice deficient for p53 are developmentally normal but susceptible to spontaneous tumours. *Nature* 356 (6366):215-21.

Dorfman, R., A. Sandford, C. Taylor, B. Huang, D. Frangolias, Y. Wang, R. Sang, L. Pereira, L. Sun, Y. Berthiaume, L. C. Tsui, P. D. Pare, P. Durie, M. Corey, and J. Zielenski. 2008. Complex two-gene modulation of lung disease severity in children with cystic fibrosis. *J Clin Invest* 118 (3):1040-9.

Drazen, J. M., C. N. Yandava, L. Dube, N. Szczerback, R. Hippensteel, A. Pillari, E. Israel, N. Schork, E. S. Silverman, D. A. Katz, and J. Drajesk. 1999. Pharmacogenetic association between ALOX5 promoter genotype and the response to anti-asthma treatment. *Nat Genet* 22 (2):168-70.

Drysdale, C. M., D. W. McGraw, C. B. Stack, J. C. Stephens, R. S. Judson, K. Nandabalan, K. Arnold, G. Ruano, and S. B. Liggett. 2000. Complex promoter and coding region beta 2-adrenergic receptor haplotypes alter receptor expression and predict in vivo responsiveness. *Proc Natl Acad Sci U S A* 97 (19):10483-8.

Durham, Will. 7-12-2001. Huge Genetic Variation Found in Human Beings. *Reuters.*

Editorial. 2006. The end of the beginning? *Nat Rev Drug Discov* 5 (9):705.

Elledge, R. M., and W. H. Lee. 1995. Life and death by p53. *Bioessays* 17 (11):923-30.

Enard, W., P. Khaitovich, J. Klose, S. Zollner, F. Heissig, P. Giavalisco, K. Nieselt-Struwe, E. Muchmore, A. Varki, R. Ravid, G. M. Doxiadis, R. E. Bontrop, and S. Paabo. 2002. Intra- and interspecific variation in primate gene expression patterns. *Science* 296 (5566):340-3.

Ennever, F. K., T. J. Noonan, and H. S. Rosenkranz. 1987. The predictivity of animal bioassays and short-term genotoxicity tests for carcinogenicity and non-carcinogenicity to humans. *Mutagenesis* 2 (2):73-8.

Ernst, E. 1999. Second thoughts about safety of St John's wort. *Lancet* 354 (9195):2014-6.

Ernst, E., J. I. Rand, J. Barnes, and C. Stevinson. 1998. Adverse effects profile of the herbal antidepressant St. John's wort (Hypericum perforatum L.). *Eur J Clin Pharmacol* 54 (8):589-94.

Eubank, S., and J. D. Farmer. 1990. Introduction to Chaos and Randomness. In *1989 Lectures in Complex Systems. Santa Fe Institute Studies in the Sciences of Complexity, Lectures*, edited by E. Jen and A. Wesley.

Evans, D. A., K. A. Manley, and McKusick V. A. 1960. Genetic control of isoniazid metabolism in man. *Br Med J* 2 (5197):485-91.

Evans, W. E., and H. L. McLeod. 2003. Pharmacogenomics—drug disposition, drug targets, and side effects. *N Engl J Med* 348 (6):538-49.

Ewald, P W. 1994. *The Evolution of Infectious Disease*. Oxford: Oxford University Press.

Farrar, Steve. 2003. Key basic research study proves fatally ambiguous. *Times Higher Education Supplement*, 21 November.

FDA News. *FDA Issues Advice to Make Earliest Stages Of Clinical Drug Development More Efficient*. USFDA 2006 [cited. Available from http://www.fda.gov/bbs/topics/news/2006/NEW01296.html

Ferrari, G., J. Ottinger, C. Place, S. M. Nigida, Jr., L. O. Arthur, and K. J. Weinhold. 1993. The impact of HIV-1 infection on phenotypic and functional parameters of cellular immunity in chimpanzees. *AIDS Res Hum Retroviruses* 9 (7):647-56.

Fiultz, P N. 1991. Human immunodeficiency virus infection of chimpanzees: An animal model for asymptomatic HIV carriers and vaccine efficacy. In *AIDS Research Reviews*, edited by W. Koff, F. Wong-Staal and R. Kennedy. New York: Marcel Dekker.

Flaveny, C., R. K. Reen, A. Kusnadi, and G. H. Perdew. 2008. The mouse and human Ah receptor differ in recognition of LXXLL motifs. *Arch Biochem Biophys* 471 (2):215-23.

Florkin, and Schoffeneils. 1970. In *Chemical Zoology*, edited by M. Florkin and B. T. Scheer: Academic Press.

Follow the yellow brick road. 2003. *Nat Rev Drug Discov* 2 (3):167.

Fomchenko, E. I., and E. C. Holland. 2006. Mouse models of brain tumors and their applications in preclinical trials. *Clin Cancer Res* 12 (18):5288-97.

Forni, G, P Caretto, P Ferraiorni, M Bosco, and M Giovarelli. 1990. The Necessity of Animal Experimentation in Tumor Immunology. In *The Importance of Animal Experiments for Safety and Biomedical Research*, edited by S. Garattini and D. W. v. Bekkum. Dordrecht Kluwer Academic Publishers.

Foster, J. R. 2005. Spontaneous and drug-induced hepatic pathology of the laboratory beagle dog, the cynomolgus macaque and the marmoset. *Toxicol Pathol* 33 (1):63-74.

Fraga, M. F., E. Ballestar, M. F. Paz, S. Ropero, F. Setien, M. L. Ballestar, D. Heine-Suner, J. C. Cigudosa, M. Urioste, J. Benitez, M. Boix-Chornet, A. Sanchez-Aguilera, C. Ling, E. Carlsson, P. Poulsen, A. Vaag, Z. Stephan, T. D. Spector, Y. Z. Wu, C. Plass, and M. Esteller. 2005. Epigenetic differences arise during the lifetime of monozygotic twins. *Proc Natl Acad Sci U S A* 102 (30):10604-9.

Frankos, V. H. 1985. FDA perspectives on the use of teratology data for human risk assessment. *Fundam Appl Toxicol* 5 (4):615-25.

Frantz, S. 2004. Vaccine approach for Alzheimer's disease revisited. *Nat Rev Drug Discov* 3 (9):726-7.

Freeman, Matthew, and Daniel St Johnston. 2008. Wherefore DMM? *Disease Models & Mechanisms* 1 (1):6-7.

Fugh-Berman, A. 2000. Herb-drug interactions. *Lancet* 355 (9198):134-8.

Fujita, K., H. Kodaira, T. Kuwabara, and H. Kobayashi. 2008. Pharmacokinetics and metabolism of KW-4490, a selective phosphodiesterase 4 inhibitor: difference in excretion of KW-4490 and acylglucuronide metabolites between rats and cynomolgus monkeys. *Xenobiotica* 38 (5):511-26.

Fultz, P. N. 1993. Nonhuman primate models for AIDS. *Clin Infect Dis* 17 Suppl 1:S230-5.

Futuyma, D. 1998. *Evolutionary Biology*. 3rd ed. Sunderland: Sinauer Associates.

Gad, SC. 1990. Model Selection in Toxicology: Principles and Practice. *International Journal of Toxicology* 9 (3):291-302.

———. 2007. Preface. In *Animal Models in Toxicology*, edited by S. Gad: CRC Press.

Gallagher, Richard. 2008. Why the philosophy of science matters. *The Scientist* 22 (10):15.

Gallo, M A, and J Doull. 1993. History and scope of toxicology. In *Casarett and Doull's Toxicology*, edited by M. O. Amdur, J. D. Doull and C. D. Klaasen. New York: McGraw-Hill.

Garattini, S. 1990. The necessity of animal experimentation. In *The Importance of Animal Experiments for Safety and Biomedical Research*, edited by S. Garattini and v. Bekkum. Dordrecht Kluwer Academic Publishers.

Garattini, S. 1985. Toxic effects of chemicals: difficulties in extrapolating data from animals to man. *Crit Rev Toxicol* 16 (1):1-29.

Gardner, M. B., and P. A. Luciw. 1989. Animal models of AIDS. *FASEB J* 3 (14):2593-606.

General Accounting Office. 2001. GAO-01-286R Drugs Withdrawn From Market. Washington, DC: US General Accounting Office.

Gerhart, John, and Marc Kirschner. 1997. *Cells, Embryos, and Evolution: Toward a Cellular and Developmental Understanding of Phenotypic Variation and Evolutionary Adaptability* Blackwell Publishers.

Giere, Ronald N. 1991. *Understanding Scientific Reasoning*. 3rd ed: Harcourt College Publishing.

Gilbert, Scott F, and Richard M Burian. 2006. Developmental Genetics. In *Keywords and Concepts in Evolutionary Developmental Biology*, edited by B. K. Hall and W. M. Olson: Harvard University Press.

Giles, J. 2006. Animal experiments under fire for poor design. *Nature* 444 (7122):981.

Giot, L., J. S. Bader, C. Brouwer, A. Chaudhuri, B. Kuang, Y. Li, Y. L. Hao, C. E. Ooi, B. Godwin, E. Vitols, G. Vijayadamodar, P. Pochart, H. Machineni, M. Welsh, Y. Kong, B. Zerhusen, R. Malcolm, Z. Varrone, A. Collis, M. Minto, S. Burgess, L. McDaniel, E. Stimpson, F. Spriggs, J. Williams, K. Neurath, N.

Ioime, M. Agee, E. Voss, K. Furtak, R. Renzulli, N. Aanensen, S. Carrolla, E. Bickelhaupt, Y. Lazovatsky, A. DaSilva, J. Zhong, C. A. Stanyon, R. L. Finley, Jr., K. P. White, M. Braverman, T. Jarvie, S. Gold, M. Leach, J. Knight, R. A. Shimkets, M. P. McKenna, J. Chant, and J. M. Rothberg. 2003. A protein interaction map of Drosophila melanogaster. *Science* 302 (5651):1727-36.

Gold, L. S., T. H. Slone, and B. N. Ames. 1998. What do animal cancer tests tell us about human cancer risk?: Overview of analyses of the carcinogenic potency database. *Drug Metab Rev* 30 (2):359-404.

Gold, L. S., T. H. Slone, N. B. Manley, and L. Bernstein. 1991. Target organs in chronic bioassays of 533 chemical carcinogens. *Environ Health Perspect* 93:233-46.

Gold, L. S., T. H. Slone, B. R. Stern, N. B. Manley, and B. N. Ames. 1992. Rodent carcinogens: setting priorities. *Science* 258 (5080):261-5.

Goldberg, A. M., and T. Hartung. 2006. Protecting more than animals. *Sci Am* 294 (1):84-91.

Goldstein, D. B., and G. L. Cavalleri. 2005. Genomics: understanding human diversity. *Nature* 437 (7063):1241-2.

Gondo, Y. 2008. Trends in large-scale mouse mutagenesis: from genetics to functional genomics. *Nat Rev Genet* 9 (10):803-10.

Gonzalez, E., H. Kulkarni, H. Bolivar, A. Mangano, R. Sanchez, G. Catano, R. J. Nibbs, B. I. Freedman, M. P. Quinones, M. J. Bamshad, K. K. Murthy, B. H. Rovin, W. Bradley, R. A. Clark, S. A. Anderson, J. O'Connell R, B. K. Agan, S. S. Ahuja, R. Bologna, L. Sen, M. J. Dolan, and S. K. Ahuja. 2005. The influence of CCL3L1 gene-containing segmental duplications on HIV-1/AIDS susceptibility. *Science* 307 (5714):1434-40.

Gonzalez, F. J., and S. Kimura. 2001. Understanding the role of xenobiotic-metabolism in chemical carcinogenesis using gene knockout mice. *Mutat Res* 477 (1-2):79-87.

Gottlieb, Gilbert. 2006. Behavioral Development and Evolution. In *Keywords and Concepts in Evolutionary Developmental Biology*, edited by B. K. Hall and W. M. Olson: Harvard University Press.

Grant, Jonathan, Liz Green, and Barbara Mason. 2003. From Bedside to Bench: Comroe and Dripps Revisited. . In *HERG Research Report No. 30* Health Economics Research Group. Brunel University, Uxbridge, Middlesex UB8 3PH, UK. .

Greaves, M. 2000. *Cancer: The Evolutionary Legacy*. Oxford: Oxford University Press.

Greek, R, and J Greek. 2002. *Specious Science*. New York: Continuum Int.

Greek, Ray, and Jean Greek. 2000. *Sacred Cows and Golden Geese: The Human Cost of Experiments on Animals*. New York: Continuum Int.

Guengerich, F. P. 1997. Comparisons of catalytic selectivity of cytochrome P450 subfamily enzymes from different species. *Chem Biol Interact* 106 (3):161-82.

Gura, T. 1997. Cancer Models: Systems for identifying new drugs are often faulty. *Science* 278 (5340):1041-2.

———. 2000. Cancer research. Caution raised about possible new drug. *Science* 288 (5467):786-7.

Hacia, J. G. 2001. Genome of the apes. *Trends Genet* 17 (11):637-45.

Hahn, W. C., and R. A. Weinberg. 2002. Modelling the molecular circuitry of cancer. *Nat Rev Cancer* 2 (5):331-41.

Hallgrimsson, Benedikt. 2006. Variation. In *Keywords and Concepts in Evolutionary Developmental Biology (Harvard University Press Reference Library*, edited by B. K. Hall and W. M. Olson: Harvard University Press.

Hansen, S., and R. G. Leslie. 2006. TGN1412: scrutinizing preclinical trials of antibody-based medicines. *Nature* 441 (7091):282.

Harare, D. M., C. Rake, C McKee, M., and H. B. MacPhillamy. 1943. The toxicity of penicillin as prepared for clinical use. *Am J. M. Sc* 206:642-52.

Harris, W E. 1997. *Low dose risk assessment*. Winnipeg: Wuerz.

Haseman, J. K. 2000. Using the NTP database to assess the value of rodent carcinogenicity studies for determining human cancer risk. *Drug Metab Rev* 32 (2):169-86.

Hau, J. 2003. Animal Models. In *Handbook of Laboratory Animal Science. Animal Models*, edited by J. Hau and G. K. van Hoosier Jr: CRC Press.

HealthDay News. *Gene Tied to Infants' Lung Maturation*. HealthDay News December 4, 2008 [cited. Available from http://news.yahoo.com/s/hsn/20081205/hl_hsn/genetiedtoinfantslungmaturation;_ylt=AqV.HpaenyHTbzw3wVdoEeS3j7AB.

Hengstler, J. G., B. Van der Burg, P. Steinberg, and F. Oesch. 1999. Interspecies differences in cancer susceptibility and toxicity. *Drug Metab Rev* 31 (4):917-70.

Hepp-Raymond. 1988. In *Neurosciences (Comparative Primate Biology)*: John Wiley & Sons.

Herbst, A. L., H. Ulfelder, and D. C. Poskanzer. 1971. Adenocarcinoma of the vagina. Association of maternal stilbestrol therapy with tumor appearance in young women. *N Engl J Med* 284 (15):878-81.

Heywood, R. 1990. Clinical Toxicity--Could it have been predicted? Post-marketing experience. In *Animal Toxicity Studies: Their Relevance for Man*, edited by CE Lumley and S. Walker. Lancaster: Quay.

Hinde, Robert. 1987. Animal-Human Comparisons. In *The Oxford Companion to the Mind*, edited by R. L. Gregory. Oxford: Oxford University Press.

Hochachka, Peter W., and George N. Somero. 2002. *Biochemical Adaptation: Mechanism and Process in Physiological Evolution*: Oxford University Press.

Hodgson, J. 2001. ADMET—turning chemicals into drugs. *Nat Biotechnol* 19 (8):722-6.

Hoffman, GR. 1993. Genetic Toxicology. In *Casarett and Doull's Toxicology*, edited by A. Amdour, J. Doull and C. Klaasen. New York: McGraw-Hill.

Holden, C. 2003. Race and medicine. *Science* 302 (5645):594-6.

———. 2005. Sex and the suffering brain. *Science* 308 (5728):1574.

Home Office. 2008. Statistics of Scientific Procedures on Living Animals Great Britain 2007. London: The Stationery Office.

Hood, L. 2003. Leroy Hood expounds the principles, practice and future of systems biology. *Drug Discov Today* 8 (10):436-8.

Hopkin, Michael. 2006. Can super-antibody drugs be tamed? *Nature* 440 (7086):855-6.

Horrobin, D. F. 2003. Modern biomedical research: an internally self-consistent universe with little contact with medical reality? *Nat Rev Drug Discov* 2 (2):151-4.

Houdebine, L. M. 2007. Transgenic animal models in biomedical research. *Methods Mol Biol* 360:163-202.

HSUS. *Number animals used in research. Animal use statistics.* 2008 [cited December 21. Available from http://www.hsus.org/web-files/PDF/ARI/awreport2006-1.pdf.

Hudson, M. 2008. The Home Office Statistics for 2007 - mutant mice and fishy tales. *Altern Lab Anim* 36 (6):695-704.

Huff, J., M. F. Jacobson, and D. L. Davis. 2008. The limits of two-year bioassay exposure regimens for identifying chemical carcinogens. *Environ Health Perspect* 116 (11):1439-42.

Hughes, B. 2009. The comparative effectiveness challenge. *Nat Rev Drug Discov* 8 (4):261-3.

Hutchinson, J. N., and W. J. Muller. 2000. Transgenic mouse models of human breast cancer. *Oncogene* 19 (53):6130-7.

IARC. 2004. *IARC monographs programme on the evaluation of carcinogenic risks to humans.* . IARC [cited 01-01-04 2004]. Available from http://monographs.iarc.fr.

———. 1972-1992. *IARC Monographs on the Evaluation of Carcinogenic Risks to Humans.* Vol. 1-55. Lyon: IARC.

———. 1972-2001. IARC Working group on the evaluation of carcinogenic risks to humans. Lyon.

———. 1974. Some aromatic amines, hydrazine and related substances, n-nitroso compounds and miscellaneous alkylating agents. In *IARC monograph on the evaluation of carcinogenic risks to humans.* Lyon.

———. 1993. *Summaries & Evaluations. Vol 56. CAFFEIC ACID. CAS No.: 331-39-5. Chem. Abstr. Name: 3-(3,4-Dihydroxyphenyl)-2-propenoic acid.* . Vol. 56.

Igarashi, T. 1994. The duration of toxicity studies required to support repeated dosing in clinical investigation—A toxicologists opinion. In *CMR Workshop: The Timing of Toxicological Studies to Support Clinical Trials*, edited by N. M. C Parkinson, C Lumley, SR Walker. Boston/UK: Kluwer.

Igarashi, T., S. Nakane, and T. Kitagawa. 1995. Predictability of clinical adverse reactions of drugs by general pharmacology studies. *J Toxicol Sci* 20 (2):77-92.

Igarashi, Y. Report from the Japanese Pharmaceutical Manufacturers Association 1994 Seiyakukyo data.

Interlandi, Jeneen. 2007. Chemo control drugs target epigenetic changes in cancer cells. *Scientific American* April:24.

Janofsky, Michael. 1993. On Cigarettes, Health and Lawyers. *New York Times*, December 6.

Jeong, H., B. Tombor, R. Albert, Z. N. Oltvai, and A. L. Barabasi. 2000. The large-scale organization of metabolic networks. *Nature* 407 (6804):651-4.

Johnson, D E, and G H Wolfgang. 2000. Predicting human safety: screening and computational approaches. *Drug Discovery Today* 5 (10):445-54.

Johnston, M. I. 2000. The role of nonhuman primate models in AIDS vaccine development. *Mol Med Today* 6 (7):267-70.

Jones, S. A., L. B. Moore, J. L. Shenk, G. B. Wisely, G. A. Hamilton, D. D. McKee, N. C. Tomkinson, E. L. LeCluyse, M. H. Lambert, T. M. Willson, S. A. Kliewer, and J. T. Moore. 2000. The pregnane X receptor: a promiscuous xenobiotic receptor that has diverged during evolution. *Mol Endocrinol* 14 (1):27-39.

Jordan, V. C., and S. P. Robinson. 1987. Species-specific pharmacology of antiestrogens: role of metabolism. *Fed Proc* 46 (5):1870-4.

Kaiser, Jocelyn. 2008. AIDS RESEARCH: Review of Vaccine Failure Prompts a Return to Basics. *Science* 320 (5872):30-31.

Kamb, A. 2005. What's wrong with our cancer models? *Nat Rev Drug Discov* 4 (2):161-5.

Kaplowitz, N. 2005. Idiosyncratic drug hepatotoxicity. *Nat Rev Drug Discov* 4 (6):489-99.

Kass, JH, and MF Huerta. 1988. The Subcortical Visual System of Primates. In *Neurosciences (Comparative Primate Biology)*, edited by H. Steklis and J. Erwin: John Wiley & Sons.

Kauffman, Stuart. 1990. In *Theoretical Biology: Epigenetic and Evolutionary Order from Complex Systems*, edited by B. Goodwin and P. Saunders: Edinburgh University Press.

Kauffman, Stuart A. 1993. *The Origins of Order: Self-Organization and Selection in Evolution* Oxford University Press.

Kehrer, James P. 2001. In *American Chemical Society Short Course: The role of toxicology in drug discovery. In association with Drug Discovery Technology*. Boston, MA.

Keller, S. J., and M. K. Smith. 1982. Animal virus screens for potential teratogens. I. Poxvirus morphogenesis. *Teratog Carcinog Mutagen* 2 (3-4):361-74.

Kelley, Sean K., Louise A. Harris, David Xie, Laura DeForge, Klara Totpal, Jeanine Bussiere, and Judith A. Fox. 2001. Preclinical Studies to Predict the Disposition of Apo2L/Tumor Necrosis Factor-Related Apoptosis-Inducing Ligand in Humans: Characterization of in Vivo Efficacy, Pharmacokinetics, and Safety. *J Pharmacol Exp Ther* 299 (1):31-38.

Khor, T. O., S. Ibrahim, and A. N. Kong. 2006. Toxicogenomics in drug discovery and drug development: potential applications and future challenges. *Pharm Res* 23 (8):1659-64.

King, M. C., and A. C. Wilson. 1975. Evolution at two levels in humans and chimpanzees. *Science* 188 (4184):107-16.

Kirschner, Marc W, and John C Gerhardt. 2006. *The Plausability of Life*: Yale University Press.

Kitano, H. 2002. Computational systems biology. *Nature* 420 (6912):206-10.

———. 2002. Systems biology: a brief overview. *Science* 295 (5560):1662-4.

Klaassen, C D, and D L Eaton. 1993. Principles of Toxicology. In *Casarett and Doull's Toxicology*, edited by M. O. Amdur, J. Doull and C. Klaassen. New York: McGraw-Hill.

Knapp, RG, and MC Miller. 1992. *Clinical Epidemiology and Biostatistics (National Medical Series for Independent Study)*. 1st ed: Harwal Pub Co.

Kneteman, N. M., and D. F. Mercer. 2005. Mice with chimeric human livers: who says supermodels have to be tall? *Hepatology* 41 (4):703-6.

Knight, A., J. Bailey, and J. Balcombe. 2006. Animal carcinogenicity studies: 1. Poor human predictivity. *Altern Lab Anim* 34 (1):19-27.

Koch, J. A., and R. M. Ruprecht. 1992. Animal models for anti-AIDS therapy. *Antiviral Res* 19 (2):81-109.

Koppanyi, T., and M. A. Avery. 1966. Species differences and the clinical trial of new drugs: a review. *Clin Pharmacol Ther* 7 (2):250-70.

Kornberg, A. 1995. Science in the stationary phase. *Science* 269 (5232):1799.

Kotani, Y., Y. Nishimura, H. Maeda, and M. Yokoyama. 1999. Beta2-adrenergic receptor polymorphisms affect airway responsiveness to salbutamol in asthmatics. *J Asthma* 36 (7):583-90.

Kuivenhoven, J. A., J. W. Jukema, A. H. Zwinderman, P. de Knijff, R. McPherson, A. V. Bruschke, K. I. Lie, and J. J. Kastelein. 1998. The role of a common variant of the cholesteryl ester transfer protein gene in the progression of coronary atherosclerosis. The Regression Growth Evaluation Statin Study Group. *N Engl J Med* 338 (2):86-93.

LaFee, Scott. 2005. Crucial or cruel? *San Diego Union Tribune*, March 16.

LaFollette, H, and N Shanks. 1995. Two Models of Models in Biomedical Research. *Philosophical Quarterly* 141-60.

LaFollette, Hugh, and Niall Shanks. 1994. Animal Experimentation: The Legacy of Claude Bernard. *International Studies in the Philosophy of Science* 8 (3):195-210.

———. 1996. *Brute Science: Dilemmas of animal experimentation*. London and New York: Routledge.

Lahav, Noam. 1999. *Biogenesis: Theories of Life's Origin*. Oxford: Oxford University Press.

Lam, Lui. 1997. *Introduction to Nonlinear Physics*: Springer-Verlag.

Langley, Gill. *Next of Kin*. BUAV 2006 [cited. Available from http://www.buav.org/downloads/pdf/BUAV_Report-Next_of_Kin.pdf.

Langmuir, A. D. 1971. New environmental factor in congenital disease. *N Engl J Med* 284 (15):912-3.

Langston, J. W., P. Ballard, J. W. Tetrud, and I. Irwin. 1983. Chronic Parkinsonism in humans due to a product of meperidine-analog synthesis. *Science* 219 (4587):979-80.

Lauer, B., G. Tuschl, M. Kling, and S. O. Mueller. 2008. Species-specific toxicity of diclofenac and troglitazone in primary human and rat hepatocytes. *Chem Biol Interact*.

Lazarou, J., B. H. Pomeranz, and P. N. Corey. 1998. Incidence of adverse drug reactions in hospitalized patients: a meta-analysis of prospective studies. *JAMA* 279 (15):1200-5.

Leaf, C. 2004. Why we are losing the war on cancer. *Fortune*, March 22.

Ledford, H. 2008. Human genes are multitaskers. *Nature* 456 (7218):9.

———. 2008. Translational research: the full cycle. *Nature* 453 (7197):843-5.

Lee, A., W. D. Ngan Kee, and T. Gin. 2002. A quantitative, systematic review of randomized controlled trials of ephedrine versus phenylephrine for the management of hypotension during spinal anesthesia for cesarean delivery. *Anesth Analg* 94 (4):920-6, table of contents.

Lehninger, Albert L., David L. Nelson, and Michael M. Cox. 1993. *Principles of Biochemistry: With an Extended Discussion of Oxygen-Binding Proteins.* Worth Publishers.

Lerer, Bernard. 2002. *Pharmacogenetics of Psychotropic Drugs.* Cambridge University Press.

Levi, Primo. 1984. *The Periodic table.* Schocken.

Lewontin, Richard. 1995. *Human Diversity.* W.H. Freeman & Company.

———. 1995. Primate Models of Human Traits. In *Perspectives on Medical Research: Aping Science,* edited by C. o. A. M. i. B. Research: Medical Research Modernization Committee.

———. 2000. *The Triple Helix: Gene, Organism and Environment.* Harvard University Press.

Li, Wen-Hsiung. 1997. *Molecular Evolution:* Sinauer Associates

Liggitt, H. D., and G. M. Reddington. 1992. Transgenic animals in the evaluation of compound efficacy and toxicity: will they be as useful as they are novel? *Xenobiotica* 22 (9-10):1043-54.

Lin, J. H. 2008. Applications and limitations of genetically modified mouse models in drug discovery and development. *Curr Drug Metab* 9 (5):419-38.

Litchfield, J. T., Jr. 1962. Symposium on clinical drug evaluation and human pharmacology. XVI. Evaluation of the safety of new drugs by means of tests in animals. *Clin Pharmacol Ther* 3:665-72.

Lord Rayleigh. 1942. *The Life of Sir J.J. Thomson.* Cambridge University Press.

Lu, Z. X., J. Peng, and B. Su. 2007. A human-specific mutation leads to the origin of a novel splice form of neuropsin (KLK8), a gene involved in learning and memory. *Hum Mutat* 28 (10):978-84.

Machin, G. A. 1996. Some causes of genotypic and phenotypic discordance in monozygotic twin pairs. *Am J Med Genet* 61 (3):216-28.

Mankoff, S. P., C. Brander, S. Ferrone, and F. M. Marincola. 2004. Lost in Translation: Obstacles to Translational Medicine. *J Transl Med* 2 (1):14.

Manson, J, and D Wise. 1993. Teratogens. In *Casarett and Doull's Toxicology.*

Markowitch. 1988. Prefrontal Cortex in Comparative Primate Biology In *Neurosciences (Comparative Primate Biology),* edited by H. Steklis and J. Erwin: John Wiley & Sons.

Marshall, E. 2003. First check my genome, doctor. *Science* 302 (5645):589.

Martini, G., and M. V. Ursini. 1996. A new lease of life for an old enzyme. *Bioessays* 18 (8):631-7.

Marwick, C. 2000. Promising vaccine treatment for alzheimer disease found. *JAMA* 284 (12):1503-5.

Masubuchi, Y. 2006. Metabolic and non-metabolic factors determining troglitazone hepatotoxicity: a review. *Drug Metab Pharmacokinet* 21 (5):347-56.

Maugh II, Thomas H., and Jia-Rui Chong. 2006. AIDS Stalks Humans as HIV Research Slows to a Crawl. *LA Times,* June 4.

Mayr, Ernst. 1988. *Toward a New Philosophy of Biology.* Harvard University Press.

———. 1989. How biology differs from the physical sciences. In *Evolution at a Crossroads: The New Biology and the New Philosophy of Science,* edited by D. J. Depew and B. H. Weber. Cambridge, Mass: MIT Press.

———. 1998. *This Is Biology: The Science of the Living World.* Belknap Press.

———. 2002. *What evolution is.* Basic Books.

McLean, A. 1991. Concluding Remarks. In *Animals and Alternatives in Toxicology: Present Status and Future Prospects* edited by M. Balls, J. Bridges and J. Southee: Wiley-VCH.

McLeod, H. L., S. C. Pritchard, J. Githang'a, A. Indalo, M. M. Ameyaw, R. H. Powrie, L. Booth, and E. S. Collie-Duguid. 1999. Ethnic differences in thiopurine methyltransferase pharmacogenetics: evidence for allele specificity in Caucasian and Kenyan individuals. *Pharmacogenetics* 9 (6):773-6.

Medawar, Peter. 1984. *Pluto's Republic: Incorporating The Art of the Soluble and Induction and Intuition in Scientific Thought*: Oxford University Press.

Meijers, J. M., G. M. Swaen, and L. J. Bloemen. 1997. The predictive value of animal data in human cancer risk assessment. *Regul Toxicol Pharmacol* 25 (2):94-102.

Menard, S., E. Tagliabue, M. Campiglio, and S. M. Pupa. 2000. Role of HER2 gene overexpression in breast carcinoma. *J Cell Physiol* 182 (2):150-62.

Mepham, T. B., R. D. Combes, M. Balls, O. Barbieri, H. J. Blokhuis, P. Costa, R. E. Crilly, T. de Cock Buning, V. C. Delpire, M. J. O'Hare, L. M. Houdebine, C. F. van Kreijl, M. van der Meer, C. A. Reinhardt, E. Wolf, and A. M. van Zeller. 1998. The Use of Transgenic Animals in the European Union: The Report and Recommendations of ECVAM Workshop 28. *Altern Lab Anim* 26 (1):21-43.

Miller, SA. 1993. Food Additives and Contaminants. In *Casarett and Doull's Toxicology*, edited by M. O. Amdur, J. Doull and C. Klaassen. New York: McGraw-Hill.

Milovanovic, Selma. 2008. Melbourne scientists hit on cancer treatment. *The Age*.

Minelli, Alessandro. 2006. *The Development of Animal Form: Ontogeny, Morphology, and Evolution* Cambridge University Press.

Mohan, T R K. 1998. Bifurcations and Chaos in a model biochemical reaction pathway. *Journal of Bifurcation and Chaos* 8 (2):381-94.

Monro, A M, and J S MacDonald. 1998. Evaluation of the carcinogenic potential of pharmaceuticals. Opportunities arising from the International Conference on Harmonisation. *Drug Saf* 18 (5).

Monro, A. 1996. Are lifespan rodent carcinogenicity studies defensible for pharmaceutical agents? *Exp Toxicol Pathol* 48 (2-3):155-66.

Moore, L B, D J Parks, S A Jones, R K Bledsoe, T G Consler, J B Stimmel, B Goodwin, C Liddell, S G Blanchard, T M Willson, J L Collins, and S. A. Kliewer. 2000. Orphan Nuclear Receptors Constitutive Androstane Receptor and Pregnane X Receptor Share Xenobiotic and Steroid Ligands. . *The Journal of Biological Chemistry* 275:15122-15126. .

Moore, L. B., B. Goodwin, S. A. Jones, G. B. Wisely, C. J. Serabjit-Singh, T. M. Willson, J. L. Collins, and S. A. Kliewer. 2000. St. John's wort induces hepatic drug metabolism through activation of the pregnane X receptor. *Proc Natl Acad Sci U S A* 97 (13):7500-2.

Moriguchi, T., H. Motohashi, T. Hosoya, O. Nakajima, S. Takahashi, S. Ohsako, Y. Aoki, N. Nishimura, C. Tohyama, Y. Fujii-Kuriyama, and M. Yamamoto. 2003. Distinct response to dioxin in an arylhydrocarbon receptor (AHR)-humanized mouse. *Proc Natl Acad Sci U S A* 100 (10):5652-7.

Moss, R. 2006. Our futile war on cancer. *New Scientist* (December 16):19.

Mueller, Gerd B, and Lennart Olsson. 2003. Epigenesis and Epigenetics. In *Keywords and Concepts in Evolutionary Developmental Biology*, edited by B. K. Hall and W. M. Olson: Harvard University Press.

Mukerjee, Madhusree. 2004. Book Review of *Speaking for the Animals Scientific American* August:96-7.

Murray, Andrew, and Tim Hunt. 1993. *The cell cycle: An introduction*: Oxford University Press.

Nara, P., W. Hatch, J. Kessler, J. Kelliher, and S. Carter. 1989. The biology of human immunodeficiency virus-1 IIIB infection in the chimpanzee: in vivo and in vitro correlations. *J Med Primatol* 18 (3-4):343-55.

Newman, L. M., E. M. Johnson, and R. E. Staples. 1993. Assessment of the effectiveness of animal developmental toxicity testing for human safety. *Reprod Toxicol* 7 (4):359-90.

News Bulletin. 1997. In aftermath of New England Journal of Medicine article, critic Bailar declares the defeat of cancer treatment. *The Oncologist* 4 (2):276-79.

Nijhout, H Frederik. 2003. The Importance of Context in Genetics. *American Scientist* 91 (5):416-23.

Nishimura, H., and S. Miyamoto. 1969. Teratogenic effects of sodium chloride in mice. *Acta Anat (Basel)* 74 (1):121-4.

Nonneman, A.J., and M.L. Woodruff. 1994. Animal Models and the Implications of Their Use. In *Toxin-Induced Models of Neurological Disorders*, edited by A. J. Nonneman and M. L. Woodruff: Springer.

Norman, Frank. *London Blog* 2008 [cited April 22. Available from http://network.nature.com/hubs/london/blog/2008/04/22/more-skeptics-than-you-could-shake-a-crystal-pendant-at.

Northrup, E. 1957. *Science looks at smoking: A new inquiry into the effects of smoking on your health.*. New York: Coward-McCann.

Ober, Carole, Dagan A. Loisel, and Yoav Gilad. 2008. Sex-specific genetic architecture of human disease. *Nat Rev Genet* 9 (12):911-922.

Olson, H., G. Betton, D. Robinson, K. Thomas, A. Monro, G. Kolaja, P. Lilly, J. Sanders, G. Sipes, W. Bracken, M. Dorato, K. Van Deun, P. Smith, B. Berger, and A. Heller. 2000. Concordance of the toxicity of pharmaceuticals in humans and in animals. *Regul Toxicol Pharmacol* 32 (1):56-67.

Oltvai, Z. N., and A. L. Barabasi. 2002. Systems biology. Life's complexity pyramid. *Science* 298 (5594):763-4.

Oser, B. L. 1981. The rat as a model for human toxicological evaluation. *J Toxicol Environ Health* 8 (4):521-42.

Overmier, J Bruce, and Marilyn E Carroll. 2001. Basic Issues in the Use of Animals in Health Research. In *Animal Research and Human Health*, edited by M. E. Carroll and J. B. Overmier: American Psychological Association.

Palfreyman, M G, V Charles, and J Blander. 2002. The importance of using human-based models in gene and drug discovery. *Drug Discovery World* Fall:33-40.

Parascandola, J. 1995. The emergence of pharmaceutical science.. *Pharmacy in History*. 37:68-75.

Parham, Peter. 1994. The rise and fall of great class I genes. *Semin Immunol* 6 (6):373=82.

————. 2000. *The Immune System.* Routledge.

Partridge, E. A., R. A. D'Souza, E. M. Lenz, S. M. Smith, J. Clarkson-Jones, and D. W. Roberts. 2008. Disposition and metabolism of the colchicine derivative [14C]-ZD6126 in rat and dog. *Xenobiotica* 38 (4):399-421.

Patton, W. 1993. *Mouse and Man.* Oxford: Oxford University Press.

Paul, J R. 1971. *A History of Poliomyelitis.* New Haven: Yale University Press.

Peller, Sigsmund. 1952. *Cancer in Man.* New York: International Universities Press.

Pennisi, E. 2000. A mouse chronology. *Science* 288 (5464):248-57.

Perry, G. H., F. Yang, T. Marques-Bonet, C. Murphy, T. Fitzgerald, A. S. Lee, C. Hyland, A. C. Stone, M. E. Hurles, C. Tyler-Smith, E. E. Eichler, N. P. Carter, C. Lee, and R. Redon. 2008. Copy number variation and evolution in humans and chimpanzees. *Genome Res* 18 (11):1698-710.

Petroski, Henry. 1994. *Design Paradigms: Case Histories of Error and Judgment in Engineering.* Cambridge University Press.

————. 1996. *Invention by Design; How Engineers Get from Thought to Thing.* Harvard University Press.

Pigliucci, M. 2003. Species as family resemblance concepts: the (dis-)solution of the species problem? *Bioessays* 25 (6):596-602.

Piscitelli, S. C., A. H. Burstein, D. Chaitt, R. M. Alfaro, and J. Falloon. 2000. Indinavir concentrations and St John's wort. *Lancet* 355 (9203):547-8.

Pound, P., S. Ebrahim, P. Sandercock, M. B. Bracken, and I. Roberts. 2004. Where is the evidence that animal research benefits humans? *BMJ* 328 (7438):514-7.

Price, Peter W. 1996. *Biological Evolution.* Harcourt.

Ptashne, M, and A Gann. 2002. *Genes & Signals.* Cold Springs Harbor Laboratory Press.

Quine, WV. 2005. *Quiddities" An Intermittently Philosophical Dictionary.* Cambridge: The Belknap Press of Harvard University Press.

Rall, D. P. 2000. Laboratory animal tests and human cancer. *Drug Metab Rev* 32 (2):119-28.

Rangarajan, A., and R. A. Weinberg. 2003. Opinion: Comparative biology of mouse versus human cells: modelling human cancer in mice. *Nat Rev Cancer* 3 (12):952-9.

Ravipati, G., W. S. Aronow, H. Lai, J. Shao, A. J. DeLuca, M. B. Weiss, A. L. Pucillo, K. Kalapatapu, C. E. Monsen, and R. N. Belkin. 2008. Comparison of sensitivity, specificity, positive predictive value, and negative predictive value of stress testing versus 64-multislice coronary computed tomography angiography in predicting obstructive coronary artery disease diagnosed by coronary angiography. *Am J Cardiol* 101 (6):774-5.

Redon, R., S. Ishikawa, K. R. Fitch, L. Feuk, G. H. Perry, T. D. Andrews, H. Fiegler, M. H. Shapero, A. R. Carson, W. Chen, E. K. Cho, S. Dallaire, J. L. Freeman, J. R. Gonzalez, M. Gratacos, J. Huang, D. Kalaitzopoulos, D. Komura, J. R. MacDonald, C. R. Marshall, R. Mei, L. Montgomery, K. Nishimura, K. Okamura, F. Shen, M. J. Somerville, J. Tchinda, A. Valsesia, C. Woodwark, F. Yang, J. Zhang, T. Zerjal, L. Armengol, D. F. Conrad, X. Estivill, C. Tyler-Smith, N. P. Carter, H. Aburatani, C. Lee, K. W. Jones, S. W. Scherer, and M.

E. Hurles. 2006. Global variation in copy number in the human genome. *Nature* 444 (7118):444-54.

Relling, M. V., M. L. Hancock, G. K. Rivera, J. T. Sandlund, R. C. Ribeiro, E. Y. Krynetski, C. H. Pui, and W. E. Evans. 1999. Mercaptopurine therapy intolerance and heterozygosity at the thiopurine S-methyltransferase gene locus. *J Natl Cancer Inst* 91 (23):2001-8.

Reville, William. 2008. Why animal testing offers the best chance for answers. *Irish Times*.

Rigamonti, E., G. Chinetti-Gbaguidi, and B. Staels. 2008. Regulation of macrophage functions by PPAR-alpha, PPAR-gamma, and LXRs in mice and men. *Arterioscler Thromb Vasc Biol* 28 (6):1050-9.

Ritchie, M. D., L. W. Hahn, N. Roodi, L. R. Bailey, W. D. Dupont, F. F. Parl, and J. H. Moore. 2001. Multifactor-dimensionality reduction reveals high-order interactions among estrogen-metabolism genes in sporadic breast cancer. *Am J Hum Genet* 69 (1):138-47.

Robles, A. I., and L. Varticovski. 2008. Harnessing genetically engineered mouse models for preclinical testing. *Chem Biol Interact* 171 (2):159-64.

Rosenzweig, M L. 1995. *Species diversity in space and time*: Cambridge University Press.

Ross, S. A., P. J. McCaffery, U. C. Drager, and L. M. De Luca. 2000. Retinoids in embryonic development. *Physiol Rev* 80 (3):1021-54.

Rueger, Alexander, and W David Sharp. 1998. Idealization and Stabiulity: A Perspective From Nonlinear Dynamics. In *Poznan Studies in the Philosophy of Science and the Humanities. Idealization IX: Idealization in Contemporary Physics*, edited by N. Shanks. Amsterdam: Rodopi.

Ruelle, David. 1994. *Dynamical Zeta Functions for Piecewise Monotone Maps of the Interval (Crm Monograph, Vol 4)* American mathematical Society.

Runner, M. N. 1967. Comparative pharmacology in relation to teratogenesis. *Fed Proc* 26 (4):1131-6.

Ruschitzka, F., P. J. Meier, M. Turina, T. F. Luscher, and G. Noll. 2000. Acute heart transplant rejection due to Saint John's wort. *Lancet* 355 (9203):548-9.

Salmon, WC. 1998. Rational Prediction. In *Philosophy of Science*, edited by M. Curd and J. Cover: Norton.

Salsburg, D. 1983. The lifetime feeding study in mice and rats--an examination of its validity as a bioassay for human carcinogens. *Fundam Appl Toxicol* 3 (1):63-7.

Sankar, U. 2005. The Delicate Toxicity Balance in Drug Discovery. *The Scientist* 19 (15):32.

Schaeffeler, E., C. Fischer, D. Brockmeier, D. Wernet, K. Moerike, M. Eichelbaum, U. M. Zanger, and M. Schwab. 2004. Comprehensive analysis of thiopurine S-methyltransferase phenotype-genotype correlation in a large population of German-Caucasians and identification of novel TPMT variants. *Pharmacogenetics* 14 (7):407-17.

Schardein, J. 2000. *Chemically Induced Birth defects*. 3rd ed: Informa HealthCare.

Schardein, James L. 1992. *Chemically Induced Birth Defects*. 2nd ed: Marcel Dekker.

Schardein, JL. 1976. *Drugs as Teratogens*: CRC Press.

———. 1985. *Chemically Induced Birth Defects*: Marcel Dekker.

Schardein, JL, and OT Macina. 2007. *Human Developmental Toxicants*: CRC Press.

Schnabel, J. 2008. Neuroscience: Standard model. *Nature* 454 (7205):682-5.

Schneierson, S. S., and E. Perlman. 1956. Toxicity of penicillin for the Syrian hamster. *Proc Soc Exp Biol Med* 91 (2):229-30.

Scialli, Anthony R. 1991. *A Clinical Guide to Reproductive and Developmental Toxicology*: CRC Press.

Science Daily. *Alzheimer's Research Using Animal Models Significantly Increases Understanding Of The Disease* (December 15). Science Daily 2008 [cited. Available from http://www.sciencedaily.com/releases/2008/12/081215075119.htm.

Shanks, N., R. Greek, and J. Greek. 2009. Are animal models predictive for humans? *Philos Ethics Humanit Med* 4 (1):2.

Shanks, N., and R. A. Pyles. 2007. Evolution and medicine: the long reach of "Dr. Darwin". *Philos Ethics Humanit Med* 2:4.

Shanks, Niall. 2001. Modeling Biological Systems: The Belousov-Zhabotinski Reaction. *Foundations of Chemistry* 3 (1):33-53.

———. 2002. *Animals and Science*: ABC Clio.

———. 2004. *God, the Devil, and Darwin: A Critique of Intelligent Design Theory*: Oxford University Press.

Shanks, Niall, Ray Greek, Nathan Nobis, and Jean Greek. 2007. Animals and Medicine: Do Animal Experiments Predict Human Response? *Skeptic* 13 (3):44-51.

Shanks, Niall, and Karl Joplin. 1999. Redundant Complexity: A Critical Analysis of Intelligent Design in Biochemistry. *Philosophy of Science* 66:268-82.

Shapiro, KJ. 2004. Animal Model Research. The Apples and Oranges Quandry. *ATLA* 32 (Suupl 1):405-09.

Shapiro, S. D. 2007. Transgenic and gene-targeted mice as models for chronic obstructive pulmonary disease. *Eur Respir J* 29 (2):375-8.

Shepard, Thomas H. 1995. *Catalog of Teratogenic Agents*: Johns Hopkins University Press.

Shima, N., M. Katagi, H. Kamata, K. Zaitsu, T. Kamata, M. Nishikawa, A. Miki, H. Tsuchihashi, T. Sakuma, and N. Nemoto. 2008. Urinary excretion of the main metabolites of 3,4-methylenedioxymethamphetamine (MDMA), including the sulfate and glucuronide of 4-hydroxy-3-methoxymethamphetamine (HMMA), in humans and rats. *Xenobiotica* 38 (3):314-24.

Shubick, P. 1980. Statement of the Problem. In *Human Epidemiology and Animal Laboratory Correlations in Chemical Carcinogenesis*, edited by F. Coulston and P. Shubick: Ablex Pub. .

Sidhu, RK. 1983. Corticosteroids in pregnancy. In *Drugs and Pregnancy: Human Teratogenesis and Related Problems*, edited by Hawkins: Churchill Livingston.

Sigma Xi. 1992. Sigma Xi Statements of the Use of Animals in Research. *American Scientist* 80 (1):73-76.

Simmons. 1988. In *Neurosciences (Comparative Primate Biology)*: John Wiley & Sons.

Simon, V. 2005. Wanted: women in clinical trials. *Science* 308 (5728):1517.

Singer, SJ. 2003. *The Splendid Feast of Reason*: University of California Press.

Sipes, I G, and A. J. Gandolfi. 1993. Biotransformation of Toxicants In *Casarett and Doull's Toxicology*, edited by A. Amdour, J. Doull and C. Klaassen. New York: McGraw-Hill.

Sistare, F. D., and J. J. DeGeorge. 2007. Preclinical predictors of clinical safety: opportunities for improvement. *Clin Pharmacol Ther* 82 (2):210-4.

Skolnick, A, and Ray Greek. 2005. Debate on animals in research. http://curedisease.com/images/debate%20transcript.pdf.

Sloan, D. A., D. M. Fleiszer, G. K. Richards, D. Murray, and R. A. Brown. 1983. Increased incidence of experimental colon cancer associated with long-term metronidazole therapy. *Am J Surg* 145 (1):66-70.

Sludden, J., S. C. Hardy, M. R. VandenBranden, S. A. Wrighton, and H. L. McLeod. 1998. Liver dihydropyrimidine dehydrogenase activity in human, cynomolgus monkey, rhesus monkey, dog, rat and mouse. *Pharmacology* 56 (5):276-80.

Smith, M. W., M. Dean, M. Carrington, C. Winkler, G. A. Huttley, D. A. Lomb, J. J. Goedert, T. R. O'Brien, L. P. Jacobson, R. Kaslow, S. Buchbinder, E. Vittinghoff, D. Vlahov, K. Hoots, M. W. Hilgartner, and S. J. O'Brien. 1997. Contrasting genetic influence of CCR2 and CCR5 variants on HIV-1 infection and disease progression. Hemophilia Growth and Development Study (HGDS), Multicenter AIDS Cohort Study (MACS), Multicenter Hemophilia Cohort Study (MHCS), San Francisco City Cohort (SFCC), ALIVE Study. *Science* 277 (5328):959-65.

Smith, O W, G van S Smith, and D Hurwitz. 1946. Increased excretion of pregnanediol in pregnancy from diethylstilbestrol with special reference to the prevention of late pregnancy accidents. . *Am J Obst Gyn* 51:411-15.

Smith, Peter. 1998. *Explaining Chaos*. Cambridge University Press.

Sober, Elliott. 1993. *Philosophy of Biology*. Westview Press.

Spriet-Pourra., C, and M Auriche. 1994. Drug Withdrawal from Sale. New York.

St Jude. *Gene therapy corrects sickle cell disease in laboratory study*. Press release from St Jude's Children's Research Hospital December 3, 2008 [cited. Available from http://www.eurekalert.org/pub_releases/2008-12/sjcr-gtc120308.php.

Staples, R. E., and D. E. Holtkamp. 1963. Effects of Parental Thalidomide Treatment on Gestation and Fetal Development. *Exp Mol Pathol* 26:81-106.

Steinberg, Douglas. 2002. Companies Halt First Alzheimer Vaccine Trial. At issue: What inflamed patients' brains? . *The Scientist* 16 (7):22.

Stump, D. S., and S. VandeWoude. 2007. Animal models for HIV AIDS: a comparative review. *Comp Med* 57 (1):33-43.

Szathmáry, E., F. Jordan, and C. Pal. 2001. Molecular biology and evolution. Can genes explain biological complexity? *Science* 292 (5520):1315-6.

Tannenbaum, S. R. 1997. Comparative metabolism of tamoxifen and DNA adduct formation and in vitro studies on genotoxicity. *Semin Oncol* 24 (1 Suppl 1):S1-81-S1-6.

Tateishi, T., M. Chida, N. Ariyoshi, Y. Mizorogi, T. Kamataki, and S. Kobayashi. 1999. Analysis of the CYP2D6 gene in relation to dextromethorphan O-demethylation capacity in a Japanese population. *Clin Pharmacol Ther* 65 (5):570-5.

Taylor, K., N. Gordon, G. Langley, and W. Higgins. 2008. Estimates for worldwide laboratory animal use in 2005. *Altern Lab Anim* 36 (3):327-42.

Teeling-Smith, George. A. 1980. *Question of Balance: the Benefits and Risks of Pharmaceutical Innovation*. London: Office of Health Economics.

Templeton, Alan R. 1989. The Meaning of Species and Speciation:A Genetic Perspective. In *Speciation and Its Consequences*, edited by D. Otte and J. A. Endler. Sunderland, Mass: Sinauei.

Thayer, H S. 2005. *Newton's Philosophy of Nature: Selections from His Writings*. Dover Publications.

Tomatis, L, and L Wilbourn. 2003. Evaluation of carcinogenic risk to humans: the experience of IARC. In *New Frontiers in Cancer Causation*, edited by Iversen. Washington, DC: Taylor and Francis.

Tong, A. H., G. Lesage, G. D. Bader, H. Ding, H. Xu, X. Xin, J. Young, G. F. Berriz, R. L. Brost, M. Chang, Y. Chen, X. Cheng, G. Chua, H. Friesen, D. S. Goldberg, J. Haynes, C. Humphries, G. He, S. Hussein, L. Ke, N. Krogan, Z. Li, J. N. Levinson, H. Lu, P. Menard, C. Munyana, A. B. Parsons, O. Ryan, R. Tonikian, T. Roberts, A. M. Sdicu, J. Shapiro, B. Sheikh, B. Suter, S. L. Wong, L. V. Zhang, H. Zhu, C. G. Burd, S. Munro, C. Sander, J. Rine, J. Greenblatt, M. Peter, A. Bretscher, G. Bell, F. P. Roth, G. W. Brown, B. Andrews, H. Bussey, and C. Boone. 2004. Global mapping of the yeast genetic interaction network. *Science* 303 (5659):808-13.

Topol, E. J. 2004. Failing the public health--rofecoxib, Merck, and the FDA. *N Engl J Med* 351 (17):1707-9.

Translational research: getting the message across. 2008. *Nature* 453 (7197):839.

Turbow, M. M., W. H. Clark, and J. A. Dipaolo. 1971. Embryonic abnormalities in hamsters following intrauterine injection of 6-aminonicotinamide. *Teratology* 4 (4):427-31.

Ulrich, A. B., J. Standop, B. M. Schmied, M. B. Schneider, T. A. Lawson, and P. M. Pour. 2002. Species differences in the distribution of drug-metabolizing enzymes in the pancreas. *Toxicol Pathol* 30 (2):247-53.

Utidjian, M. 1988. In *Perspectives in Basic and Applied Toxicology*, edited by B. Ballantyne: Butterworth-Heinemann.

Van Regenmortel, M. H. 2004. Reductionism and complexity in molecular biology. Scientists now have the tools to unravel biological and overcome the limitations of reductionism. *EMBO Rep* 5 (11):1016-20.

VandeWoude, S., and C. Apetrei. 2006. Going wild: lessons from naturally occurring T-lymphotropic lentiviruses. *Clin Microbiol Rev* 19 (4):728-62.

Varki, A., and T. K. Altheide. 2005. Comparing the human and chimpanzee genomes: searching for needles in a haystack. *Genome Res* 15 (12):1746-58.

Varki, A., D. H. Geschwind, and E. E. Eichler. 2008. Explaining human uniqueness: genome interactions with environment, behaviour and culture. *Nat Rev Genet* 9 (10):749-63.

Vineis, P., and R. Melnick. 2008. A Darwinian perspective: right premises, questionable conclusion. A commentary on Niall Shanks and Rebecca Pyles' "evolution and medicine: the long reach of "Dr. Darwin"". *Philos Ethics Humanit Med* 3:6.

Wake, D B. 2003. Homology and Homoplasy In *Keywords and Concepts in Evolutionary Developmental Biology*, edited by B. K. Hall and W. M. Olson: Harvard University Press.

Walker, B. E. 1989. Animal models of prenatal exposure to diethylstilboestrol. *IARC Sci Publ* (96):349-64.

Walker, Mark D, Joyce K Nelson, and John C Bernard. 2007. Primates: toxicology. In *Animal Models in Toxicology*, edited by S. Gad: CRC Press.

Walker, R. M., and T. F. McElligott. 1981. Furosemide induced hepatotoxicity. *J Pathol* 135 (4):301-14.

Waltz, E. 2005. Price of mice to plummet under NIH's new scheme. *Nat Med* 11 (12):1261.

Wang, C., T. K. Bammler, and D. L. Eaton. 2002. Complementary DNA cloning, protein expression, and characterization of alpha-class GSTs from Macaca fascicularis liver. *Toxicol Sci* 70 (1):20-6.

Ward, R. M., and T. P. Green. 1988. Developmental pharmacology and toxicology: principles of study design and problems of methodology. *Pharmacol Ther* 36 (2-3):309-34.

Watkins, D. I., D. R. Burton, E. G. Kallas, J. P. Moore, and W. C. Koff. 2008. Nonhuman primate models and the failure of the Merck HIV-1 vaccine in humans. *Nat Med* 14 (6):617-21.

Weatherall, D. 2006. The use of nonhuman primates in research 2006. London.

Weatherall, M. 1982. An end to the search for new drugs? *Nature* 296:387-90.

Webster, W. S., and J. A. Freeman. 2003. Prescription drugs and pregnancy. *Expert Opin Pharmacother* 4 (6):949-61.

Weinshilboum, R. 2003. Inheritance and drug response. *N Engl J Med* 348 (6):529-37.

Weiss, Rick. March 2, 2002. Alzheimer's Vaccine Permanently Shelved *Washington Post*.

West-Eberhard, Mary Jane. 2003. *Developmental Plasticity and Evolution*: Oxford University Press.

Wikipedia. 2008. *Animal testing* 2008 [cited December 21 2008]. Available from http://en.wikipedia.org/wiki/Animal_testing.

Wikipedia Commons. *Roessler attractor* 2009 [cited. Available from http://en.wikipedia.org/wiki/File:Roessler_attractor.png.

———. 2009. *Scale Free Network* 2009 [cited May 30 2009]. Available from http://images.google.com/imgres?imgurl=http://upload.wikimedia.org/wiki pedia/commons/7/77/Scale-free_network_sample.png&imgrefurl=http://co mmons.wikimedia.org/wiki/File:Scale-free_network_sample.png&usg=__4Ux rl6XTzIZd644cLQfxXc3IBR8=&h=410&w=870&sz=34&hl=en&start=3&u m=1&tbnid=_sFoEGyMM7_I4M:&tbnh=68&tbnw=145&prev=/images%3F q%3Dscale%2Bfree%2Bnetwork%26hl%3Den%26client%3Dsafari%26rls%3 Den-us%26sa%3DG%26um%3D1.

Wilbourn, J., L. Haroun, E. Heseltine, J. Kaldor, C. Partensky, and H. Vainio. 1986. Response of experimental animals to human carcinogens: an analysis based upon the IARC Monographs programme. *Carcinogenesis* 7 (11):1853-63.

Wilkins, A S. 2001. *The Evolution of Developmental Pathways*: Sinauer Associates.

Williams, G, and J Weisburger. 1993. Chemical carcinogenesis. In *Casarett and Doull's Toxicology, The Basic Science of Poisons,*, edited by M. Amdur, J. Doull and C. Klassen. New York: McGraw-Hill.

Williams, James. 2008. What Makes Science 'Science'? *The Scientist* 22 (10):29.

Willson, T. M., and S. A. Kliewer. 2002. PXR, CAR and drug metabolism. *Nat Rev Drug Discov* 1 (4):259-66.

Willyard, C. 2007. Blue's clues. *Nat Med* 13 (11):1272-3.

Wilson, D S, and E Sober. 1994. Reintroducing group selection to the human behavioral sciences. *Behavioral and Brain Sciences* 17 (4):585-654.

Wogan, G. N. 1997. Review of the toxicology of tamoxifen. *Semin Oncol* 24 (1 Suppl 1):S1-87-S1-97.

Wray, Gregory A. 2001. DEVELOPMENT: Resolving the Hox Paradox. *Science* 292 (5525):2256-2257.

Xceleron. [cited. Available from http://www.xceleron.com/metadot/index.pl.

Yates, C. R., E. Y. Krynetski, T. Loennechen, M. Y. Fessing, H. L. Tai, C. H. Pui, M. V. Relling, and W. E. Evans. 1997. Molecular diagnosis of thiopurine S-methyltransferase deficiency: genetic basis for azathioprine and mercaptopurine intolerance. *Ann Intern Med* 126 (8):608-14.

Young, Malcolm. 2008. Prediction v Attrition *Drug Discovery World* (Fall):9-12.

Zbinden, G. 1993. The concept of multispecies testing in industrial toxicology. *Regul Toxicol Pharmacol* 17 (1):85-94.

Zucker, Arthur. 1995. *Introduction to the Philosophy of Science.* Prentice Hall.

INDEX

2

2-naphthylamine, 271

A

Acetaminophen, 271, 283
adenovirus, 349
ADMET, 223, 274, 331
African, 153, 231-232
Agarwal and Moorchung, 315
Ahn, 82, 353
AIDS, 119, 153, 320, 342, 344, 347-349
Alberts, 194, 195, 197, 199, 216-217, 221, 236
Alternative splicing, 151, 154
Altman, 256
Alzheimer's, 304, 320, 344, 372
Amedis Pharmaceuticals, 274
American Medical Association, 62, 368
AN1792, 333
AN-1792, 333
angina, 235
animal rights movement, 23
Antarctic notothenioids, 31
Arctic cod, 31
Asians, 343
attractors, 91-94

B

Bailar, JC, 388-389
Bains, William, 223, 274
basic biological research, 30, 116, 118, 359, 381
basic research, 114, 267, 369-370, 376-378, 380-383, 385, 387, 389
Begley, Sharon, 380-381, 383, 385
Bernard, Claude, 60-69, 80, 111, 117, 131, 136-137, 209, 220-221, 263, 378
birth defects, 121, 276, 281-284, 288, 291
Black, 234
blacks, 231, 233
Blakemore, Colin, 369
body-as-machine, 40
Botting and Morrison, 369
brain, 334, 382
Braun and Johnson, 348

Brenner, Sydney, 384
Brigandt, 31
bushbaby, 290

C

C9D loop, 334
caffeic acid, 211, 230
Caldwell, 244
Cambridge, 18, 44
CAMs, 105-106, 108, 117-118, 129, 208, 211-212, 222, 226, 237, 246, 310, 321, 359
cancer, 286
cardiovascular diseases, 235
Carroll, Marilyn, 105
Castalia, 375-376
Caucasian, 232
Caucasians, 230-231, 341
causal determinism, 51, 64-65, 107
causal functional disanalogies, 312
CCR5, 348
CCR5-Delta32, 348
CD4, 345-349
CD4+, 345, 348-349
Center for Science in the Public Interest, 371
CF, 314-315, 368
chaos, 84, 99
Charles River, 377
Cherry picking, 258, 261
Cheung and Gonzalez, 238
Chiba, 342
chimpanzees, 59, 115, 119-120, 132, 170-173, 177, 215, 225, 321-322, 324, 326-328, 330, 342, 345, 347
chloramphenicol, 271
chloroform, 230, 275
Chloroform, 230
chlortrimazole, 241
cigarette, 36, 272-273, 388
cinchona bark, 275
clioquinol, 284
CNVs, 152
Collins, 233, 243-244
Committee on Models for Biomedical Research, 377
Committee on Toxicity Testing and Assessment of Environmental Agents, 264

ABOUT THE AUTHORS

Niall Shanks, PhD, is currently the Curtis D. Gridley Distinguished Professor in the History and Philosophy of Science at Wichita State University, and Vice-President of *Americans for Medical Advancement*. Shanks has served as President of the SWARM Division of the *American Association for the Advancement of Science* (2008-2009). Shanks' recent research interests have focused on the implications of evolutionary biology for the practice of biomedical research. His recent books include *Animals in Science* (2002), *God, the Devil and Darwin* (2004), and (with C. Ray Greek) *FAQs about the Use of Animals in Science* (2009).

◆

C. Ray Greek received his MD from the University of Alabama in Birmingham and trained in anesthesiology at the University of Missouri-Columbia and the University of Wisconsin-Madison. Along with his wife Jean Greek, DVM he has written *Sacred Cows and Golden Geese* (2000), *Specious Science* (2002), and *What Will We Use if We Don't Experiment on Animals?* (2004). Ray and Jean Greek founded Americans For Medical Advancement (www.curedisease.com), a not-for-profit dedicated to advancing medical science. Specifically, AFMA studies the use of animal models. He currently resides, with his family, in southern California.

LaVergne, TN USA
07 October 2009
160067LV00003B/56/P